HANDBOOK ON THE MORPHOLOGY OF COMMON GRASSES

Identification and Characterization of Caryopses and Seedlings

HANDBOOK ON THE MORPHOLOGY OF COMMON GRASSES

Identification and Characterization of Caryopses and Seedlings

Dhara Gandhi, Susy Albert, and Neeta Pandya

Department of Botany, Faculty of Science,
The Maharaja Sayajirao University of Baroda,
Vadodara – 390002, Gujarat, India

AAP | APPLE ACADEMIC PRESS

Apple Academic Press Inc.
3333 Mistwell Crescent
Oakville, ON L6L 0A2
Canada

Apple Academic Press Inc.
9 Spinnaker Way
Waretown, NJ 08758
USA

Exclusive worldwide distribution by CRC Press, a member of Taylor & Francis Group

No claim to original U.S. Government works

Printed in the United States of America on acid-free paper

International Standard Book Number-13: 978-1-77188-249-1 (Hardcover)

International Standard Book Number-13: 978-1-77188-250-7 (eBook)

Library and Archives Canada Cataloguing in Publication

Gandhi, Dhara, author
Handbook on the morphology of common grasses : identification and characterization of caryopses and seedlings / Dhara Gandhi, Susy Albert, and Neeta Pandya.

Includes bibliographical references and index.
Issued in print and electronic formats.
ISBN 978-1-77188-249-1 (bound).--ISBN 978-1-77188-250-7 (pdf)
1. Grasses--Seedlings. 2. Grasses--Morphology. 3. Grasses--Identification. 4. Seedlings--Morphology. 5. Seedlings--Identification. 6. Caryopses--Morphology. 7. Caryopses--Identification. I. Albert, Susy, author II. Pandya, Neeta author III. Title.

QK495.G74G36 2016 584'.9 C2016-900858-4 C2016-900859-2

Library of Congress Cataloging-in-Publication Data

Names: Gandhi, Dhara, author. | Albert, Susy, author. | Pandya, Neeta, author.
Title: Handbook on the morphology of common grasses : identification and characterization of caryopses and seedlings / authors: Dhara Gandhi, Susy Albert, and Neeta Pandya.
Description: 1st ed. | Waretown, NJ : Apple Academic Press, [2016] | Includes bibliographical references and index.
Identifiers: LCCN 2016004886 (print) | LCCN 2016007325 (ebook) | ISBN 9781771882491 (hardcover : alk. paper) | ISBN 9781771882507 () Subjects: LCSH: Grasses--Identification--Handbooks, manuals, etc. Classification: LCC QK495.G74 G355 2016 (print) | LCC QK495.G74 (ebook) | DDC 584/.9--dc23
LC record available at http://lccn.loc.gov/2016004886

Apple Academic Press also publishes its books in a variety of electronic formats. Some content that appears in print may not be available in electronic format. For information about Apple Academic Press products, visit our website at www.appleacademicpress.com and the CRC Press website at www.crcpress.com

CONTENTS

Preface .... *vii*
Introduction .... *ix*
About the Authors .... *xi*

GENERAL INTRODUCTION **1**
Identification Criteria of Grasses **2**
Grass Biology **2**
Guide to Taxonomic Ranks **4**
Keywords **6**
References **6**

SECTION 1: CARYOPSES **9**
1.1. Historical Preview **10**
1.2. List of Grass Caryopses Characterized **13**
1.3. Morphometric Features **15**
1.4. Light and Scanning Electron Microscopic Features **84**
1.5. Identification Key **144**
1.6. Cluster Analysis **150**
Keywords **156**
References **157**

SECTION 2: SEEDLING **159**
2.1. Introduction **160**
2.2. List of Grass Seedling Characterized **162**
2.3. Materials and Methods **165**
2.4. Illustrated Terminologies **166**

2.5. Illustrated Seedling Description **200**
2.6. Identification Key **251**
2.7. Cluster Analysis **257**
Keywords **259**
References **259**
Appendix: Synonyms **261**
Index *303*

PREFACE

Grasslands are the largest and most important natural ecosystems of the world, in which grasses, sedges and other herbaceous legumes are dominant. The grass family is one of the largest and most diverse families in the plant kingdom, covering 70% land surface of the globe and showing high adaptability with respect to their environments and their ability to coexist with grazing animals and man.

Grasses are of great economic value. They provide human beings and their domestic animals with the main necessities of life, add diversity to the landscape, and stability to the ground surface. They also have ornamental and amenity value. Generally, identification of grasses are done on the characteristic features of the flower and vegetative features. Grasses, especially in the vegetative phase are easily confused with the sedges (Cypreaceae) and rushes (Juncaceae) and are difficult to differentiate.

In this book, 100 common grasses (palatable and unpalatable) growing in the grasslands have been characterized on the basis of the vegetative characters of the seedling. A key to the identification of the grasses at its seedling stage has been provided. This will help the foresters and other researchers to identify the grass at their early stages of development and remove any unwanted species at its earlier stages.

Generally grasses are propogated by their vegetative means. Even taxonomists generally avoid to study the seeds of grasses, for example, caryopses. Characteristic features of caryopses serve as an important feature of identification. In the present work, the morphometric and micromorphological study of the caryopses has been conducted, and on the basis of these features, an identification key has been provided.

The book has been divided into two sections. The first section deals with characteristic features of the caryopses. It includes light and scanning electron microscopic features and a diagnostic key to the identification of the species. The second section deals with grass seedling morphology and a key to the identification of the species on the basis of early vegetative features. Each of the sections include an introduction of the topic, material and methods and results supplemented with microphotographs representing the features of identification.

This handbook indeed is the first of its kind that includes so many grass species that can be authentically identified with the help of pictorial diagnostic features of the seedlings and caryopses. The identifying features are solely on the basis of morphological, micromorphological, and morphometric characters.

INTRODUCTION

Seeds are vital part of life on earth. They are fundamental part for plant reproduction. Seeds have various uses and play an important role in the diet of human and domestic and wild animals. Seeds provide numerous morphological characters and can be used for taxonomic purposes. In this book, morphometric studies of grass seeds, generally termed as caryopses, have been conducted. Light and scanning electron microscopic features have been described and supplemented with photographs of identifying features.

Apart from this, seedlings are also important part from which the species can be identified based on the diagnostic features. A seedling is helpful in assessing the natural regeneration of an ecosystem and is of great importance to the forest planners. Recognition of plants by their vegetative characters is essential in the development of a sound pasture-improvement program. Seedlings of tree species are very easily identifiable, while with grasses and herbaceous plants it is very difficult to identify at the seedling stages, particularly in the one-to-two-leaf seedling growth stage. For the perfect identification, the flowering condition of grasses is needed.

This book is a pictorial resource guide to the identification of 98 different grasses in their early growth stage. The majority of these grasses are cosmopolitan, and thus this information will be of wide use. Terms used to describe a grass seedling have been depicted with help of photographs. Different diagnostic features of the seedling such as growth habit, type of vernation, node, internode, leaf lamina, leaf tip, leaf sheath, ligule, auricle and collar have been used for their identification. The description and photographs will enable a user to successfully and very easily identify these species in the field environment.

Identification of these seedlings would be very easy as it would match with the documented features. Identifying the seedling with the help of photographs is always more authentic rather than with the help of diagrams. The contents include photographs of seedlings of each species where features used in identification are clearly exposed. An identification key and easy-to-use pictorial field guide for identifying the common grass seedlings have been developed and included. The key to the identification is also supplemented with dendrograms. The diagnostic features of the seedlings represented will be a useful tool in the field of forestry, agriculture, taxonomy, and other related scientific studies.

ABOUT THE AUTHORS

Dhara Gandhi is a researcher working in the Department of Botany of the Faculty of Science at The Maharaja Sayajirao University of Baroda in Vadodara, India. She has several years' experience in research. She worked as a project fellow in a research project funded by the Gujarat Forest Department, India. Her specializations are plant taxonomy and plant anatomy, with her main area of research concentrated on the diversity and characterization of grasses. She has published several research papers.

Susy Albert, PhD, is currently a professor in the Department of Botany, The Maharaja Sayajirao University of Baroda in Vadodara, India. She completed her PhD in developmental plant anatomy from that university under the guidance of renowned plant anatomist professor J. J. Shah. She has 15 years of research and teaching experience and has published 50 research papers in reputed journals. She is presently guiding four students for their PhD programs. Her main area of research work is on fungal wood degradation and its biotechnological applications mainly for biopulping in the paper industry. Other research areas of interests include morphometrics and developmental anatomy of monocots, mainly grasses, and some dicots.

Neeta R. Pandya, PhD, is working as professor in the Department of Botany at The Maharaja Sayajirao University of Baroda in Vadodara, India. She has 22 years of research and teaching experience. Her subject specialization is in plant ecology, phytochemistry, and medicobotany. She is teaching plant physiology, ecology, analytical techniques, and phytochemistry at undergraduate and postgraduate levels.

GENERAL INTRODUCTION

CONTENTS

Identification Criteria of Grasses 2
Grass Biology 2
Habit 3
Inflorescence 3
Panicle 4
Raceme 4
Spike 4
Guide to Taxonomic Ranks 4
Family 4
Subfamily 5
Tribe 5
Genus 5
Species 6
Keywords 6
References 6

Grasses comes from Old High German word *"gras"* – generally used to describe any suitable for livestock grazing and they evolved in late cretaceous era (Stromberg, 2011; Prasad et al., 2011). They are one of the largest groups of the plants. Grasses are uniformly distributed on all continents and in all climatic zones. They are cosmopolitan in distribution with important centers of diversity in Brazil, Central North America, Southern-east Africa, and Australia and occupy an enormous. It is surprise, but fact is that the grass species (Gramineae members) are more in Arctic regions than the other family. Even they grow in range of habitats (Clayton and Renvoize, 1986; Osborne et al., 2011), for example, grow in marshy area, desert, water, water logged area, etc. Grass species are ubiquitous on earth, occurring in ecosystems on every continent and when they form grasslands, they have a major influence on climate through the cycling of carbon and water between soils and the atmosphere.

IDENTIFICATION CRITERIA OF GRASSES

Grasses are very important group of plants not only to human beings but also to animals. Grass species are most of the world's major food crops, including wheat, barley, oats, rice, maize, millets, sugarcane and pasture species. Grasses are very difficult to identify and generally they are identified on the basis of their inflorescences type, for example, on the basis of their reproductive parts. The other feature was the falling of lemma and palea after maturity. Rather than this identification of them are difficult and no more literatures available.

GRASS BIOLOGY

Grasses are most dominated and evoluted plants of all plant kingdoms. They are economically very important. Use of grasses, as food resources or as fodder has led to extensive breeding programs and improvement in pasture land. Grasses are also very important use as hay, so that various species cultivated in hay fields and for pastures. Many tall grasses, such as Bamboos are used as good construction material and plumbing as well as good raw material for the paper and various other products. Tuft and ornamental grasses are used and appreciated for their durability and beauty throughout the work. Apart from that grasses are used as a green fodder which the most important factor responsible for the success of animal husbandry. Natural grasslands play an important role in supplying fodder to the animals.

Grasses have unique features so they can differentiate easily form the other families. They are cosmopolitan in distribution. The culm is either erect or prostrate. It consists of nodes and internodes. Few grasses have stolons or rhizomes. Leaves have mainly two parts: leaf blade and leaf sheath. At the junction of these two parts there is a ligule present, which is either membranous or hairy or sometime absent. Rather than ligule other appendages may also be present, and auricle and the zone present at the base of leaf blade is called collar. Leaf blade may be lanceolate, linear, ovate or oblong. Its base is rounded, cuneate or cordate and the margin wavy. There is a distinct midrib and on holding the leaf against the light, four or five small veins come in to view. All veins run parallel from the base to the apex. Leaf sheaths protect the meristems. On the basis of leaf anatomy grasses are divided into two physiological groups: C_3 and C_4 plants. C_4 type grasses usually have kranz anatomy. Radiate mesophyll is characteristics of C_4 grasses, but C_4 photosynthesis has multiple independent origins in the Poaceae (Kellogg, 2000; Giussani et al., 2001).

The grass inflorescence appears at the free ends of branches. Length, type, shape varies from species to species. It may be open, contracted, cylindrical or ovate. Spikelet is a unit of inflorescence and depicts variations in glumes, lemma, palea, lodicules, stamens and pistil. It may be sessile or pedicellate. In Andropogoneae the spikelets are born in pairs, one is sessile, which is fertile one and other is pedicellate, which is sterile. Glumes are bracts of spikelet. Shape, size, texture, nature of glumes are used in species identification. Lemma and palea length, shape, width, texture, keeled or not were noticed. Inside the lemma and palea stamens an ovary are present. The lemma and palea are sometime adherent to the surface of the caryopsis.

HABIT

Grasses are either annual or biennial or perennial. Annual means they complete their life cycle, starting from germination to dispersal of seed and death of the mother plant in a year and biennial plants complete their life cycle in two years and perennial are very long lived, they take more than two years for complete their life cycle. In annuals, seed setting is good so that majorly they reproduce by sexually while perennials seed setting is poor so that they are reproduce by vegetative. Out of 100 studied grasses, 40 were perennial and 60 were annual.

INFLORESCENCE

Inflorescence is a portion of a flowering culm upward from the node at the base of the uppermost leaf (Abercrombie et al., 1960; Benson, 1979; Gould, 1968; Smith, 1977). In grasses, true flower parts are inconsequential in identification. The grass inflorescence is a highly modified and complex structure with a different ontogeny. Based on the spikelet arrangement will be used for description of the inflorescence type. Basically three type of inflorescence are there: the panicle, the raceme and the spike.

Panicle

The spikelets are born on branches from the central axis of the inflorescence. Panicles are the most common grass inflorescence. It has two different forms: (i) spreading panicles having varying branch lengths; and (ii) compact panicles have short panicle braches.

Raceme

The spikelets are attached directly to the rachis by a single stalk. There are two forms: (i) digitate raceme, and (ii) multiple racemes.

Spike

The sessile spikelets are attached directly to the central axis of the inflorescence. It may in three forms: (i) solitary spikes have one rachis of spikelets; (ii) digitate spikes have more than one rachis of spikelets; and (iii) multiple spikes have more than one rachis and they form from various points.

GUIDE TO TAXONOMIC RANKS

FAMILY

The grass family, scientifically known as Poaceae or Gramineae is a 5th largest family after Asteraceae, Leguminosae, Orchidaceae and Rubiaceae of flowering plants in the world (Tzevlev and Michaelova, 1989). The family Gramineae distinguished by characteristic morphological features including sheathing, ligulate leaves with distinctive epidermal features and flowers in spikelets, glumes, palea, lemma and caryopses. The grasses are not closely related to other families of monocotyledons.

Grasses range from tiny inconspicuous herbs less than an inch to the giant bamboos that grow up to 130 feet tall. It is difficult to calculate the exact number of species of family Poaceae; however, according to Tzvelev, Poaceae consists of 11,000 species (Osborne, 2010) belonging to 898 genera (1989).

Grasses and their values have been recognized since time immemorial as the present day cereal crops are the cultivated varieties of their wild ancestors. The grasses show high adaptability with respect to changing environments, ability to coexist with grazing animals and with man. They have endless variations with distinct life forms. Grasses grow, reproduce and die back in one short season. Great emphasis is perforce on reproduction, for example, seed production. The plant body is just a few thin leaves; one or two stems but the inflorescence that weight as much as the rest of the plant producing large number of seeds. Clearly, these are organisms whose success lies in their ability to grow when conditions are right. Since the proper conditions for growth and reproduction may be limited to a few short weeks, the grasses have evolved to reproduce as quickly as possible and this makes them one of the most successful terrestrial life forms on the earth (Gosavi, 2010, PhD thesis).

SUBFAMILY

The two main subfamilies in the Poaceae based on the Shah's classification (1978). One is Panicoideae and other one is Pooideae. These two subfamilies classified on the basis of the inflorescence characters.

TRIBE

A tribe is defined on the basis of major differences in spikelet and inflorescence structure and compares to the subfamily group. The tribes are highly natural groups.

GENUS

It is a group of one or more species with features or ancestry (or both) in common. It is the principal category of taxa intermediate in rank between family and species in the nomenclatural hierarchy.

SPECIES

It is a group or populations of individuals sharing common features and/or ancestral generally the smallest groups that can be readily and consistently recognized, often a group of individuals capable of interbreeding and producing fertile offspring.

KEYWORDS

- **Grasses**
- **Habit**
- **Habitat**
- **Inflorescence**
- **Morphology**
- **Species**

REFERENCES

1. Aberuombk, M.; Hkkman, C. J.; Johnson, M. L. *A Dictionary of Biology*. Penguin Books, Harmondworth, England. 1960, 251 p.
2. Benson, L. *Plant Classification*. Znded, D. C. Heath and Co., Lexington, Mass. 1979, 901 p.
3. Clayton, W. D.; Renvoize, S. A. Genera Graminum. *Kew Bulletin Additional Series* 1986, 13, 1–389.
4. Giussani, L. M.; Cota-Sanchez, J. H.; Zuloaga, F. O.; Kellogg, E. A. A molecular phylogeny of the grass subfamily Panicoideae (Poaceae) shows multiple origins of C4 photosynthesis. *American Journal of Botany* 2001, 88, 1993–2012.
5. Gosavi, K. V. C. *Study on Lithophytic grasses of Maharashtra*. PhD thesis, Shivaji University, Kolhapur, 2010.

6. Gould, F. W. *Grass Systematics*. McGraw-Hill, New York. 1968, 382 p.
7. Kellogg, E. A. The grasses: A case study in macroevolution. *Annual Review of Ecology and Systematics* 2000, 31, 217–238.
8. Osborne, C. P.; Visser, V.; Chapman, S.; Baker, A.; Freckleton, R. P.; Salamin, N.; Simpson, D.; Uren, V. GrassPortal: an online ecological and evolutionary data facility. [WWW document]. http://www.grassportal.org. 2011.
9. Prasad, V.; Strömberg, A. E.; Alimohammadian, H.; Sahni, A. Dinosaur coprolites and the early evolution of grasses and grazers. *Science* 2005, 310, 1177–1180.
10. Shah, G. L. *Flora of Gujarat State*. Vol. II. Sardar Patel University: Vidhyanagar, 1978.
11. Smith, J. P. *Vascular Plant Families*. Mad River Press, Eureka, Calif. 1977, 320 p.
12. Strömberg, C. A. E. Evolution of grasses and grassland ecosystems. *Annual Review of Earth and Planetary Sciences* 2011, 39, 517–544.
13. Tzvelev, N. N. The system of grasses (Poaceae) and their evolution. *The Botanical Review* 1989, 55(3), 141–203.
14. Tzvelev, N. N.; Michaelova, V. V. The system of grasses (Poaceae) and their evolution. *Botanical Review* 1989, 55(3), 141–204.
15. www.fsl.orst.edu
16. courses.eeb.utoronto.ca

SECTION 1

CARYOPSES

CONTENTS

1.1 Historical Preview 10
1.2 List of Grass Caryopses Characterized 13
1.3 Morphometric Features 15
1.3.1 The Spikelet and Floret 15
1.3.2 Callus 15
1.3.3 Characteristic Features of Caryopses 17
1.3.3.1 Structure 17
1.3.3.2 Presentation of Data 17
1.3.3.3 Measurements 17
1.3.3.4 Shape 18
1.3.3.5 Embryo 18
1.3.3.6 Embryo Type 30
1.3.3.7 Scutellum Shape 30
1.3.3.8 Hilum 32
1.3.3.9 Hilum Visibility 33
1.3.3.10 Ventral Furrow 33
1.3.3.11 Dorsal/Lateral Striations 33
1.3.3.12 Caryopses Length 34
1.3.3.13 Caryopsis Breadth 34
1.3.3.14 Narrowness and Length: Breadth Ratio 35
1.3.3.15 Compression and Thickness: Breadth Ratio 35
1.3.3.16 Embryo Size (%) 36
1.3.3.17 Sculpturing 37

1.4 Light and Scanning Electron Microscopic Features 84
1.5 Identification Key 144
1.6 Cluster Analysis 150
Keywords 156
References 157

Grasses are very difficult to identify and generally they are identified on the basis of their inflorescences. The grass flower mature into the fruit called caryopsis. Generally Gramineae fruits, also called as grains, can usually be considered also as seeds. Morphological both micro and macro morphological features are known to serve as a diagnostic character specific to the species and is of great taxonomic value.

1.1 HISTORICAL PREVIEW

Caryopses the fruit in grasses show diverse characteristic features. But these characters have been relatively little used by plant taxonomists. Terrel and Peterson (1993) studied the caryopsis morphology of tribe Triticeae. The purpose of their study was to evaluate the morphological structure of the caryopses within Triticeae on the basis of which the tribe Triticeae could be divided into two major subtribes and one monogeneric tribe. Tribe Chlorideae, has been dealt with to address phylogenetic relationship, mainly on the basis of morphological and molecular data (Hilu and Wright, 1982; Hilu and Alice, 2001). Liu et al. (2005b) studied 58 species representing 45 genera of tribe Chlorideae and revealed that Chlorideae allows recognition of three major types of caryopsis on the basis of differences in ventral surface and hilum morphology. Hoagland and Paul compared and differentiated caryopsis of *Oryza sativa* varieties including the weed red rice and several cultivars (1978). Osman et al. (2012) studied the fruit morphology of 33 annual grasses species from Egypt belonging to 24 genera and 11 tribes. The study was based on light and SEM characters; fruit shape, color and surface topography. Osman et al. (2012) diagnosed the seeds majorly on the basis of the three characters: fruit shape and seed surface topography. Zhang et al. (2014) studied the genus *Themeda* and its allied spathaceous genera with an aim to discuss the taxonomic value of caryopsis characters at suprageneric and interspecific level and evaluate the evolutionary tendency of caryopsis. They concluded that micromorphological characters had limited taxonomic value at suprageneric level, while the caryopsis shape, embryo proportion and sculpturing pattern were valuable features at interspecific level.

The three identification guides to grasses are: (i) Udelgard Körber-Grohne's guide to identification of waterlogged European grass caryopses (Körber-Grohne, 1964) on basis of subtle bran and hilum characters. (ii) Mordechai Kislev's Computerized Identification Guide to Near Eastern Grass Seeds mainly based on dimensions of caryopses (Kislev et al., 1997). (iii) Identification guide for Near Eastern Grass Seeds (Nesbitt, 2006) in which 122 genera and 324 species are studied based on presence of genera in the Near East and variability in the shapes are known.

The importance of microstructural pattern analysis of the seed coat observed under stereoscopic and scanning electron microscopy, as a reliable approach for resolving taxonomic problems, has been well recognized (Bogdan, 1966; Heywood, 1971; Barthlott, 1981; Wang Guo and Li, 1986; Koul et al., 2000). These characters, especially the microstructural features of the seed surface, can be useful for assessing phenetic relationships and delimiting taxa at various levels (Hufford, 1995; Karcz et al., 2000). Most systematists agree that data concerning the macro- and microstructure of seeds are very significant for the classification of Angiosperm taxa. Seed morphology has been proven to be a valuable feature in many systematic studies, particularly with the application of SEM (Hill 1976; Mathews and Levins, 1986). Heywood drew attention to the importance and impact of Scanning Electron Microscope in the study of systematic problems by using this technique (1971). Seed SEM characters are relatively consistent for a plant species and may thus prove useful in distinguishing different species and also in grouping them under definite categories (Agrawal, 1984). Few Scanning Electron Microscopic (SEM) studies have been concerned with fine structural differences in taxonomic and morphological features of closely related species, especially within groups of plants of the identical species (Qing Liu et al.,

2005, Joshi et al., 2008). Seed surface are also characterized by a secondary sculpture (Barhlott, 1984). Closely related species often have greater similarities compared to species belonging to different genera and families, which makes differentiation within a genus a difficult task. In such cases, SEM plays a major role (Joshi et al., 2008). During the past decade, SEM studies of small seeds have contributed significantly to the systematic of various angiosperms, for example, Whiffin and Tomb (1972) on Melastomataceae, Hill (1976) on *Mentzelia, Seavey*, Chuang and Heckarel (1972) on *Cordylanthus*, etc.

In this book, 100 grass species listed below have been characterized on the basis of the morphometirc variations in the features.

1.2 LIST OF GRASS CARYOPSES CHARACTERIZED

1. *Coix lachryma-jobi* L.
2. *Chionachne koenigii* (Spr.) Thw.
3. *Andropogon pumilus* Roxb.
4. *Apluda mutica* L.
5. *Arthraxon lanceolatus* (Roxb.) Hochst.
6. *Bothriochloa pertusa* (L.) A. Camus
7. *Capillipedium hugelii* (Hack.) A. Camus
8. *Chrysopogon fulvus* (Spreng.) Chiov.
9. *Cymbopogon martini* (Roxb.) W. Watson
10. *Dichanthium annulatum* (Forssk.) Stapf
11. *Dihcanthium caricosum* (L.) A.Camus
12. *Dimeria orinthopoda* Trin.
13. *Heteropogon contortus* var. *contortus* sub var. *typicus* Blatt. and McCann
14. *Heteropogon contortus* var. *contortus* sub var. *genuinus* Blatt. and McCann
15. *Heteropogon ritcheii* (Hook.f.) Blatt. and McCann
16. *Heteropogon triticeus* (R.Br.) Stapf ex Craib
17. *Ischaemum indicum* (Houtt.) Merr.
18. *Ischaemum molle* Hook.f.
19. *Ischaemum pilosum* (Willd.) Wight
20. *Ischaemum rugosum* Salisb.

21. *Iseilema laxum* Hack.
22. *Ophiuros exaltatus* (L.) Kuntze
23. *Rottboellia exaltata* Linn. f.
24. *Saccharum spontanum* L.
25. *Sehima ischaemoides* Forssk.
26. *Sehima nervosum* (Rottler) Stapf
27. *Sehima sulcatum* (Hack.) A.Camus
28. *Sorghum halepense* (L.) Pers.
29. *Sorghum purpureo-sericeum* (A.Rich.) Schweinf. and Asch.
30. *Thelepogn elegans* Roth
31. *Themeda cymbaria* Hack.
32. *Themeda laxa* A. Camus
33. *Themeda triandra* Forssk.
34. *Themeda quadrivalvis* (L.) Ketz.
35. *Triplopogon ramosissimus* (Hack.) Bor
36. *Vetivaria zizanoides* (L.) Nash
37. *Alloteropsis cimicina* (L.) Stapf
38. *Brachiaria eruciformis* (Sm.) Griseb.
39. *Brachiaria distachya* (L.) Stapf
40. *Brachiaria ramosa* (L.) Stapf
41. *Brachiaria reptans* (L.) C. A. Gardner and C. E. Hubb.
42. *Cenchrus biflorus* Roxb.
43. *Cenchrus ciliaris* L.
44. *Cenchrus setigerus* Vahl
45. *Cenchrus preurii* (Kunth) Maire
46. *Digitaria ciliaris* (Retz.) Koeler
47. *Digitaria microbachne* (Persl) Hern.
48. *Digitaria granularis* (Trin.) Henrard
49. *Echinochloa colona* (L.) Link
50. *Echinochloa crusgalli* (L.) P. Beauv.
51. *Echinochloa stagnina* (Retz.) P. Beauv.
52. *Eriochloa procera* (Retz.) C. E. Hubb.
53. *Oplismenus burmanii* (Retz.) P. Beauv.
54. *Oplismenus composites* (L.) P. Beauv.
55. *Panicum antidotale* Retz.
56. *Panicum miliaceum* L.

57. *Panicum trypheron* Schult.
58. *Paspalidium flavidum* (Retz.) A. Camus
59. *Paspalidium geminatum* (Forssk.) Stapf
60. *Paspalum scrobiculatum* L.
61. *Pennisetum setosum* (Sw.) Rich.
62. *Setaria glauca* (L.) P. Beauv.
63. *Setaria tomentosa* (Roxb.) Kunth
64. *Setaria verticillata* (L.) P. Beauv.
65. *Aeluropus lagopoides* (L.) Thwaites
66. *Isachne globosa* (Thunb.) Kuntze
67. *Aristida adscensionis* L.
68. *Aristida funiculata* Trin. and Rupr.
69. *Perotis indica* (L.) O. Ketz
70. *Chloris barbata* Sw.
71. *Chloris Montana* Roxb.
72. *Chloris virgata* Sw.
73. *Cynadon dactylon* (L.) Pers.
74. *Melanocenchris jaequemontii* Jaub. and Spach
75. *Oropetium villosulum* Stapf ex Bor
76. *Schoenefeldia gracilis* Kunth
77. *Tetrapogon tenellus* (Roxb.) Chiov.
78. *Tetrapogon villosus* Desf.
79. *Dactyloctenium aegyptium* (L.) Willd.
80. *Dactyloctenium sindicum* Boiss.
81. *Dactyloctenium giganteum* Fischer and Schweick.
82. *Desmostachya bipinnata* (L.) Stapf
83. *Dinebra retroflexa* (Vahl) Panz.
84. *Eleusine indica* (L.) Gaertn.
85. *Eleusine verticillata* Roxb.
86. *Eragrostiella bifaria* (Vahl) Bor
87. *Eragrostis cilianensis* (All.) Janch.
88. *Eragrostis ciliaris* (L.) R.Br.
89. *Eragrostis japonica* (Thunb.) Trin
90. *Eragrostis nutans* (Retz.) Nees ex Steud.
91. *Eragrostis pilosa* (L.) P. Beauv.
92. *Eragrostis tenella* (Linn.) P.Beauv. ex Roem.

93. *Eragrostis tremula* Hochst. ex Steud.
94. *Eragrostis unioloides* (Retz.) Nees ex Steud.
95. *Eragrostis viscosa* (Retz.) Trin.
96. *Sporobolus coromardelianus* (Retz.) Kunth
97. *Sporobolus diander* (Retz.) P. Beauv.
98. *Sporobolus indicus* (L.) R.Br.
99. *Urochondra setulosa* (Trin.) C. E. Hubb.
100. *Tragus biflorus* (Roxb.) Schult. (illegimate name)

1.3 MORPHOMETRIC FEATURES

1.3.1 THE SPIKELET AND FLORET

Spikelet is a most obvious unit of the inflorescence (Figure 1). The central axis of the spikelet is the rachilla. It comprises of a pair of glumes, which encloses at its base one or more florets. Floret is the main reproductive unit of the spikelet a reduced or modified flower enclosed by two bracts, the larger outer lemma and the smaller inner palea. The number of florets per spikelet varies widely among the grass species.

Glumes, Lemma and palea are analogous to leaf (Figure 2) having midrib, which varies from obscure to prominent sometimes extending into the awn and veins, which vary in number, location and prominence.

1.3.2 CALLUS

Callus is hard projection at the base of the floret, spikelet or inflorescence segment, indicating a disarticulation point. It is a proximal end of the diaspore (i.e., A reproductive plant part, such as a seed, fruit, or spore, that is modified for dispersal). Generally, it is restricted to those grasses where a diaspore is either floret or spikelet. In genera like *Eragrostis, Sporobolus,* caryopses itself behave as a diaspore while in grasses *Cenchrus* and *Sorghum* group of spikelet behaves as a diaspore and grasses like *Rottbelia,* inflorescence breaks into segments with embedded spikelets. Callus shape

FIGURE 1 Grass intact spikelet (adapted from www.fsl.orst.edu).

FIGURE 2 Grass floret (adapted from courses.eeb.utoronto.ca).

is useful for the separating genus and species. It has different shapes and it may be hard, sharp and penetrating, blunt, etc.

1.3.3 *CHARACTERISTIC FEATURES OF CARYOPSES*

1.3.3.1 Structure

Angiospserm fruit consists of fruit wall and seed. The grass fruit is termed a caroypsis because the pericarp is tightly bounded to the seed wall. In grass caroypses, the pericarp and testa are adherent and must be studied together (Nesbitt, 2006). The distinctiveness extends to the caroypses with their abundant starchy endosperm and laterally placed embryos.

Caroypsis is similar to an achene because both are dry indehiscent fruits. The pericarp is adnate to the seed coat in the caroypsis. Even though the pericarp is free in some grasses, these fruits are not considered achenes, but rather the free pericarp represents a modification of the caroypsis. Identification of grass caryopses (grains or seeds) is complicated by a large number of genera and species in the family, by considerable overlapping in caryopses characteristics of different tribes and genera and by fact that the seeds may be either enclosed in lemma and palea of the floret or naked. Grass seed morphological features and surface patterns have been used in many studies to identify and compare taxa and genera (Benerjee et al., 1981; Bogdan 1965; Colledge 1988; Hillman 1916; Jensen 1957; Matsutani 1986 and Nesbitt 2006, Terrel and Peterson 1993).

1.3.3.2 Presentation of Data

The generic and species name of the studied grasses have been represented according to their tribes. Morphometric data of 100 studied species are presented in Table 1. Based on light microscopic data using Minitab Ver. 16 (Statistical software) cluster analysis was conducted and is represented in Figure 48. Photographs of Light microscopy and scanning electron microscopy of all species are represented in Figures 4–47.

1.3.3.3 Measurements

Measurements carried out as per Nesbitt (2006) are averages (n=15–20). Dimensions are measured parallel to the embryonic axis. Length of caryopses (L) was measured (in mm) parallel to the middle vertical axis included embryo tip, either in dorsal or ventral view. Breadth of caryopses (B) was the maximum width (in mm) on the horizontal axis measured either in dorsal or ventral view. Thickness of caryopses (T) was the maximum width (in mm) measured at right angles to the breadth and in the same horizontal plane, such that T≤B. The length to breadth ratio (L:B) was calculated as the length of caryopses divided by breadth and multiplied by 10. The thickness to breadth ratio (T:B) was calculated as the thickness of caryopses divided by breadth and multiplied by 100. The length of the embryo (from embryo tip to scutellum/endosperm boundary) was calculated as a % of caryopses length (Embryo %). Hilum % was calculated as the length of the hilum for linear hila (measured from base to tip) and for basal and subbasal hila (from base of caryopses to end of hilum) and calculated as a % of caryopses length. All the dimensions details are observed through Stereo Microscope.

All the characteristic morphological and morphometric features are represented in Table 1.

1.3.3.4 Shape

It was difficult to identify the species only on the basis of the shape. Shape of the caryopses ranged from ovate, obovate, oblong, ovoid, linear to fusiform. Species of *Heteropogon* showed a linear shaped caryopses while only *Aristida* sp. has acicular shaped caryopses among the studied species. Shape of *Eragrostis* genus ranges from oblong to ovoid to obovate to orbicular. *E. cilianensis* have orbicular shape while *E. tremula* have round to oblong shape.

1.3.3.5 Embryo

Caryopses can be mainly divided into two parts: embryo and endosperm. Embryo is the basic unit of the caryopses, form where a new emergence or seedling develop and it is rich in protein and oil. Endosperm has a reserved starchy food material. These two parts are separated by means of scutellum. The role of scutellum is to assist the breakdown of starchy endosperm

TABLE 1 Morphometric Features of Caryopses

Sr. No.	Plant name	Length (mm)	Breadth (mm)	Thickness (mm)	L:B Ratio	T:B Ratio	Embryo		Hilum		Embryo %	Hilum %
							Length (mm)	Breadth (mm)	Length (mm)	Breadth (mm)		
Group: Panicoideae												
Tribe: Maydeae												
1.	*Chionachne koenigii*	3.34 ± 0.16	2.21 ± 0.14	1.31 ± 0.12	15.08 ± 0.94	59.24 ± 5.66	2.71 ± 0.32	0.55 ± 0.08	1.01 ± 0.09	0.59 ± 0.05	81.22 ± 6.16	30.41 ± 1.49
2.	*Coix lachryma-jobi*	4.50 ± 0.31	3.51 ± 0.40	2.99 ± 0.58	12.81 ± 1.43	85.14 ± 10.50	3.85 ± 0.44	3.62 ± 0.49	2.41 ± 0.29	0.93 ± 0.12	85.62 ± 11.59	53.50 ± 4.54
Tribe: Andropogoneae												
3.	*Andropogon pumilus*	2.73 ± 0.10	0.40 ± 0.05	0.31 ± 0.03	68.53 ± 7.49	77.77 ± 16.45	1.37 ± 0.09	0.38 ± 0.03	0.25 ± 0.04	0.15 ± 0.03	49.97 ± 3.39	8.99 ± 1.80
4.	*Apluda mutica*	2.20 ± 0.87	0.48 ± 0.36	0.78 ± 0.23	46.11 ± 24.68	162.99 ± 53.14	1.64 ± 0.53	0.75 ± 0.23	0.33 ± 0.11	0.25 ± 0.08	74.58 ± 20.66	14.84 ± 10.09
5.	*Arthraxon lanceolatus*	2.29 ± 0.17	0.48 ± 0.05	0.41 ± 0.03	48.11 ± 5.18	86.36 ± 12.52	1.22 ± 0.20	0.43 ± 0.06	0.22 ± 0.06	0.18 ± 0.03	53.21 ± 5.04	9.40 ± 3.01
6.	*Bothriochloa pertusa*	1.68 ± 0.11	0.75 ± 0.08	0.25 ± 0.01	22.26 ± 3.53	32.64 ± 3.29	1.16 ± 0.09	0.67 ± 0.06	0.24 ± 0.03	0.22 ± 0.02	69.20 ± 5.26	14.38 ± 0.91
7.	*Capillipedium hugelii*	1.39 ± 0.14	0.62 ± 0.12	0.29 ± 0.06	22.53 ± 3.27	46.39 ± 17.42	0.87 ± 0.08	0.45 ± 0.07	0.24 ± 0.05	0.26 ± 0.06	62.76 ± 4.23	17.12 ± 4.43
8.	*Chrysopogon fulvus*	4.89 ± 0.48	0.68 ± 0.08	0.97 ± 0.05	71.89 ± 7.74	142.22 ± 17.25	2.54 ± 0.66	0.68 ± 0.08	0.45 ± 0.08	0.18 ± 0.04	51.91 ± 10.48	9.13 ± 2.60
9.	*Cymbopogon martinii*	2.52 ± 0.11	0.94 ± 0.04	0.43 ± 0.03	26.73 ± 1.79	45.81 ± 2.75	1.68 ± 0.11	0.69 ± 0.04	0.35 ± 0.07	0.25 ± 0.04	66.75 ± 4.24	14.04 ± 2.85

TABLE 1 Continued

Sr. No.	Plant name	Length (mm)	Breadth (mm)	Thickness (mm)	L:B Ratio	T:B Ratio	Embryo		Hilum		Embryo %	Hilum %
							Length (mm)	Breadth (mm)	Length (mm)	Breadth (mm)		
10.	*Dicanthium annulatum*	1.92 ± 0.09	0.76 ± 0.04	0.45 ± 0.03	25.32 ± 1.21	59.59 ± 5.14	1.18 ± 0.15	0.54 ± 0.04	0.22 ± 0.03	0.23 ± 0.02	61.42 ± 6.76	11.47 ± 1.41
11.	*Dicanthium caricosum*	1.88 ± 0.15	0.90 ± 0.05	0.41 ± 0.01	20.90 ± 0.79	46.07 ± 3.44	1.16 ± 0.11	0.66 ± 0.09	0.30 ± 0.03	0.26 ± 0.03	61.47 ± 2.50	15.76 ± 1.49
12.	*Dimeria orinthopoda*	2.33 ± 0.14	0.19 ± 0.03	0.19 ± 0.03	123.17 ± 20.38	102.65 ± 21.73	0.81 ± 0.06	0.18 ± 0.02	0.13 ± 0.02	0.21 ± 0.26	34.99 ± 3.87	5.60 ± 0.98
13.	*Heteropogon contortus var. contortus sub var. typicus*	3.89 ± 0.22	0.39 ± 0.06	0.42 ± 0.03	100.89 ± 12.76	109.38 ± 15.26	2.39 ± 0.20	0.30 ± 0.02	0.32 ± 0.01	0.29 ± 0.01	61.40 ± 3.00	8.13 ± 0.59
14.	*Heteropogon contortus var. contortus sub var. genuinus*	4.28 ± 0.17	0.66 ± 0.02	0.46 ± 0.05	65.06 ± 2.56	70.52 ± 6.21	2.58 ± 0.32	0.39 ± 0.03	0.36 ± 0.03	0.28 ± 0.01	60.34 ± 8.19	8.29 ± 1.00
15.	*Heteropogon ritchiei*	4.48 ± 0.03	0.39 ± 0.04	0.37 ± 0.03	113.18 ± 9.26	94.44 ± 7.37	2.69 ± 0.02	0.29 ± 0.02	0.28 ± 0.03	0.18 ± 0.01	60.06 ± 6.15	6.25 ± 8.73
16.	*Heteropogon triticeus*	10.20 ± 1.11	2.78 ± 0.17	2.46 ± 0.10	36.64 ± 3.00	88.21 ± 3.04	6.86 ± 0.59	2.07 ± 0.13	1.60 ± 0.21	1.34 ± 0.15	67.20 ± 3.63	15.72 ± 1.93
17.	*Ischaemum indicum*	1.58 ± 0.12	0.71 ± 0.04	0.40 ± 0.02	22.02 ± 1.59	56.31 ± 4.69	1.18 ± 0.07	0.68 ± 0.05	0.28 ± 0.04	0.29 ± 0.02	74.67 ± 2.52	17.66 ± 3.49
18.	*Ischaemum molle*	2.13 ± 0.31	0.88 ± 0.06	0.70 ± 0.03	24.41 ± 2.85	79.64 ± 7.13	1.29 ± 0.19	0.74 ± 0.12	0.52 ± 0.09	0.45 ± 0.08	60.59 ± 3.26	24.48 ± 5.27

TABLE 1 Continued

Sr. No.	Plant name	Length (mm)	Breadth (mm)	Thickness (mm)	L:B Ratio	T:B Ratio	Embryo		Hilum		Embryo %	Hilum %
							Length (mm)	Breadth (mm)	Length (mm)	Breadth (mm)		
19.	*Ischaemum pilosum*	2.47 ± 0.15	0.83 ± 0.03	0.69 ± 0.05	29.67 ± 2.06	83.09 ± 7.22	1.40 ± 0.12	0.71 ± 0.03	0.33 ± 0.06	0.26 ± 0.03	56.70 ± 1.79	13.42 ± 2.38
20.	*Ischaemum rugosum*	2.20 ± 0.20	0.99 ± 0.06	0.81 ± 0.07	22.26 ± 2.78	82.29 ± 9.72	1.45 ± 0.19	0.87 ± 0.07	0.41 ± 0.03	0.39 ± 0.08	66.07 ± 4.39	18.55 ± 1.32
21.	*Iseilema laxum*	1.87 ± 0.13	0.69 ± 0.03	0.28 ± 0.01	27.18 ± 1.92	40.66 ± 1.66	1.18 ± 0.14	0.59 ± 0.03	0.31 ± 0.02	0.27 ± 0.01	63.12 ± 3.82	16.75 ± 2.64
22.	*Ophiuros exaltatus*	1.37 ± 0.03	0.66 ± 0.03	0.62 ± 0.03	20.73 ± 0.99	94.01 ± 5.13	0.92 ± 0.05	0.56 ± 0.04	0.24 ± 0.01	0.24 ± 0.02	67.00 ± 4.32	17.82 ± 0.51
23.	*Rottboellia exaltata*	3.73 ± 0.02	1.93 ± 0.02	1.75 ± 0.06	19.31 ± 0.15	90.52 ± 2.98	3.13 ± 0.11	0.74 ± 0.02	0.86 ± 0.04	0.51 ± 0.01	84.03 ± 3.19	22.95 ± 1.03 ±
24.	*Saccharum spontaneum*	1.68 ± 0.07	0.51 ± 0.02	0.28 ± 0.01	32.97 ± 0.90	55.89 ± 2.76	1.10 ± 0.02	0.31 ± 0.02	0.35 ± 0.01	0.24 ± 0.01	65.28 ± 1.59	20.97 ± 1.20
25.	*Sehima ischaemoides*	2.71 ± 0.39	0.88 ± 0.06	0.53 ± 0.11	30.98 ± 4.41	60.35 ± 10.05	1.76 ± 0.28	0.79 ± 0.13	0.39 ± 0.13	0.29 ± 0.04	64.82 ± 2.88	14.27 ± 4.08
26.	*Sehima nervosum*	3.18 ± 0.44	0.95 ± 0.05	0.57 ± 0.11	33.66 ± 2.79	60.22 ± 9.60	2.07 ± 0.35	0.81 ± 0.05	0.60 ± 0.06	0.37 ± 0.02	64.98 ± 8.33	18.94 ± 3.09
27.	*Sehima sulcatum*	4.22 ± 0.23	0.97 ± 0.15	0.39 ± 0.06	43.34 ± 5.17	39.80 ± 8.67	2.44 ± 0.24	0.97 ± 0.12	0.44 ± 0.04	0.29 ± 0.02	57.84 ± 6.00	10.46 ± 1.46
28.	*Sorghum halepense*	2.95 ± 0.32	1.64 ± 0.20	1.16 ± 0.05	18.01 ± 0.54	70.70 ± 10.15	1.94 ± 0.14	1.17 ± 0.08	0.42 ± 0.05	0.40 ± 0.07	65.70 ± 11.86	14.14 ± 2.65
29.	*Sorghum purpureo-sericeum*	4.14 ± 0.08	1.81 ± 0.13	1.21 ± 0.06	22.84 ± 1.81	66.51 ± 5.15	2.75 ± 0.15	1.52 ± 0.16	0.78 ± 0.08	0.55 ± 0.04	66.53 ± 2.53	18.80 ± 2.24

TABLE 1 Continued

Sr. No.	Plant name	Length (mm)	Breadth (mm)	Thickness (mm)	L:B Ratio	T:B Ratio	Embryo		Hilum		Embryo %	Hilum %
							Length (mm)	Breadth (mm)	Length (mm)	Breadth (mm)		
30.	*Thelepogn elegans*	3.66 ± 0.12	1.36 ± 0.06	1.03 ± 0.07	26.94 ± 1.63	76.06 ± 6.05	2.33 ± 0.16	1.23 ± 0.22	0.55 ± 0.10	0.41 ± 0.07	63.69 ± 2.61	14.96 ± 2.38
31.	*Themeda cymbaria*	2.08 ± 0.19	0.75 ± 0.06	0.46 ± 0.03	27.80 ± 4.47	61.24 ± 9.06	1.22 ± 0.10	0.63 ± 0.06	0.32 ± 0.04	0.30 ± 0.05	58.59 ± 9.38	15.48 ± 3.18
32.	*Themeda laxa*	2.82 ± 0.43	0.74 ± 0.05	0.41 ± 0.04	38.18 ± 7.32	55.96 ± 3.57	1.79 ± 0.08	0.56 ± 0.01	0.38 ± 0.04	0.27 ± 0.02	63.21 ± 11.84	13.52 ± 2.92
33.	*Themeda triandra*	3.42 ± 0.37	0.91 ± 0.06	0.69 ± 0.09	37.54 ± 5.43	75.42 ± 10.98	1.83 ± 0.30	0.60 ± 0.06	0.29 ± 0.06	0.22 ± 0.05	53.37 ± 5.12	8.46 ± 1.94
34.	*Themeda quadrivalvis*	3.5 ± 0.02	1.09 ± 0.03	0.39 ± 0.04	31.99 ± 6.36	35.47 ± 3.18	1.69 ± 0.03	0.81 ± 0.04	0.52 ± 0.24	0.31 ± 0.02	48.4 ± 5.20	14.85 ± 1.23
35.	*Triplopogon ramosissimus*	2.28 ± 0.24	0.73 ± 0.11	1.03 ± 0.12	31.17 ± 6.75	140.80 ± 31.21	1.31 ± 0.06	0.65 ± 0.02	0.33 ± 0.04	0.19 ± 0.01	57.39 ± 9.45	14.64 ± 3.19
36.	*Vetivaria zizanioides*	2.41 ± 0.09	0.60 ± 0.03	0.83 ± 0.01	40.37 ± 1.39	138.75 ± 7.00	1.64 ± 0.11	0.51 ± 0.04	0.32 ± 0.02	0.22 ± 0.01	68.13 ± 7.16	13.16 ± 1.03
Tribe: Paniceae												
37.	*Alloteropsis cimicina*	1.76 ± 0.08	1.19 ± 0.06	0.47 ± 0.08	14.85 ± 1.01	39.39 ± 6.63	1.00 ± 0.03	0.62 ± 0.05	0.39 ± 0.04	0.25 ± 0.04	56.80 ± 3.25	22.00 ± 3.49
38.	*Bracharia cruciformis*	1.07 ± 0.03	0.68 ± 0.03	0.38 ± 0.03	15.73 ± 0.80	55.12 ± 5.53	0.64 ± 0.06	0.34 ± 0.09	0.27 ± 0.05	0.24 ± 0.05	59.10 ± 5.12	25.26 ± 4.74
39.	*Bracharia distachya*	1.91 ± 0.01	1.16 ± 0.01	0.54 ± 0.01	16.53 ± 0.09	46.32 ± 0.42	1.34 ± 0.01	0.66 ± 0.01	0.66 ± 0.01	0.31 ± 0.01	70.23 ± 0.34	34.45 ± 0.27

TABLE 1 Continued

Sr. No.	Plant name	Length (mm)	Breadth (mm)	Thickness (mm)	L:B Ratio	T:B Ratio	Embryo		Hilum		Embryo %	Hilum %
							Length (mm)	Breadth (mm)	Length (mm)	Breadth (mm)		
40.	*Bracharia ramosa*	1.63 ± 0.04	1.14 ± 0.06	0.61 ± 0.02	14.31 ± 0.53	53.95 ± 3.52	1.16 ± 0.04	0.55 ± 0.02	0.53 ± 0.02	0.21 ± 0.01	71.50 ± 1.51	32.68 ± 1.91
41.	*Bracharia reptans*	1.94 ± 0.10	1.38 ± 0.09	0.62 ± 0.06	14.07 ± 0.37	44.82 ± 4.76	1.34 ± 0.05	0.58 ± 0.01	0.56 ± 0.01	0.29 ± 0.01	69.00 ± 1.96	28.66 ± 1.84
42.	*Cenchrus biflorus*	1.60 ± 0.26	1.07 ± 0.14	0.76 ± 0.13	14.99 ± 1.19	71.27 ± 8.84	1.20 ± 0.30	0.72 ± 0.08	0.30 ± 0.06	0.35 ± 0.02	74.78 ± 12.63	18.81 ± 4.32
43.	*Cenchrus ciliaris*	2.42 ± 0.07	0.96 ± 0.07	0.54 ± 0.06	25.17 ± 1.75	55.91 ± 7.90	1.37 ± 0.12	0.71 ± 0.08	0.28 ± 0.04	0.24 ± 0.03	56.60 ± 3.68	11.39 ± 1.60
44.	*Cenchrus setigerus*	1.78 ± 0.07	1.09 ± 0.04	0.68 ± 0.04	16.22 ± 0.46	62.07 ± 4.17	1.30 ± 0.09	0.76 ± 0.10	0.28 ± 0.04	0.29 ± 0.05	73.49 ± 2.50	15.77 ± 1.96
45.	*Cenchrus preurii*	1.64 ± 0.08	1.09 ± 0.08	0.60 ± 0.07	15.00 ± 0.83	54.46 ± 8.99	1.31 ± 0.07	0.71 ± 0.08	0.27 ± 0.03	0.26 ± 0.03	79.91 ± 5.11	16.62 ± 1.89
46.	*Digitaria ciliaris*	2.12 ± 0.06	0.93 ± 0.02	0.55 ± 0.04	22.83 ± 0.98	59.26 ± 5.56	0.97 ± 0.06	0.59 ± 0.02	0.55 ± 0.06	0.32 ± 0.04	46.05 ± 3.53	26.16 ± 3.17
47.	*Digitaria microbachne*	1.88 ± 0.03	0.80 ± 0.01	0.49 ± 0.03	23.55 ± 0.20	61.25 ± 3.40	0.99 ± 0.02	0.53 ± 0.01	0.45 ± 0.01	0.25 ± 0.01	52.34 ± 1.15	23.89 ± 0.34
48.	*Digitaria granularis*	0.87 ± 0.02	0.58 ± 0.03	0.43 ± 0.02	14.95 ± 0.83	73.00 ± 4.85	0.43 ± 0.04	0.23 ± 0.01	0.15 ± 0.03	0.11 ± 0.01	49.48 ± 4.93	17.76 ± 3.89
49.	*Echinochloa colonum*	1.52 ± 0.10	1.07 ± 0.09	0.56 ± 0.04	14.21 ± 0.77	52.20 ± 7.81	1.10 ± 0.10	0.63 ± 0.05	0.44 ± 0.03	0.33 ± 0.02	72.24 ± 2.31	29.03 ± 3.32
50.	*Echinochloa crusgalli*	1.83 ± 0.08	1.40 ± 0.05	0.70 ± 0.05	13.06 ± 0.76	50.09 ± 5.25	1.51 ± 0.07	0.81 ± 0.06	0.44 ± 0.06	0.34 ± 0.01	82.96 ± 1.76	24.19 ± 4.21

TABLE 1 Continued

Sr. No.	Plant name	Length (mm)	Breadth (mm)	Thickness (mm)	L:B Ratio	T:B Ratio	Embryo		Hilum		Embryo %	Hilum %
							Length (mm)	Breadth (mm)	Length (mm)	Breadth (mm)		
51.	*Echinochloa stagnina*	2.67 ± 0.14	2.32 ± 0.07	1.54 ± 0.06	11.50 ± 0.36	66.59 ± 3.68	2.07 ± 3.68	1.13 ± 0.08	0.92 ± 0.10	0.90 ± 0.10	77.70 ± 4.90	34.45 ± 3.39
52.	*Eriochloa procera*	1.75 ± 0.05	0.88 ± 0.05	0.46 ± 0.01	20.05 ± 1.27	52.62 ± 2.47	1.04 ± 0.05	0.52 ± 0.02	0.43 ± 0.01	0.25 ± 0.03	59.51 ± 3.38	24.67 ± 0.83
53.	*Oplismenus burmannii*	1.63 ± 0.03	0.83 ± 0.02	0.70 ± 0.02	19.69 ± 0.66	85.17 ± 3.76	0.77 ± 0.09	0.47 ± 0.08	0.71 ± 0.09	0.18 ± 0.02	47.57 ± 5.78	43.53 ± 5.50
54.	*Oplismenus compositus*	1.98 ± 0.09	0.89 ± 0.04	0.3 ± 0.02	22.25 ± 2.44	33.71 ± 5.29	1.18 ± 0.03	0.58 ± 0.04	0.38 ± 0.05	0.14 ± 0.02	59.49 ± 2.46	19.19 ± 6.16
55.	*Panicum antidotale*	1.65 ± 0.03	0.95 ± 0.04	0.73 ± 0.01	17.43 ± 0.60	76.67 ± 3.01	0.96 ± 0.07	0.73 ± 0.06	0.35 ± 0.03	0.27 ± 0.05	58.27 ± 5.23	21.10 ± 2.12
56.	*Panicum trypheron*	1.70 ± 0.03	1.24 ± 0.06	0.73 ± 0.02	13.73 ± 0.50	58.78 ± 3.62	1.27 ± 0.06	0.61 ± 0.01	0.54 ± 0.02	0.27 ± 0.04	74.37 ± 2.65	31.74 ± 0.92
57.	*Panicum miliaceum*	2.04 ± 0.03	1.62 ± 0.07	0.92 ± 0.07	12.58 ± 0.60	56.93 ± 6.70	1.63 ± 0.05	0.93 ± 0.07	0.70 ± 0.07	0.24 ± 0.04	79.97 ± 1.85	34.56 ± 2.96
58.	*Paspalidium flavidum*	1.34 ± 0.05	1.19 ± 0.04	1.02 ± 0.06	11.30 ± 0.53	86.04 ± 6.32	1.12 ± 0.07	0.60 ± 0.03	0.51 ± 0.02	0.26 ± 0.05	83.62 ± 7.66	37.72 ± 2.03
59.	*Paspalidium geminatum*	1.25 ± 0.03	1.15 ± 0.05	0.47 ± 0.01	10.91 ± 0.50	40.77 ± 2.10	0.65 ± 0.01	0.47 ± 0.02	0.36 ± 0.02	0.23 ± 0.04	52.00 ± 1.16	28.45 ± 1.12
60.	*Paspalum scrobiculatum*	2.17 ± 0.06	2.06 ± 0.02	1.34 ± 0.08	10.53 ± 0.29	65.22 ± 3.93	1.12 ± 0.05	0.90 ± 0.10	0.94 ± 0.04	0.27 ± 0.02	51.61 ± 2.56	43.55 ± 2.72
61.	*Pennisetum setosum*	1.65 ± 0.11	0.59 ± 0.03	0.24 ± 0.01	28.16 ± 2.46	41.07 ± 1.73	0.76 ± 0.06	0.40 ± 0.03	0.28 ± 0.03	0.24 ± 0.01	46.19 ± 5.27	17.08 ± 1.93

TABLE 1 Continued

Sr. No.	Plant name	Length (mm)	Breadth (mm)	Thickness (mm)	L:B Ratio	T:B Ratio	Embryo		Hilum		Embryo %	Hilum %
							Length (mm)	Breadth (mm)	Length (mm)	Breadth (mm)		
62.	*Setaria glauca*	1.59 ± 0.09	1.00 ± 0.04	0.51 ± 0.05	15.90 ± 0.48	51.10 ± 5.36	1.18 ± 0.20	0.62 ± 0.06	0.28 ± 0.03	0.23 ± 0.02	74.19 ± 11.28	17.43 ± 1.32
63.	*Setaria tomentosa*	1.29 ± 0.10	0.90 ± 0.08	0.50 ± 0.03	14.40 ± 1.37	55.97 ± 2.24	0.88 ± 0.09	0.51 ± 0.03	0.28 ± 0.02	0.21 ± 0.01	67.75 ± 5.46	21.89 ± 1.53
64.	*Setaria verticillata*	1.54 ± 0.02	1.06 ± 0.02	0.65 ± 0.02	14.48 ± 0.39	60.71 ± 2.33	1.19 ± 0.02	0.57 ± 0.02	0.34 ± 0.03	0.25 ± 0.02	77.39 ± 2.15	21.86 ± 2.12
Group: Pooideae												
Tribe: Aeluropodeae												
65.	*Aeluropus lagopoides*	0.91 ± 0.03	0.59 ± 0.03	0.32 ± 0.03	15.41 ± 0.42	54.76 ± 4.75	0.56 ± 0.04	0.35 ± 0.03	0.29 ± 0.02	0.17 ± 0.05	61.37 ± 3.34	32.01 ± 2.69
Tribe: Isachneae												
66.	*Isachne globosa*	1.06 ± 0.04	0.93 ± 0.09	0.62 ± 0.04	11.39 ± 0.88	66.61 ± 7.90	0.26 ± 0.02	0.42 ± 0.04	0.72 ± 0.03	0.10 ± 0.01	24.68 ± 1.79	68.15 ± 4.23
Tribe: Aristideae												
67.	*Aristida adscensionis*	5.05 ± 0.60	0.26 ± 0.03	0.28 ± 0.03	192.84 ± 36.99	108.55 ± 10.53	0.76 ± 0.14	0.30 ± 0.18	0.54 ± 0.08	0.21 ± 0.06	15.02 ± 2.84	10.67 ± 2.16
68.	*Aristida funiculata*	3.77 ± 0.21	0.52 ± 0.05	0.53 ± 0.08	71.92 ± 8.06	100.99 ± 20.50	1.84 ± 0.27	0.42 ± 0.01	0.27 ± 0.04	0.17 ± 0.02	48.84 ± 6.22	7.13 ± 0.83

TABLE 1 Continued

Sr. No.	Plant name	Length (mm)	Breadth (mm)	Thickness (mm)	L:B Ratio	T:B Ratio	Embryo		Hilum		Embryo %	Hilum %
							Length (mm)	Breadth (mm)	Length (mm)	Breadth (mm)		
Tribe: Perotideae												
69.	*Perotis indica*	1.74 ± 0.06	0.29 ± 0.02	0.31 ± 0.04	60.92 ± 3.98	106.85 ± 14.93	0.58 ± 0.06	0.21 ± 0.02	0.17 ± 0.03	0.14 ± 0.01	33.40 ± 3.93	9.96 ± 2.06
Tribe: Chlorideae												
70.	*Chloris barbata*	1.42 ± 0.08	0.52 ± 0.06	0.39 ± 0.07	27.50 ± 3.11	75.85 ± 15.77	0.96 ± 0.08	0.32 ± 0.06	0.14 ± 0.03	0.12 ± 0.01	67.81 ± 6.04	9.70 ± 1.98
71.	*Choris montana*	1.45 ± 0.05	0.64 ± 0.02	0.53 ± 0.02	22.58 ± 0.42	82.92 ± 1.46	1.14 ± 0.04	0.48 ± 0.02	0.20 ± 0.02	0.13 ± 0.02	78.13 ± 1.46	13.48 ± 0.72
72.	*Chloris virgata*	1.70 ± 0.11	0.50 ± 0.05	0.52 ± 0.06	33.97 ± 4.60	103.03 ± 9.59	1.28 ± 0.10	0.44 ± 0.04	0.15 ± 0.01	0.12 ± 0.03	74.92 ± 1.70	8.62 ± 0.49
73.	*Cynadon dactylon*	1.35 ± 0.05	0.55 ± 0.06	0.71 ± 0.03	24.70 ± 3.60	130.03 ± 16.20	0.67 ± 0.09	0.34 ± 0.02	0.28 ± 0.07	0.23 ± 0.05	49.87 ± 7.41	21.11 ± 4.79
74.	*Melanocenchris jaequemontii*	1.65 ± 0.06	0.64 ± 0.02	0.26 ± 0.01	25.95 ± 0.84	41.31 ± 2.31	0.76 ± 0.07	0.42 ± 0.02	0.19 ± 0.01	0.17 ± 0.01	46.22 ± 5.50	11.25 ± 0.92
75.	*Oropetium villosulum*	0.83 ± 0.01	0.19 ± 0.02	0.22 ± 0.01	44.36 ± 4.33	116.87 ± 12.97	0.28 ± 0.02	0.13 ± 0.01	0.05 ± 0.01	0.08 ± 0.01	33.39 ± 2.15	5.44 ± 0.58
76.	*Sachoenefeldia gracilis*	1.87 ± 0.05	0.40 ± 0.01	0.56 ± 0.02	46.35 ± 0.86	139.42 ± 5.16	1.03 ± 0.06	0.32 ± 0.02	0.18 ± 0.02	0.13 ± 0.02	55.36 ± 3.62	9.37 ± 1.00
77.	*Tetrapogon tenellus*	2.32 ± 0.12	0.88 ± 0.04	1.07 ± 0.07	26.31 ± 35.53	121.04 ± 3.72	1.40 ± 0.03	0.50 ± 0.03	0.43 ± 0.03	0.24 ± 0.04	60.28 ± 2.26	18.57 ± 6.24
78.	*Tetrapogon villosus*	2.81 ± 0.18	0.80 ± 0.07	0.72 ± 0.06	35.35 ± 4.93	90.32 ± 13.47	1.63 ± 0.10	0.68 ± 0.06	0.29 ± 0.05	0.24 ± 0.01	58.12 ± 4.73	10.19 ± 1.83

TABLE 1 Continued

Sr. No.	Plant name	Length (mm)	Breadth (mm)	Thickness (mm)	L:B Ratio	T:B Ratio	Embryo		Hilum		Embryo %	Hilum %
							Length (mm)	Breadth (mm)	Length (mm)	Breadth (mm)		
Tribe: Eragrosteae												
79.	*Acrachne racemosa*	0.84 ± 0.07	0.52 ± 0.03	0.41 ± 0.06	16.12 ± 1.49	79.76 ± 13.81	0.46 ± 0.05	0.33 ± 0.03	0.11 ± 0.01	0.12 ± 0.01	54.36 ± 6.51	12.63 ± 1.51
80.	*Dactyloctenium aegyptium*	0.92 ± 0.05	0.46 ± 0.03	0.73 ± 0.05	19.87 ± 0.56	157.24 ± 17.51	0.43 ± 0.06	0.30 ± 0.05	0.15 ± 0.01	0.13 ± 0.03	46.66 ± 4.32	16.33 ± 1.63
81.	*Dactyloctenium sindicum*	1.01 ± 0.07	0.51 ± 0.04	0.70 ± 0.06	19.65 ± 1.14	137.39 ± 2.27	0.37 ± 0.05	0.29 ± 0.03	0.14 ± 0.01	0.11 ± 0.01	36.42 ± 6.70	13.87 ± 0.85
82.	*Dactyloctenium giganteum*	1.00 ± 0.04	0.52 ± 0.02	0.86 ± 0.01	19.19 ± 0.49	165.70 ± 7.46	0.44 ± 0.01	0.32 ± 0.02	0.12 ± 0.01	0.12 ± 0.01	44.17 ± 1.85	11.60 ± 1.27
83.	*Desmostachya bipinnata*	0.75 ± 0.15	0.39 ± 0.05	0.48 ± 0.05	19.08 ± 3.84	121.60 ± 15.76	0.34 ± 0.07	0.25 ± 0.02	0.10 ± 0.02	0.11 ± 0.03	44.65 ± 15.23	13.13 ± 3.88
84.	*Dinebra retroflexa*	1.34 ± 0.04	0.44 ± 0.03	0.45 ± 0.02	30.59 ± 2.25	103.75 ± 12.32	0.59 ± 0.05	0.30 ± 0.03	0.23 ± 0.04	0.18 ± 0.05	43.87 ± 4.38	17.31 ± 2.50
85.	*Eleusine indica*	1.35 ± 0.13	0.61 ± 0.03	0.58 ± 0.05	22.10 ± 2.29	95.34 ± 7.38	0.41 ± 0.02	0.35 ± 0.03	0.20 ± 0.02	0.17 ± 0.03	30.63 ± 4.35	14.90 ± 2.52
86.	*Eragrostiella bifaria*	0.66 ± 0.02	0.39 ± 0.02	0.40 ± 0.02	16.75 ± 1.15	101.37 ± 6.25	0.26 ± 0.02	0.28 ± 0.01	0.09 ± 0.01	0.08 ± 0.01	39.47 ± 2.03	13.71 ± 0.93
87.	*Eragrostis cilianensis*	0.49 ± 0.01	0.40 ± 0.02	0.41 ± 0.02	12.25 ± 0.55	102.32 ± 2.83	0.17 ± 0.03	0.17 ± 0.03	0.09 ± 0.01	0.09 ± 0.01	35.58 ± 6.24	17.73 ± 1.11
88.	*Eragrostis ciliaris*	0.47 ± 0.01	0.26 ± 0.02	0.26 ± 0.01	18.45 ± 0.72	100.70 ± 5.32	0.18 ± 0.02	0.16 ± 0.01	0.08 ± 0.01	0.07 ± 0.01	38.88 ± 5.43	17.83 ± 1.50

TABLE 1 Continued

Sr. No.	Plant name	Length (mm)	Breadth (mm)	Thickness (mm)	L:B Ratio	T:B Ratio	Embryo Length (mm)	Embryo Breadth (mm)	Hilum Length (mm)	Hilum Breadth (mm)	Embryo %	Hilum %
89.	*Eragrostis japonica*	0.47 ± 0.01	0.23 ± 0.01	0.25 ± 0.01	20.46 ± 0.79	106.89 ± 4.45	0.15 ± 0.03	0.13 ± 0.04	0.07 ± 0.01	0.07 ± 0.01	31.91 ± 6.38	15.30 ± 2.82
90.	*Eragrostis nutans*	0.53 ± 0.01	0.24 ± 0.02	0.23 ± 0.02	22.34 ± 2.16	98.64 ± 11.58	0.24 ± 0.03	0.15 ± 0.02	0.08 ± 0.01	0.06 ± 0.01	45.64 ± 5.28	14.57 ± 0.64
91.	*Eragrostis pilosa*	0.74 ± 0.05	0.34 ± 0.01	0.42 ± 0.01	21.50 ± 1.70	120.89 ± 5.05	0.46 ± 0.03	0.18 ± 0.01	0.10 ± 0.01	0.11 ± 0.01	61.75 ± 5.77	13.33 ± 1.60
92.	*Eragrostis tenella*	0.55 ± 0.02	0.30 ± 0.02	0.29 ± 0.02	18.19 ± 1.39	96.33 ± 9.81	0.26 ± 0.01	0.16 ± 0.02	0.06 ± 0.01	0.05 ± 0.01	47.82 ± 3.34	11.69 ± 2.03
93.	*Eragrostis tremula*	0.55 ± 0.03	0.38 ± 0.02	0.50 ± 0.02	14.48 ± 1.42	131.36 ± 9.38	0.27 ± 0.03	0.25 ± 0.01	0.07 ± 0.01	0.07 ± 0.01	49.82 ± 5.76	13.46 ± 0.78
94.	*Eragrostis unioloides*	0.69 ± 0.04	0.19 ± 0.01	0.40 ± 0.03	35.51 ± 3.67	204.66 ± 17.35	0.33 ± 0.04	0.17 ± 0.02	0.09 ± 0.01	0.07 ± 0.01	48.76 ± 4.35	12.55 ± 1.86
95.	*Eragrostis viscosa*	0.25 ± 0.01	0.14 ± 0.01	0.17 ± 0.01	17.33 ± 0.86	117.20 ± 6.85	0.16 ± 0.03	0.09 ± 0.01	0.06 ± 0.01	0.05 ± 0.01	64.08 ± 15.37	24.46 ± 3.38
Tribe: Sporoboleae												
96.	*Sporobolus coromardelianus*	0.96 ± 0.05	0.28 ± 0.05	0.60 ± 0.03	34.65 ± 4.93	217.12 ± 42.34	0.47 ± 0.03	0.14 ± 0.03	0.16 ± 0.01	0.10 ± 0.01	48.60 ± 1.95	16.26 ± 0.93
97.	*Sporobolus diander*	1.09 ± 0.11	0.46 ± 0.03	0.61 ± 0.03	23.46 ± 1.37	132.18 ± 11.84	0.55 ± 0.06	0.33 ± 0.02	0.18 ± 0.03	0.16 ± 0.02	50.64 ± 7.66	16.13 ± 1.07
98.	*Sporobolus indicum*	0.49 ± 0.02	0.22 ± 0.01	0.22 ± 0.01	22.47 ± 1.62	101.10 ± 8.04	0.25 ± 0.01	0.14 ± 0.01	0.04 ± 0.01	0.03 ± 0.01	50.33 ± 0.42	8.31 ± 0.85

TABLE 1 Continued

Sr. No.	Plant name	Length (mm)	Breadth (mm)	Thickness (mm)	L:B Ratio	T:B Ratio	Embryo Length (mm)	Embryo Breadth (mm)	Hilum Length (mm)	Hilum Breadth (mm)	Embryo %	Hilum %
99.	*Urochondra setulosa*	0.69 ± 0.06	0.38 ± 0.02	0.44 ± 0.03	17.98 ± 2.26	115.15 ± 11.38	0.34 ± 0.02	0.23 ± 0.01	0.14 ± 0.01	0.08 ± 0.01	50.23 ± 4.29	20.85 ± 1.29
Tribe: Zoysieae												
100.	*Tragrus biflorus*	1.50 ± 0.11	0.57 ± 0.04	0.40 ± 0.04	26.51 ± 3.37	71.60 ± 11.91	0.80 ± 0.06	0.39 ± 0.05	0.21 ± 0.01	0.19 ± 0.01	53.09 ± 5.44	14.30 ± 1.01

*L = Length, B = Breadth, T = Thickness, L:B = (Length/Breadth) × 10, T:B = (Thickness/Breadth) × 100.

and transfer of breakdown products to the embryo axis during germination. Embryo axis is smaller than the scutellum and is visible in the form of a band.

The embryos in the different species can be recognized by their differences in color, presence of furrows, ridges or other diagnostic features. In many genera the embryo is confined to the lower part of the caryopses, whereas in many others it extends about halfway up. In still others it reaches entire length extending from the base to the tip of the grain.

1.3.3.6 Embryo Type

Generally embryo axis is surrounded by the scutellum (Figure 3). Sometimes clear and prominent embryo axis is seen which is surrounded by band of scutellum and sometimes scutellum band is narrow or indistinct with an indistinct embryo axis. Based on these characters embryo could be categorized into two types:

(i) N type: clearly defined embryo axis and scutellum; and
(ii) L type: 'blob-like' embryo without clear embryo axis and invisible or very narrow scutellum.

83 species has 'N' type of embryo while only 17 species have 'L' type of embryo. In Paniceae only two species, for example, *Digitaria granularis* and *Oplismenus burmanii* have 'L' embryo type. In Pooideae, Eragrostreae and Sporoboleae all the genera have an 'N' type of embryo.

1.3.3.7 Scutellum Shape

Two types of shapes were observed:

(i) 'V' shaped; and
(ii) sickle shaped.

Among the studied species 32 species have 'V' type of scutellum and 68 species have sickle shaped scutellum. In Andropogoneae, genera *Chryosopogon, Dimeria* and *Vetiveria* have 'V' shaped scutellum while

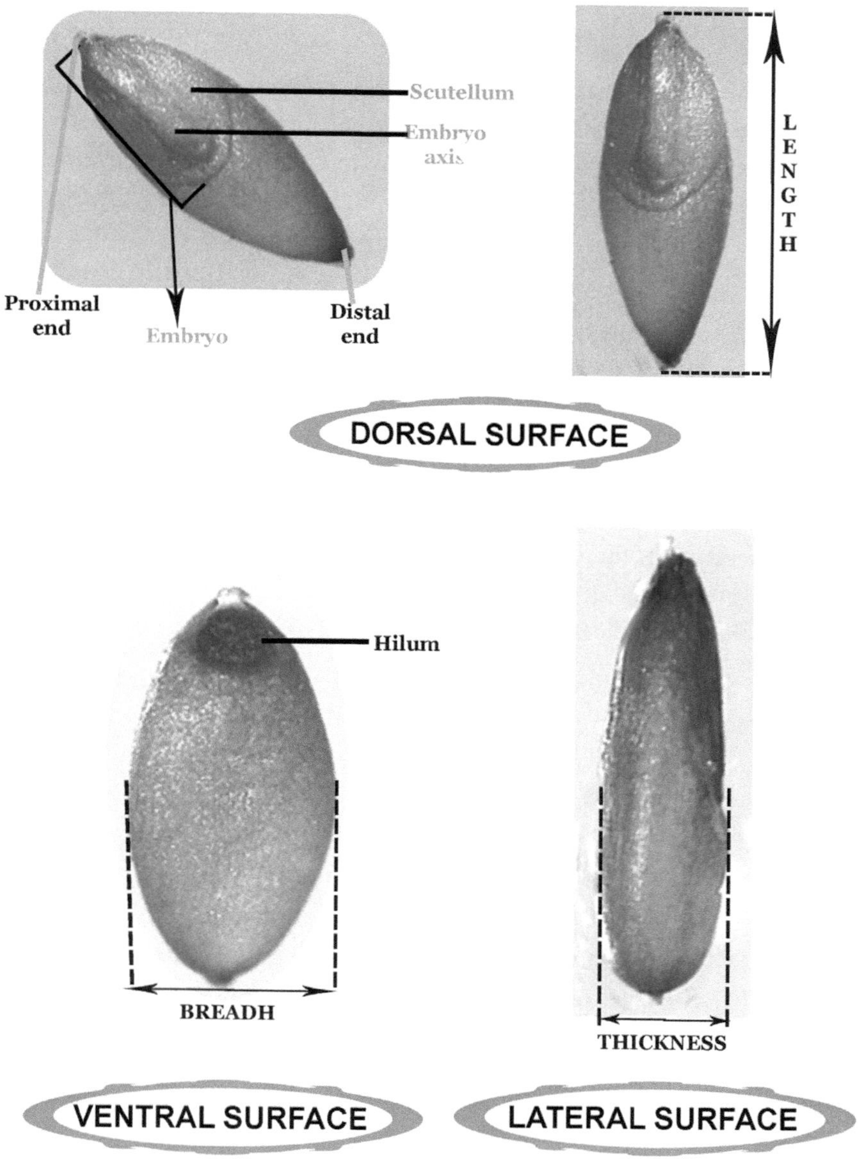

FIGURE 3 Grass caryopses features.

the rest of the species have sickle shaped scutellum while in Sporoboleae and most of Eragrosteae members Sickle shaped scutellum was observed.

1.3.3.8 Hilum

Hilum is the attachment scar of the funicle. It refers only to the scar left on the seed by the stalk of the ovule (Guest, 1966). In grasses, hilum is present on ventral surface and towards the proximal end (Figure 3).

Nesbitt divided the type of hilum into eight groups, which are follows (2006):

(i) Linear (long type): narrow, over 50% of the caryopsis length;
(ii) Linear (short type): narrow, under 50% of the caryopsis length;
(iii) Basal – oval type;
(iv) Basal – circular;
(v) Basal – fan shaped: obovate or wedge shaped, usually in a distinct depression;
(vi) Basal – 'V' shaped: a 'V' shaped thickening projecting from the bottom of the ventral side; exact morphology unclear;
(vii) Basal – linear; and
(viii) Basal – unclear.

Based on observations of the studied species hilum could be categorized on the basis of:

(I) Position of hilum:

(i) Basal
(ii) Sub basal
(iii) Linear

(a) Long (extending throughout the length of caryopses)
(b) Short (extending upto 3/4th of the caryopses length)

(II) Shape of the hilum:

(i) Oval
(ii) Fan shaped
(iii) Circular
(iv) Linear
(v) 'V' shaped

Position of the hilum varied in the species. All the Andropogoneae and Paniceae members have either basal or sub basal hilum. Among the studied species, 64 species have basal hilum and 26 species have sub basal hilum.

Species *Saccharum spontaneum, Eriochloa procera, Oplismenus burmannii, Oplismenus composites* and *Panicum miliare* have a basal linear hilum. Chlorideae and Eragrosteae also have a basal hilum. Only *Isachne globosa, Aristida adscensionis* and *Aristida funiculata* have linear hilum.

Shape of the hilum also varied among the studied species with 30 species having oval shaped hilum, most of them belonging to Paniceae. Most of the Andropogoneae members have fan shaped or 'V' shaped hilum and all the Sporoboleae genera have 'V' shaped hilum. In Eragrosteae most of the genera have a circular hilum.

1.3.3.9 Hilum Visibility

The visibility of hilum varied in the different species from being prominent to faint. The visibility of the hilum was prominent in the caryopses, which had raised and linear hila. In some of the species, for example, *Themeda, Apluda, Coix* due to the darker intensity in color, the hilum could be clearly distinguished from the remaining portion of the caryopses. In species like *Echinochloa, Setaria,* etc., the color of the hilum was faint and could not be clearly discriminated from the remaining portion.

1.3.3.10 Ventral Furrow

Generally when the hilum is basal and linear, longitudinal furrow appears. In studied species only 8 species have ventral furrow, with only one species having a linear hilum (*Isachne globosa*), 4 species (*Sehima ischaemoides, Sehima nervosum, Sehima sulcatum* and *Dinebra retroflexa*) with a basal hilum and 3 species have sub basal hilum (*Cymbopogon martini, Acrachne racemosa* and *Eleusine indica*). Paniceae and Chlorideae members show an absence of ventral furrow.

1.3.3.11 Dorsal/Lateral Striations

Striations represent the imprint on nerves of the palea or lemma. Among the studied species 55 species do not have any striations on dorsal or lateral surface and 45 species shows striations.

1.3.3.12 Caryopses Length

Length of the caryopses varied among and within all the studied tribes. There is a wide range of variation in the length varying from 0.4–10.20 mm (Figure 49). Among the studied species, caryopses of Panicoideae group showed large variation in length and it ranges from 0.87 mm – 10.20 mm. In this group, mainly three tribes were covered, namely, Maydeae, Andropogoneae and Paniceae. The length of caryopses was found to be maximum in the tribe Andropogoneae (*Heteropogon triticeus)* and minimum in the tribe Eragrosteae (*Eragrostis ciliaris, Eleusine indica*).

In Maydeae, among studied species *Coix lachrymal-jobi* showed maximum length (4.50) mm while *Chionachne Koengii* has minimum length (3.34) mm.

In Tribe Paniceae members, caryopses length varies between 0.87 mm and 2.67 mm with maximum in *Digitaria granularis* and minimum in *Echinochloa stagnina*, respectively.

Of the eight tribes studied under this group Aleuropoideae and Ischaneae has single genus with 0.91mm length (*Aleuropus lagopoides*) and 1.06 mm length (*Isachne globosa*), respectively. Aristideae tribe with two species of *Aristida* varies with *Aristida adscensionis* (5.05 mm) and *Aristida funiculata* (3.77 mm). In tribe Chlorideae and Eragrostideae the length varies between 0.83 mm and 1.35 mm.

Large variations in length aided in classifying the studied species into three groups:

(i) short (entirely <3 mm);
(ii) long (entirely > 5 mm);
(iii) medium (falling between 3.1 and 4.9 mm).

1.3.3.13 Caryopsis Breadth

Range of variation in the breadth of the caryopses in Panicoideae (0.19–3.51 mm) was more compared to that of Pooideae (0.14–0.93 mm) (Figure 50). Among Pooideae in tribe Eragrostideae the breadth varies from 0.47 mm to 0.61 mm and in tribe Sporoboleae from 0.22 mm to 0.44 mm. Aleuropoideae has 0.59 mm breadth and Ischaneae has 0.93 mm breadth while in the other four tribes it ranges between 0.14 mm and 0.88 mm.

1.3.3.14 Narrowness and Length: Breadth Ratio

Narrowness of the caryopses is evaluated by length and breadth ratio. Length is very important measurement for this ratio. According to Nesbitt depending upon the increase in numerical value of the ratio narrowness of the caryopses increases, for example, if the L:B ratio is 10 than caryopses has equal length and breadth, if L:B ratio is 20 than caryopses has twice long as broad, if L:B ratio is 40 or above than caryopses looks distinctly narrow and if L:B ratio is 100 than caryopses ten times long as broad.

Among the studied species 59 species had ratio below 25, 26 species have ratio between 25 and 40, 11 species between 40 and 100 and only 4 species have ratio above 100 (Figure 52). *Heteropogon contortus* var. *contortus* sub var. *typicus, Heteropogon ritchiei, Dimeria orinthopoda* and *Aristida adscensionis* have L:B ratio above 100 categorizing them to have caryopses ten times long as broad also indicating them to be have the most narrow seeds among the studied species.

The L:B ratio of the caryopses was found to be maximum in the tribe Andropogoneae (*Heteropogon ritchii*) and minimum in the tribe Paniceae (*Paspalum scorbiculatum*). Caryopses of Panicoideae members showed large variation in ratio (10.53–123.17) having caryopses of varying narrowness. In Maydeae the members have caryopses with more or less equal length and breadth (L:B ratio ranging between 12.81 and 15.08). Group Pooideae members had L:B ratio ranging between 11.39 and 192.84 with caryopses varying narrowness.

1.3.3.15 Compression and Thickness: Breadth Ratio

Compression is one of the valuable diagnostic features for characterization of grass caryopses. Caryopses are compressed or non-compressed. If compressed than either dorsally or laterally.

(i) If T:B ratio is < 100: Dorsally compressed;
(ii) If T:B ratio > 100: Laterally compressed;
(iii) If T:B ratio is ~ 100: Non compressed.

Among the studied species, 24 are laterally compressed, 66 species are dorsally compressed while only10 species are not compressed.

Thickness of caryopses was found to be maximum in the tribe Maydeae of Panicoideae (*Coix lachryma-jobi*) and minimum in the tribe Eragrosteae (*Eragrostis viscosa*) of pooideae group (Figure 51). Caryopses of Panicoideae members showed variation in thickness (0.25–2.99 mm) with a maximum in *Bothriochloa pertusa* and minimum in *Coix lachryma-jobi,* respectively.

Group Pooideae members showed caryopses thickness ranging between 0.17 mm – 1.07 mm with a minimum thickness in *Eragrostis viscosa* of Eragrosteae and maximum in *Tetrapogon tenellus* of chlorideae respectively.

When difference between thickness and breadth becomes more then the ratio will be come more and vice versa. The T:B ratio of the caryopses was found to be maximum in the tribe Sporoboleae (*Sporobolus coromardelianus*) of Pooideae and minimum in the tribe Andropogoneae (*Apluda* mutica) of Panicoideae (Figure 53).

Group Pooideae members showed the range of T:B ratio 41.31–217.12 varying between with minimum in *Melanocenchris jaquemontii* of tribe Chlorideae and maximum in *Sporobolus coromardelianus* of tribe Sporoboleae respectively.

Among the studied species T:B ratio is maximum in *Sporobolus coramendelianus* (217.12), *Apluda mutica* (162.99), *Sachoenefeldia gracilis* (139.42) and *Aristida funiculata* (100.99) indicating it to be distinctly flattened having breadth greater than the thickness.

1.3.3.16 Embryo Size (%)

Embryo size is related caryopses length on the basis of which embryo percentage can be calculated. Depending on the embryo percentage caryopses could be categorized into two classes:

(i) embryo % 46 or over: large embryo class; and
(ii) embryo % 45 or under: short embryo class.

Among the studied species 85 species belong to the large embryo class and 15 species belong to the short embryo class (Figure 54). Most of the Panicoideae members belong to the large embryo class except *Dimeria orinthopoda,* which has the embryo percentage 34.99% and so belonging to short embryo class. Among the Pooideae most of the Eragrosteae members belongs to the short embryo class, while all the others have a large embryo class.

Along with the maximum percentage embryo size, the percentage of area occupied by hilum was also found to be maximum in *Coix lachrymal-jobi*. The percentage embryo size was found to be minimum in the *Aristida adscensionis* (15.20 %) but the minimum percentage of area occupies by the hilum was found to be in *Aristida funiculata* (7.13%) (Figure 55).

1.3.3.17 Sculpturing

Genera *Dactyloctenium* and *Eleusine* could be easily identified with naked eye due to the prominent wrinkled sculpturing observed on the surface of the caryopses. The other species had a smooth surface without any sculpturing under light microscopy. For the light microscopy total 11 different morphological features like shape, color, texture, compressions, striations, ventral groove, scutellum shape, embryo type and class, Hilum shape and type have been taken into account along with the features noted in the different views of caryopses, for example, dorsal, ventral and lateral shown in Figures 4–13.

Scanning electron microscopic study was used to depict surface features of dorsal, ventral and lateral surfaces, features of embryo and hilum. The diagnostic features are supplemented with Figures 14–47. Most of the members of grass caryopses have reticulate type of pattern on their surface. This reticulate pattern has either undulating or straight wall. Pattern of undulations also varied: wavy or '∩' shaped, 'Ω' shaped or '∧' shaped. The undulating wall may be either smooth angled, sharp angled, broadly undulated or narrowly undulated. Between the reticulum interspace is present which is shallow, concave or sunken, depending upon the elevation of the wall. Apart from reticulate pattern, ruminate, blister or rugose patterns are also seen. Hilum of most of species have ruminate pattern with folded walls traversing in same or different directions.

Both *Acrachne racemosa* and *Eleusine indica* had compressed caryopses with sickle shaped scutellum, sub basal raised circular hilum but the class of embryo was different, for example, short in the former and large in the later species. SEM observations revealed *Acrachne racemosa* with compound reticulate pattern with blister on the surface while *Eleusine indica* had reticulate pattern with Ω shaped undulations. The two species showed close relationship with genus *Dactyloctenium* belonging to the same tribe Eragrosteae. All the three species of *Dactyloctenium* studied; had reticulate pattern with different type of undulations making them easily identifiable. In *Eragrostis* color of mature caryopses varied from light to dark brown with a smooth shining surface. Shape of the caryopses in *E. japonica, E. nutans, E. tenella, E. unioloides* and *E. viscosa* varied from oblong to ovate. A common feature noted was the absence of the ventral groove/sulcus on the caryopses. Apart from this, dorsal/lateral striations were present only in *E. cilianensis, E. tremula* and *E. unioloides* whereas other species had no striations. Moreover caryopses of *E. pilosa, E. tremula* and *E. unioloides* were compressed laterally while caryopses of the others were terete. Peterson and Sanchez reported *E. japonica, E. pilosa* and *E. ciliaris* as dorsally compressed however we found that *E. ciliaris* and *E. japonica* were terete but *E. pilosa* was laterally compressed (2007). Caryopses in most of the species were sticky (Kreitschitz et al., 2009) because of the presence of surface slime cells, giving the surfaces a shiny and translucent appearance that was difficult to observe under a light microscope but very apparent under SEM. *Eragrostis cilianensis* and *E. tremula* both had a sickle-shaped scutellum whereas all other species had a 'V' shaped scutellum. *Eragrostis pilosa* showed a contrasting feature in having a reticulate-foveate surface with thick rugae. Two unique diagnostic features observed in *E. tremula* were superimposed rows of reticulum on the lateral surface and globular slimy glands at the proximal end on the ventral surfaces. The upper reticulum was pentagonal to hexagonal with a smooth, thick, and elevated tangential wall in *E. tremula*, while the reticulum was elongated rectangular with smooth thin undulating walls in *E. pilosa*.

Tribe Paniceae members have dorsally compressed caryopses except *Oplismenus burmannii* which has uncompressed caryopses. *Panicum* and *Setaria* species have 'V' shaped scutellum with large embryo class but *Panicum* and *Setaria* has a basal and sub basal oval shaped hilum respectively. Costea et al. (2002) reported *Echinochloa colonum* to have 'W' or 'S' shaped undulation and *Echinochloa crusgalli* 'Ω' shaped undulations.

In the present work *Echinochloa colonum* had '∩' and '∧' shaped undulation on dorsal, ventral and lateral surface while *Echinochloa crusgalli* showed 'Ω' shaped undulations on ventral, and '∩' and '∧' shaped and narrow undulation with smooth angle on the dorsal and ventral surface. *Echinochloa stagnina* had reticulate pattern with indistinct undulating walls and a pitted surface.

Qing et al. (2005) studied 45 genera and 58 species of grasses from the Tribe Chlorideae and revealed that Chlorideae allows recognition of three major types of caryopses on the basis of the differences in ventral surface and hilum morphology. *Chloris* in the present study had variation in fusiform to ellipsoid shape of caryopses with smooth and shiny surface. They also had basal hilum and ventral groove. Qing et al. (2005) reported striate pattern (without pubescence) in *Chloris barbata* but present study showed a reticulate pattern in *Chloris* species.

Terminologies used to describe the features were adapted from Barthlott (1981), Murley (1951) and Koul et al. (2000). The terms used are as follows:

1. Blister: Swelling structure on the surface;
2. Convoluted: Curled, wound or twisted together or overlapped margins;
3. Crimpled: Compressed into small folds;
4. Foveate: Pitted or having depressions marked with little pits;
5. Foveolate: Marked with little shallow pits;
6. Ocellate: Having eye-like depressions, each with a raised circular border;
7. Punctate: Marked with dots/tiny spots;
8. Reticulate: Having a raised network of narrow and sharply angled lines frequently presenting a geometric appearance, each area outlined by a reticulum being an interspace;
9. Reticulate-foveate: A type intermediate between reticulate and foveate types;
10. Rugose: Wrinkled, the irregular elevations making up the wrinkles and running mostly in one direction;
11. Ruminate: Penetrated by irregular channels giving an eroded appearance and running in different directions;
12. Sulcate: Grooved or furrowed with long V-formed depressions;
13. Undulate: Having a wave like pattern;
14. Verrucate: Irregular projections or knobs.

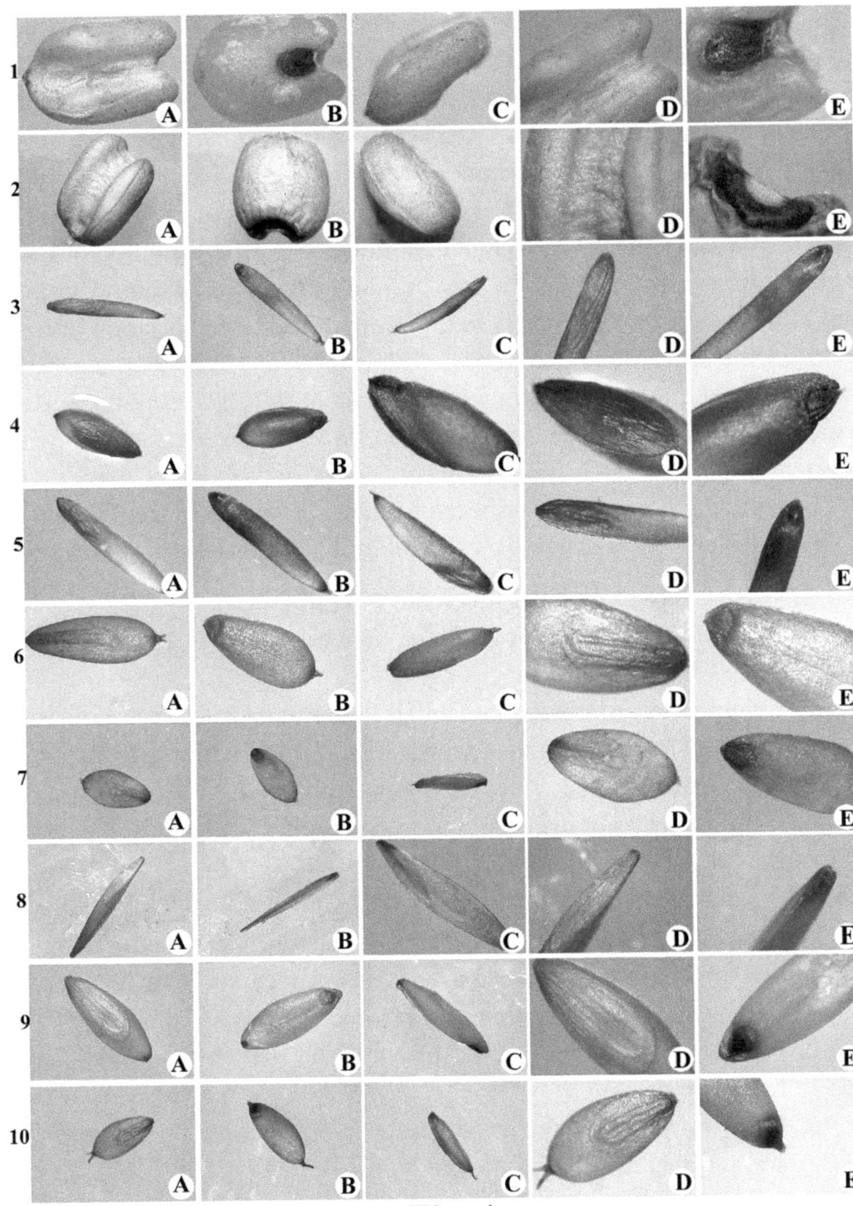

FIGURE 4 Light microscopic study of the grass caryopses (A. Dorsal surface; B. Ventral surface; C. Lateral surface; D. Embryo; E. Hilum; 1. *Chionachne koenigii*; 2. *Coix lachryma-jobi*; 3. *Andropogon pumilus*; 4. *Apluda mutica*; 5. *Arthraxon lanceolatus*; 6. *Bothriochloa pertusa*; 7. *Capillipedium hugelii*; 8. *Chrysopogon fulvus*; 9. *Cymbopogon martinii*; 10. *Dicanthium annulatum*).

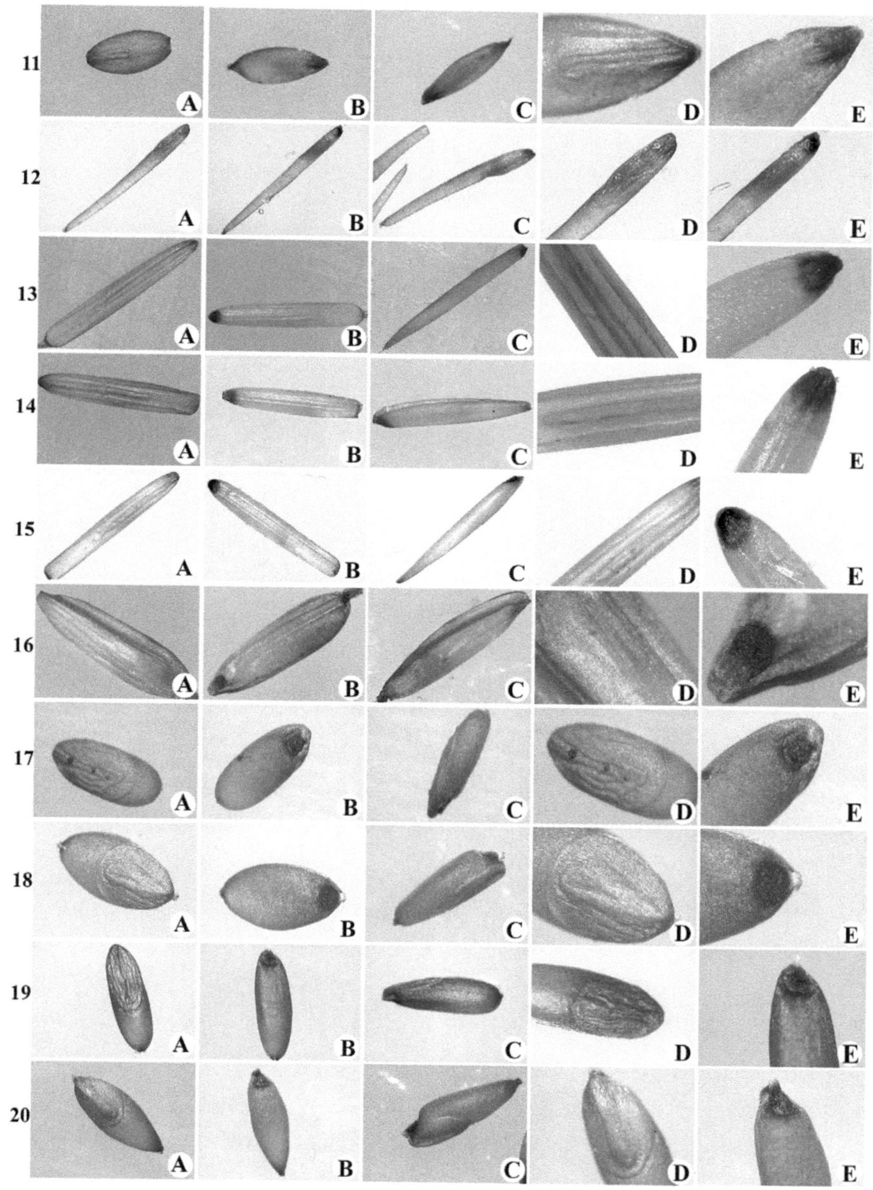

FIGURE 5 Light microscopic study of the grass caryopses (A. Dorsal surface; B. Ventral surface; C. Lateral surface; D. Embryo; E. Hilum; 11. *Dicanthium caricosum*; 12. *Dimeria orinthopoda*; 13. *Heteropogon contortus var. contortus sub var. typicus;* 14. *Heteropogon contortus var. contortus sub var. genuinus*; 15. *Heteropogon ritchiei*; 16. *Heteropogon triticeus*; 17. *Ischaemum indicum*; 18. *Ischaemum molle*; 19. *Ischaemum pilosum*; 20. *Ischaemum rugosum*).

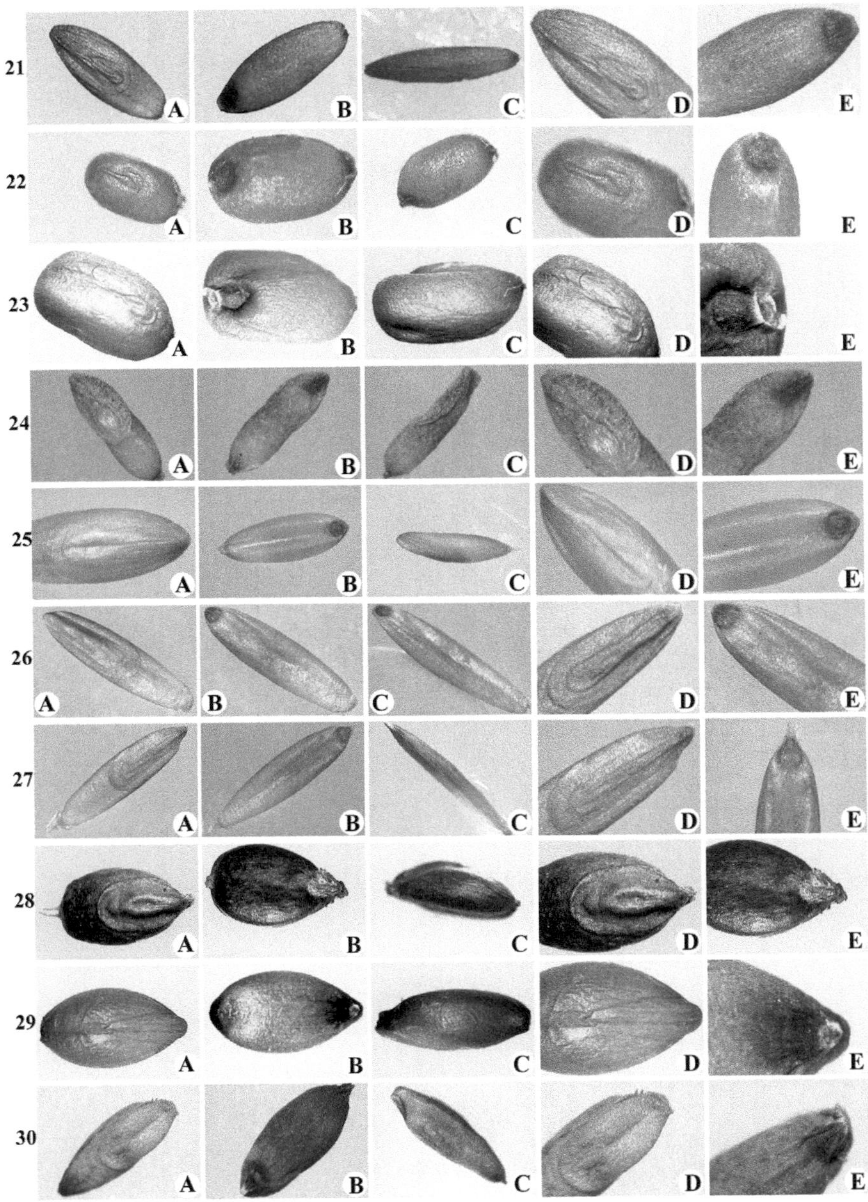

FIGURE 6 Light microscopic study of the grass caryopses (A. Dorsal surface; B. Ventral surface; C. Lateral surface; D. Embryo; E. Hilum; 21. *Iseilema laxum*; 22. *Ophiuros exaltatus*; 23. *Rottboellia exaltata*; 24. *Saccharum spontaneum*; 25. *Sehima ischaemoides*; 26. *Sehima nervosum*; 27. *Sehima sulcatum*; 28. *Sorghum halepense*; 29. *Sorghum purpureo-sericeum*; 30. *Thelepogn elegans*).

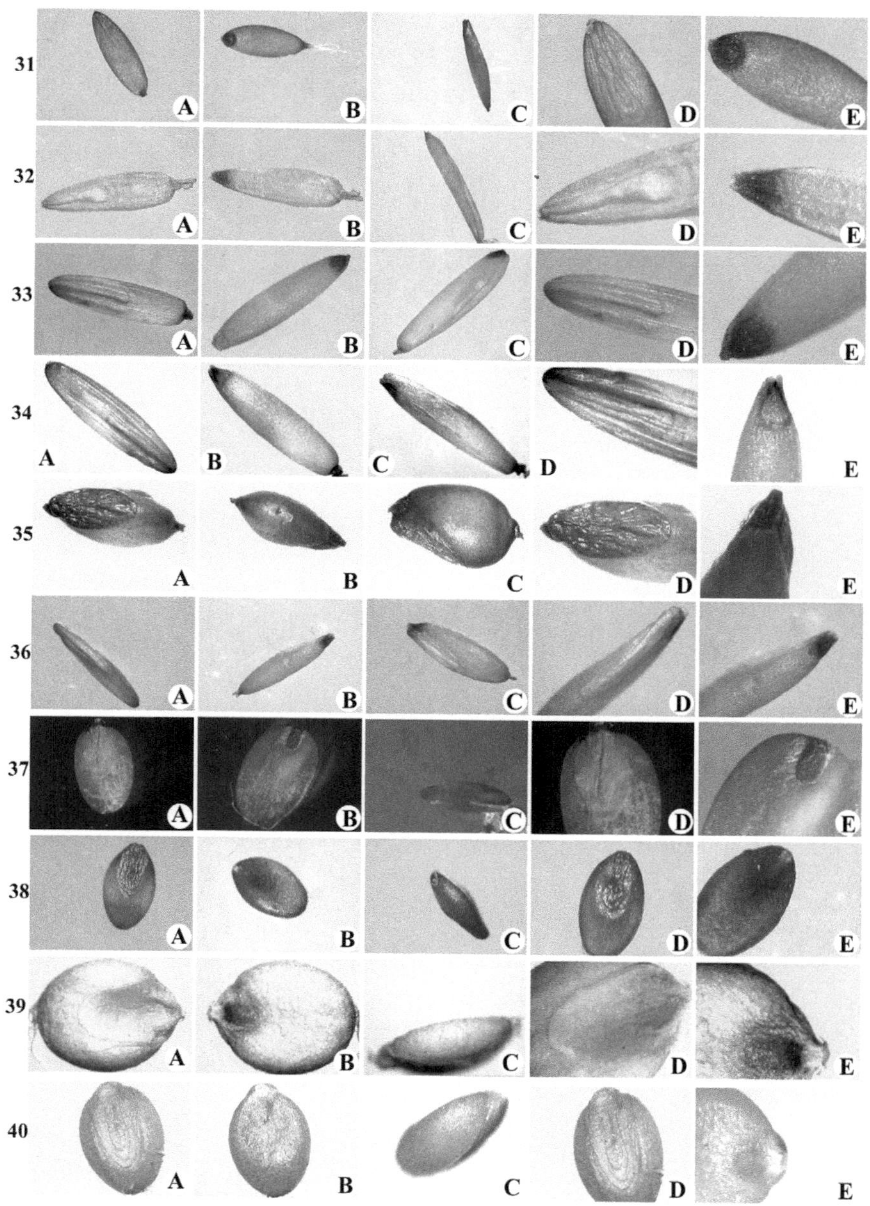

FIGURE 7 Light microscopic study of the grass caryopses (A. Dorsal surface; B. Ventral surface; C. Lateral surface; D. Embryo; E. Hilum; 31. *Themeda cymbaria*; 32. *Themeda laxa*; 33. *Themeda triandra*; 34. *Themeda quadrivalvis*; 35. *Triplopogon ramosissimus*; 36. *Vetivaria zizanioides*; 37. *Alloteropsis cimicina*; 38. *Bracharia cruciformis*; 39. *Bracharia distachya*; 40. *Bracharia ramosa*).

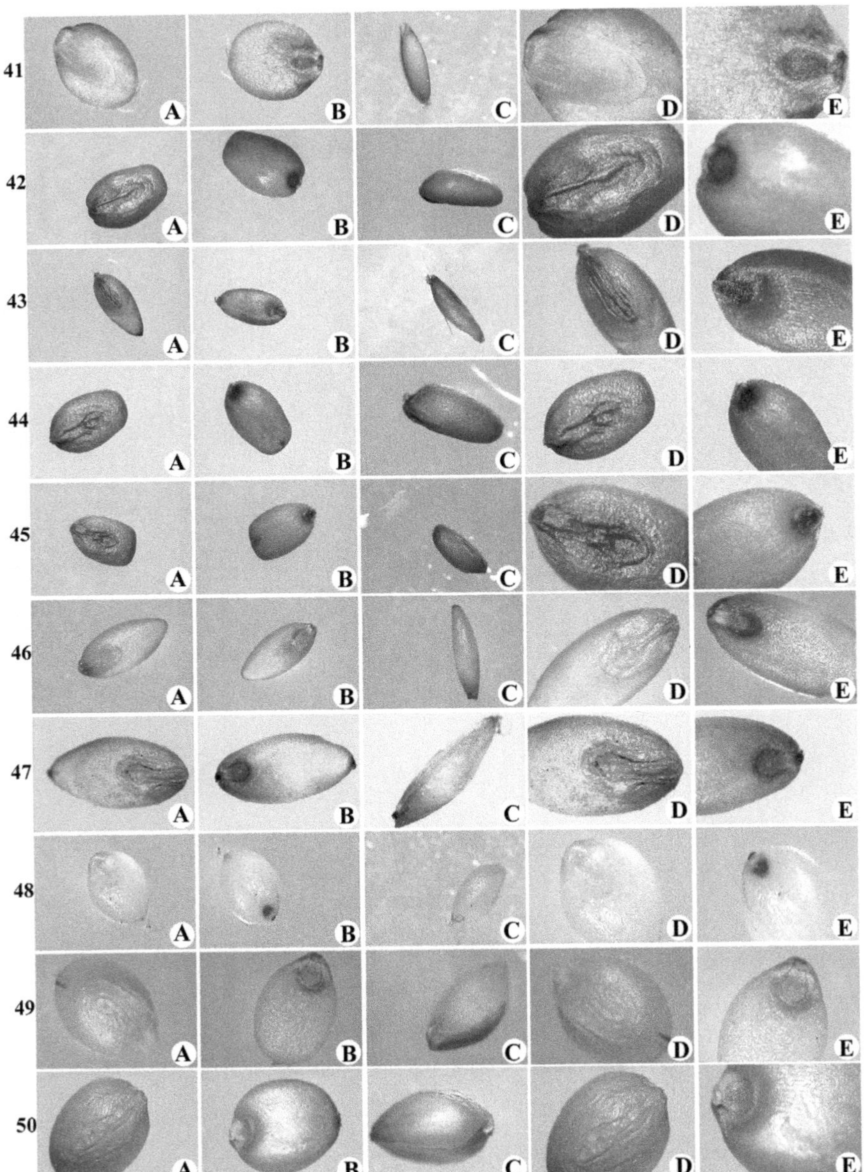

FIGURE 8 Light microscopic study of the grass caryopses (A. Dorsal surface; B. Ventral surface; C. Lateral surface; D. Embryo; E. Hilum; 41. *Bracharia reptans*; 42. *Cenchrus biflorus*; 43. *Cenchrus ciliaris*; 44. *Cenchrus setigerus*; 45. *Cenchrus preurii*; 46. *Digitaria adscendens*; 47. *Digitaria ciliaris*; 48. *Digitaria granularis*; 49. *Echinochloa colonum*; 50. *Echinochloa crusgalli*).

51 52 53 54 55 56 57 58 59 60

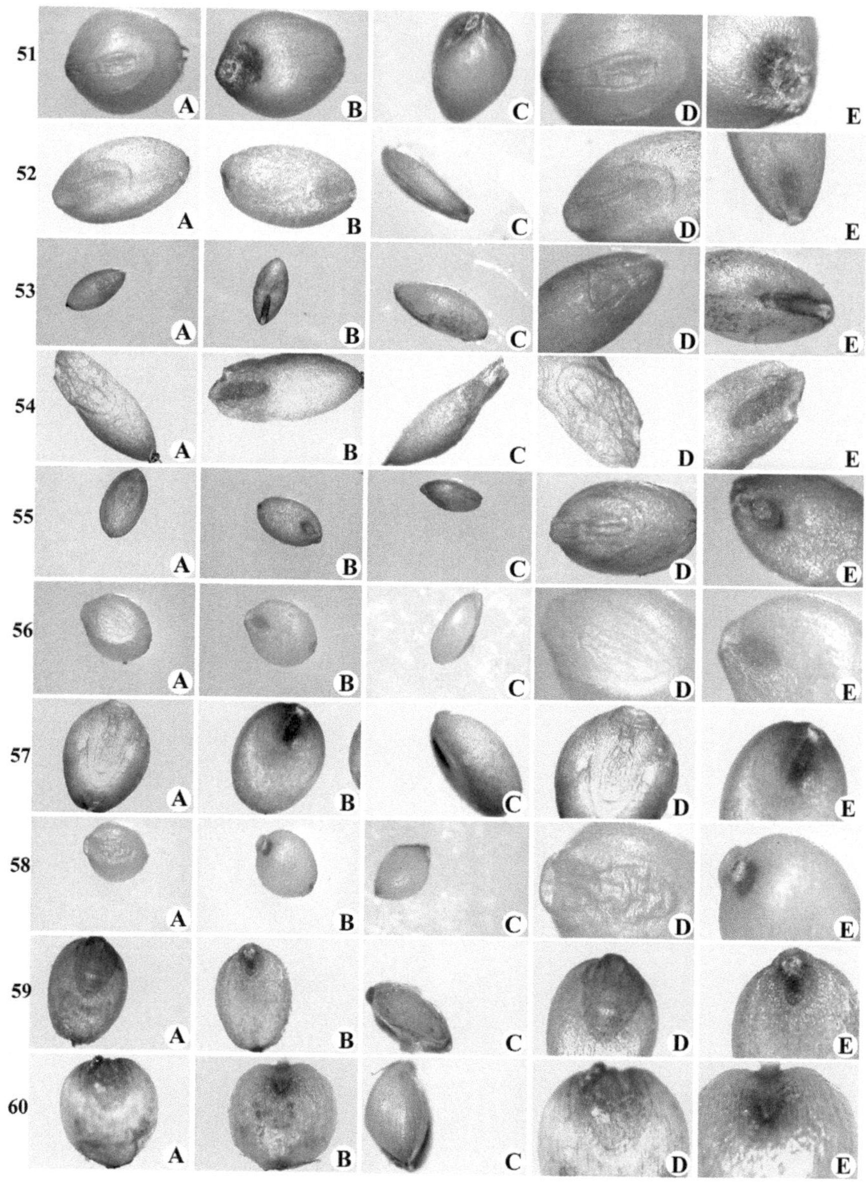

FIGURE 9 Light microscopic study of the grass caryopses (A. Dorsal surface; B. Ventral surface; C. Lateral surface; D. Embryo; E. Hilum; 51. *Echinochloa stagnina*; 52. *Eriochloa procera*; 53. *Oplismenus burmannii*; 54. *Oplismenus composites*; 55. *Panicum antidotale*; 56. *Panicum trypheron*; 57. *Panicum miliaceum*; 58. *Paspalidium flavidum*; 59. *Paspalidium geminatum*; 60. *Paspalum scrobiculatum*).

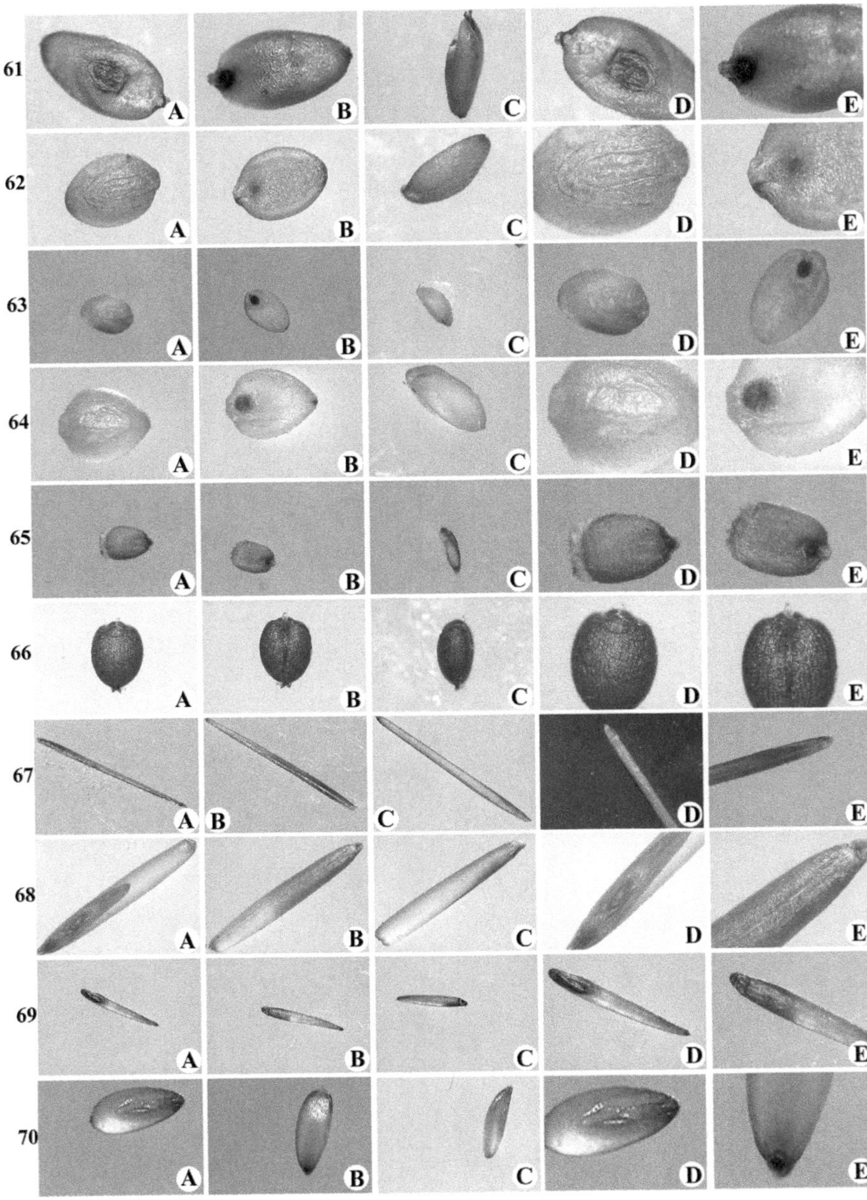

FIGURE 10 Light microscopic study of the grass caryopses (A. Dorsal surface; B. Ventral surface; C. Lateral surface; D. Embryo; E. Hilum; 61. *Pennisetum setosum*; 62. *Setaria glauca*; 63. *Setaria tomentosa*; 64. *Setaria verticillata*; 65. *Aeluropus lagopoides*; 66. *Isachne globosa*; 67. *Aristida adscensionis*; 68. *Aristida funiculate*; 69. *Perotis indica*; 70. *Chloris barbata*).

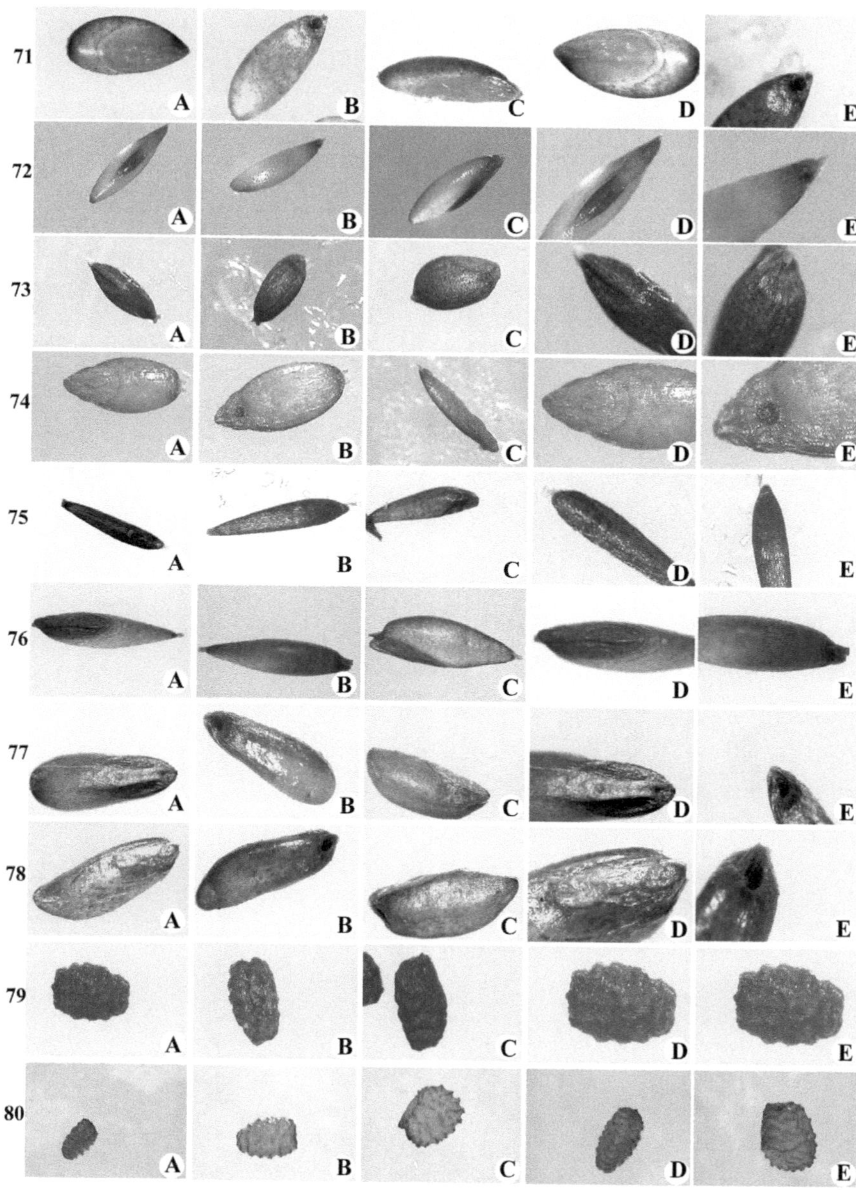

FIGURE 11 Light microscopic study of the grass caryopses (A. Dorsal surface; B. Ventral surface; C. Lateral surface; D. Embryo; E. Hilum; 71. *Choris Montana*; 72. *Chloris virgata*; 73. *Cynadon dactylon*; 74. *Melanocenchris jaequemontii*; 75. *Oropetium villosulum*; 76. *Sachoenefeldia gracilis*; 77. *Tetrapogon tenellus*; 78. *Tetrapogon villosus*; 79. *Acrachne racemosa*; 80. *Dactyloctenium aegyptium*).

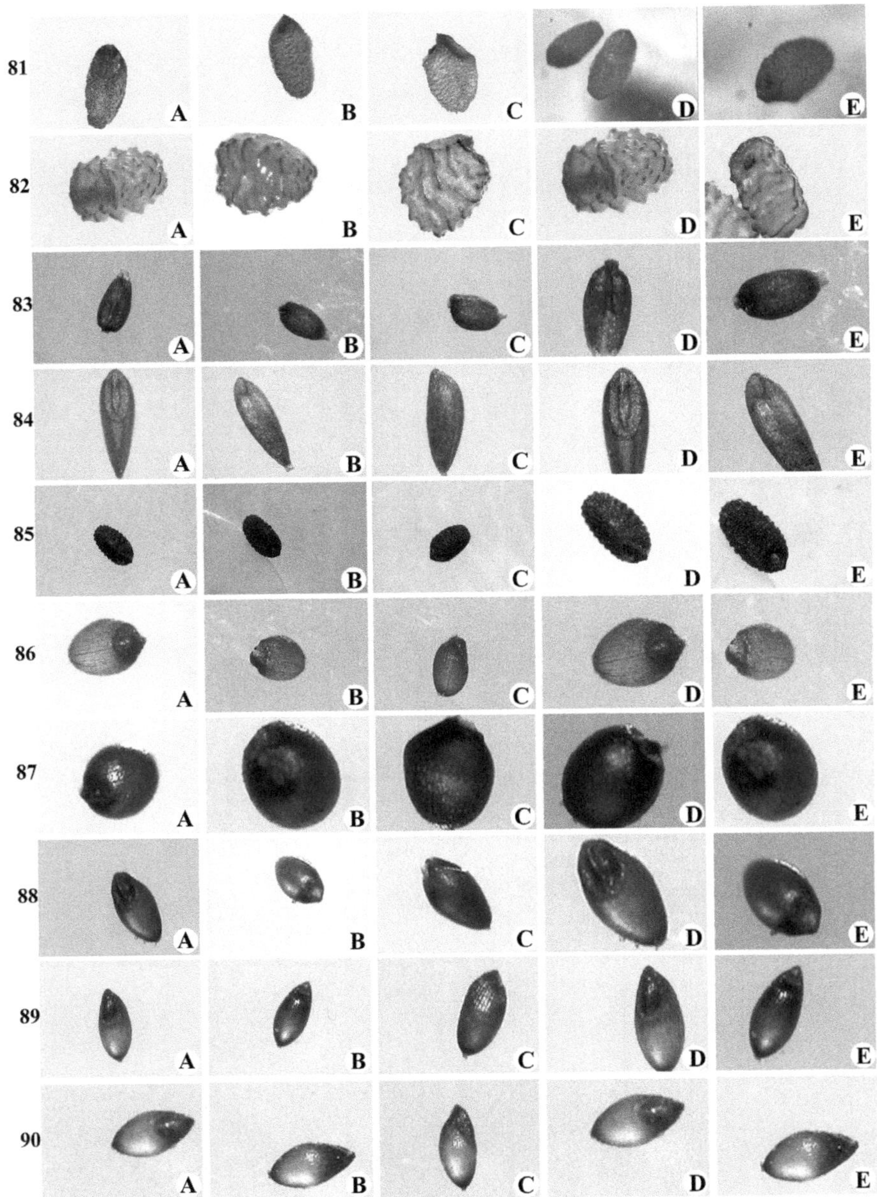

FIGURE 12 Light microscopic study of the grass caryopses (A. Dorsal surface; B. Ventral surface; C. Lateral surface; D. Embryo; E. Hilum; 81. *Dactyloctenium sindicum*; 82. *Dactyloctenium giganteum*; 83. *Desmostachya bipinnata*; 84. *Dinebra retroflexa*; 85. *Eleusine indica*; 86. *Eragrostiella bifaria*; 87. *Eragrostis cilianensis*; 88. *Eragrostis ciliaris*; 89. *Eragrostis japonica*; 90. *Eragrostis nutans*).

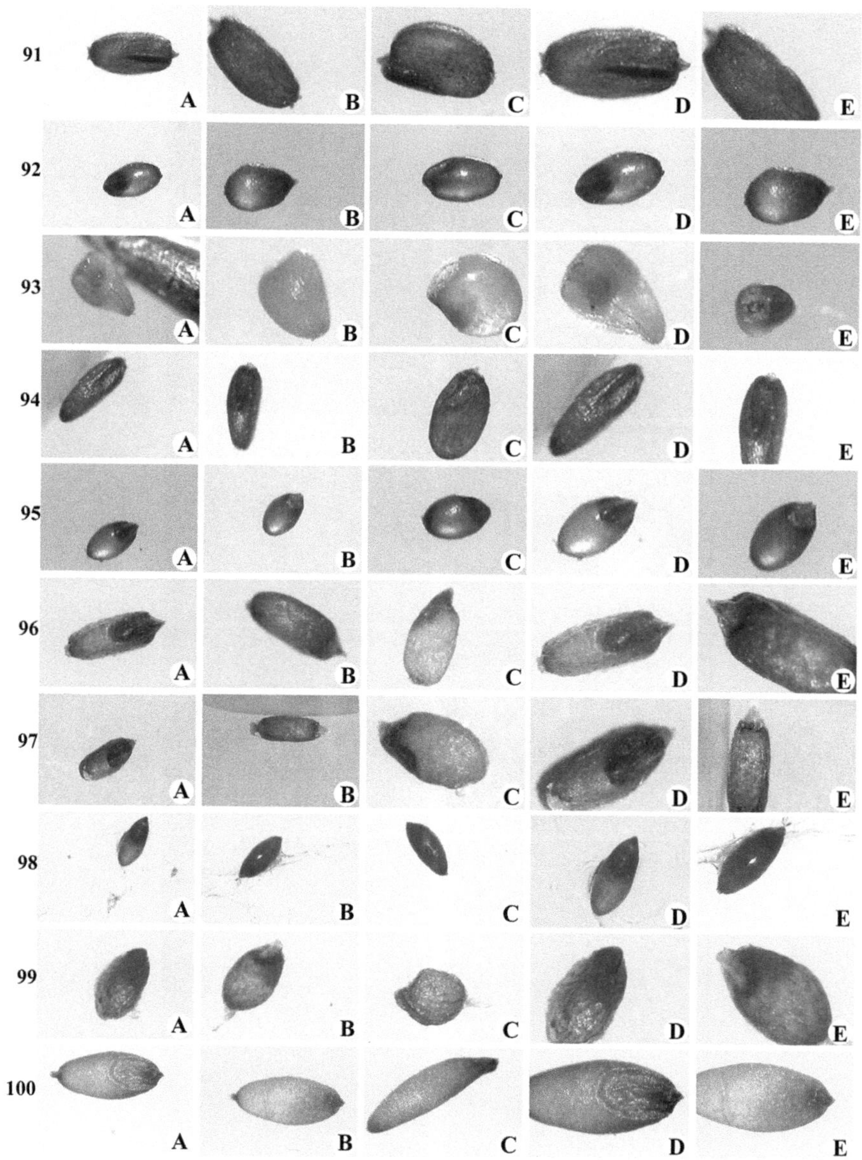

FIGURE 13 Light microscopic study of the grass caryopses (A. Dorsal surface; B. Ventral surface; C. Lateral surface; D. Embryo; E. Hilum; 91. *Eragrostis pilosa*; 92. *Eragrostis tenella*; 93. *Eragrostis tremula*; 94. *Eragrostis unioloides*; 95. *Eragrostis viscosa*; 96. *Sporobolus coromardelianus*; 97. *Sporobolus diander*; 98. *Sporobolus indicum*; 99. *Urochondra setulosa*; 100. *Tragrus biflorus*).

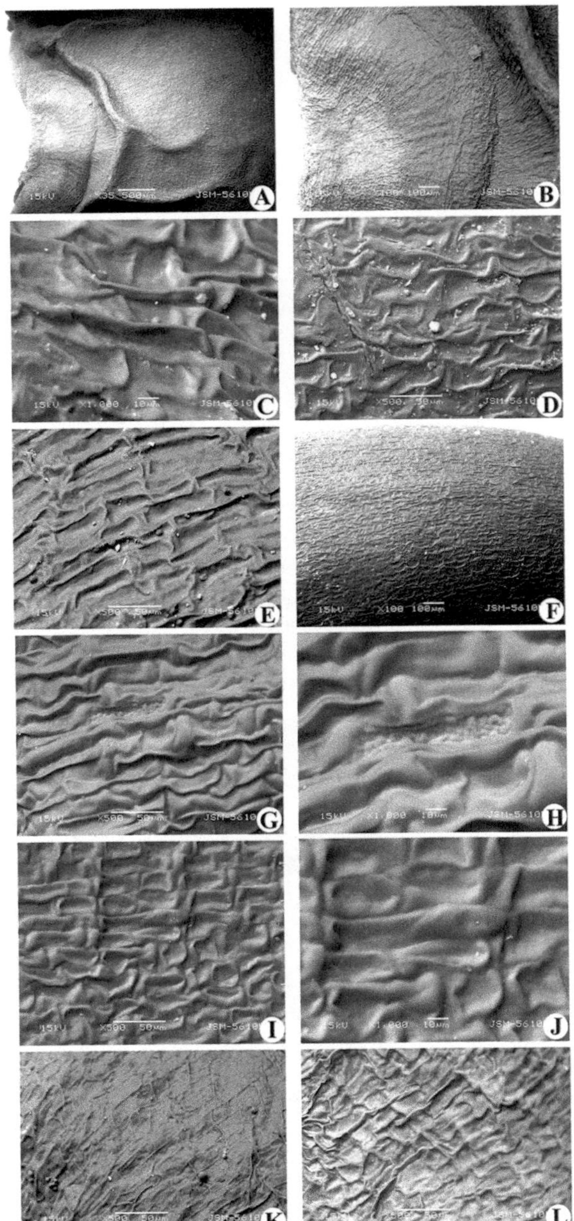

FIGURE 14 Scanning electron microscopic study of *Coix lachryma-jobi* (A, B. Ventral view (low magnification); C, D. Dorsal surface (high magnification); F. Ventral surface; G. Lateral surface; G, H. Embryo surface (proximal end); I, J. Embryo surface (distal end); K. Hilum; L. Hilum end portion).

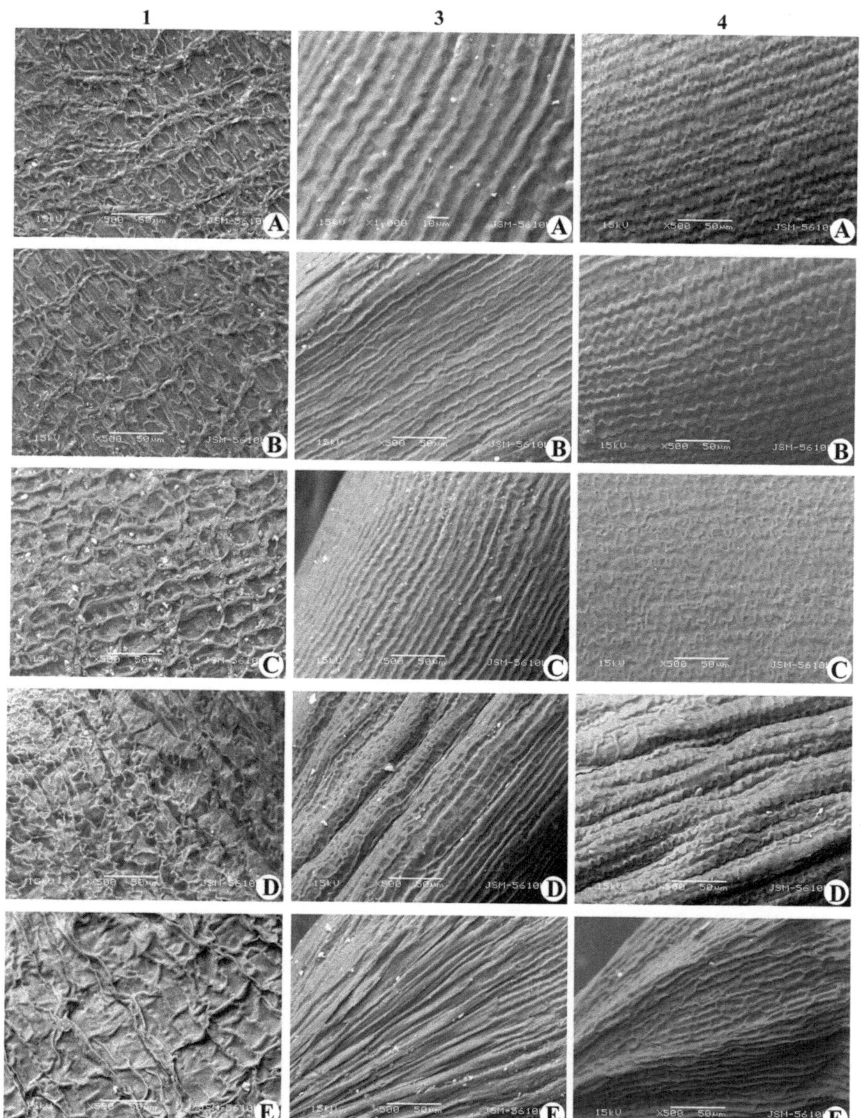

FIGURE 15 Scanning electron microscopic study of the grass caryopses (A. Dorsal surface; B. Ventral surface; C. Lateral surface; D. Embryo; E. Hilum; 2. *Chionachne koenigii*; 3. *Andropogon pumilus*; 4. *Apluda mutica*).

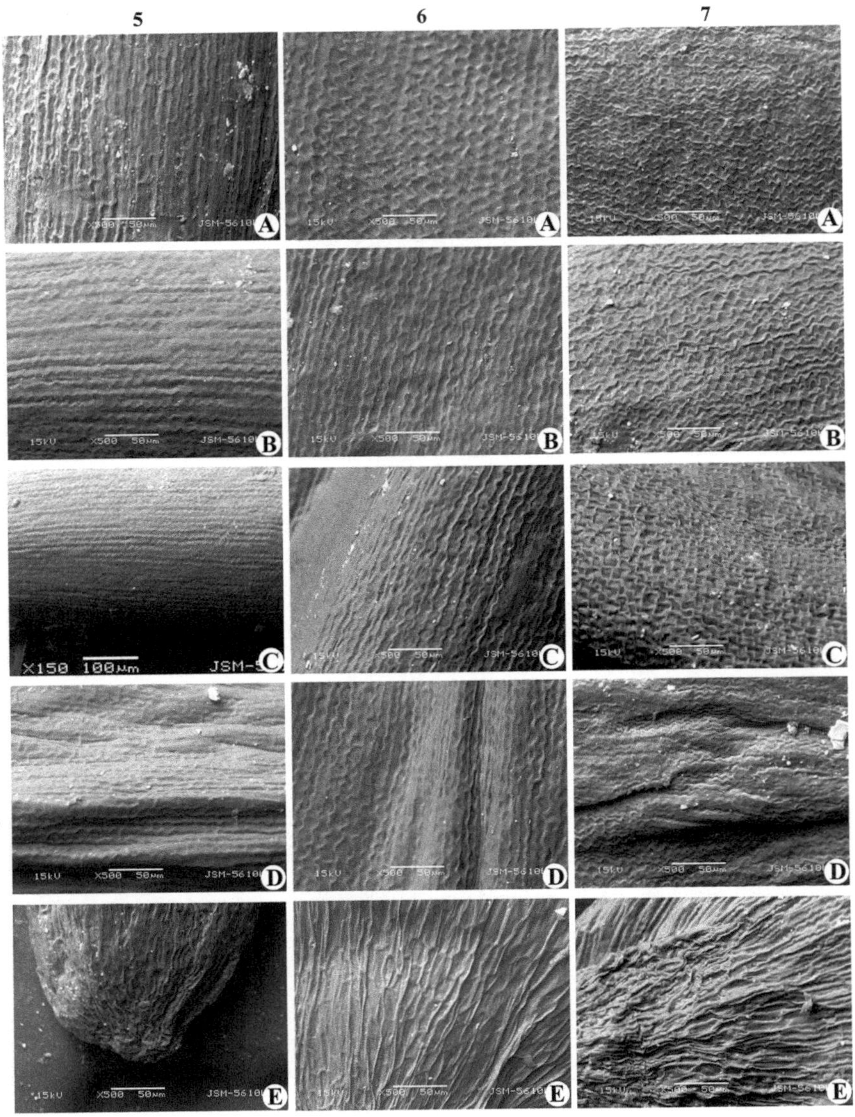

FIGURE 16 Scanning electron microscopic study of the grass caryopses (A. Dorsal surface; B. Ventral surface; C. Lateral surface; D. Embryo; E. Hilum; 5. *Arthraxon lanceolatus*; 6. *Bothriochloa pertusa*; 7. *Capillipedium hugelii*).

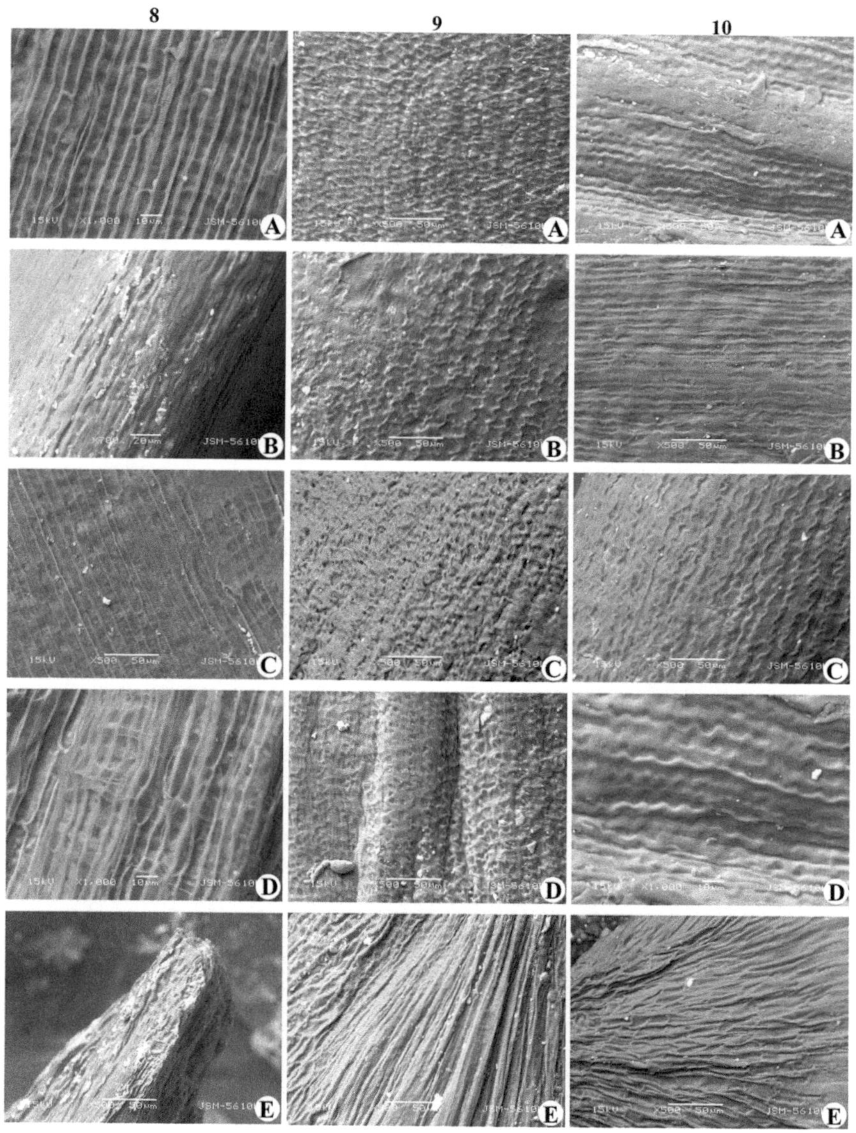

FIGURE 17 Scanning electron microscopic study of the grass caryopses (A. Dorsal surface; B. Ventral surface; C. Lateral surface; D. Embryo; E. Hilum; 8. *Chrysopogon fulvus*; 9. *Cymbopogon martini*; 10. *Dicanthium annulatum).*

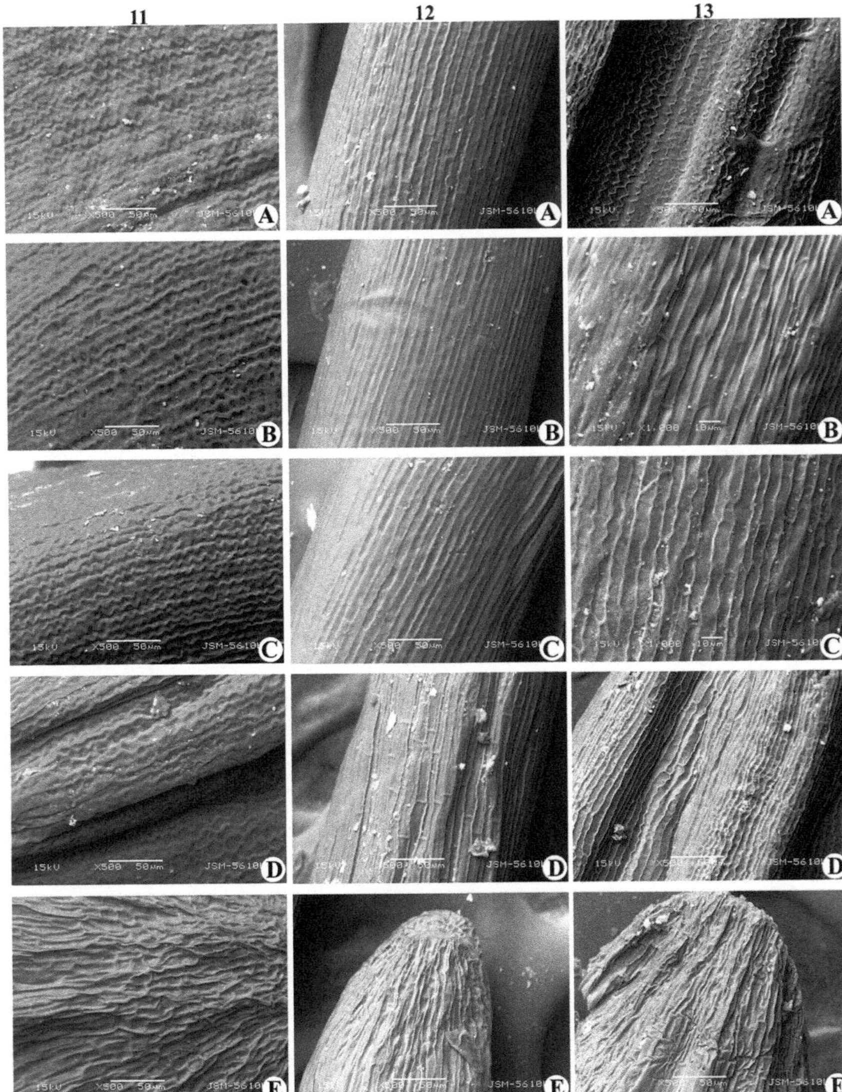

FIGURE 18 Scanning electron microscopic study of the grass caryopses (A. Dorsal surface; B. Ventral surface; C. Lateral surface; D. Embryo; E. Hilum; 11. *Dicanthium caricosum*; 12. *Dimeria orinthopoda*; 13. *Heteropogon contortus var. contortus sub var. typicus*).

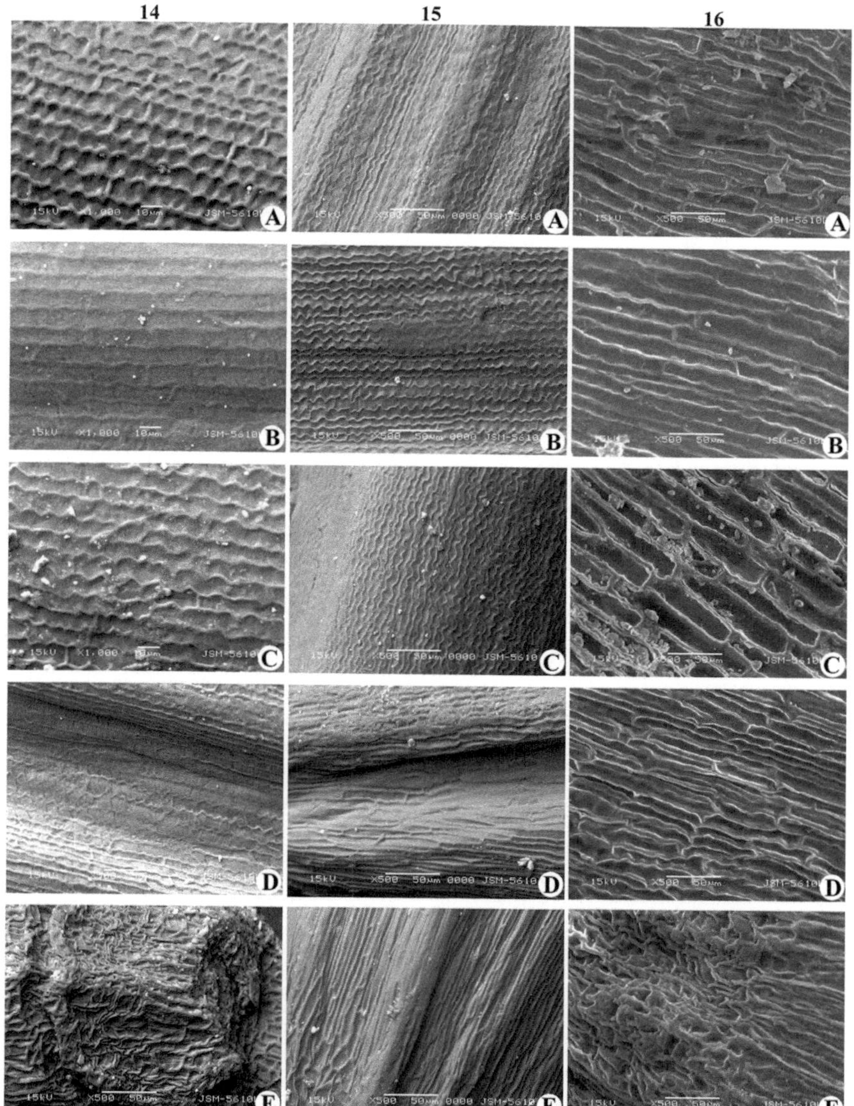

FIGURE 19 Scanning electron microscopic study of the grass caryopses (A. Dorsal surface; B. Ventral surface; C. Lateral surface; D. Embryo; E. Hilum; 14. *Heteropogon contortus var. contortus sub var. genuinus*; 15. *Heteropogon ritchiei*; 16. *Heteropogon triticeus*).

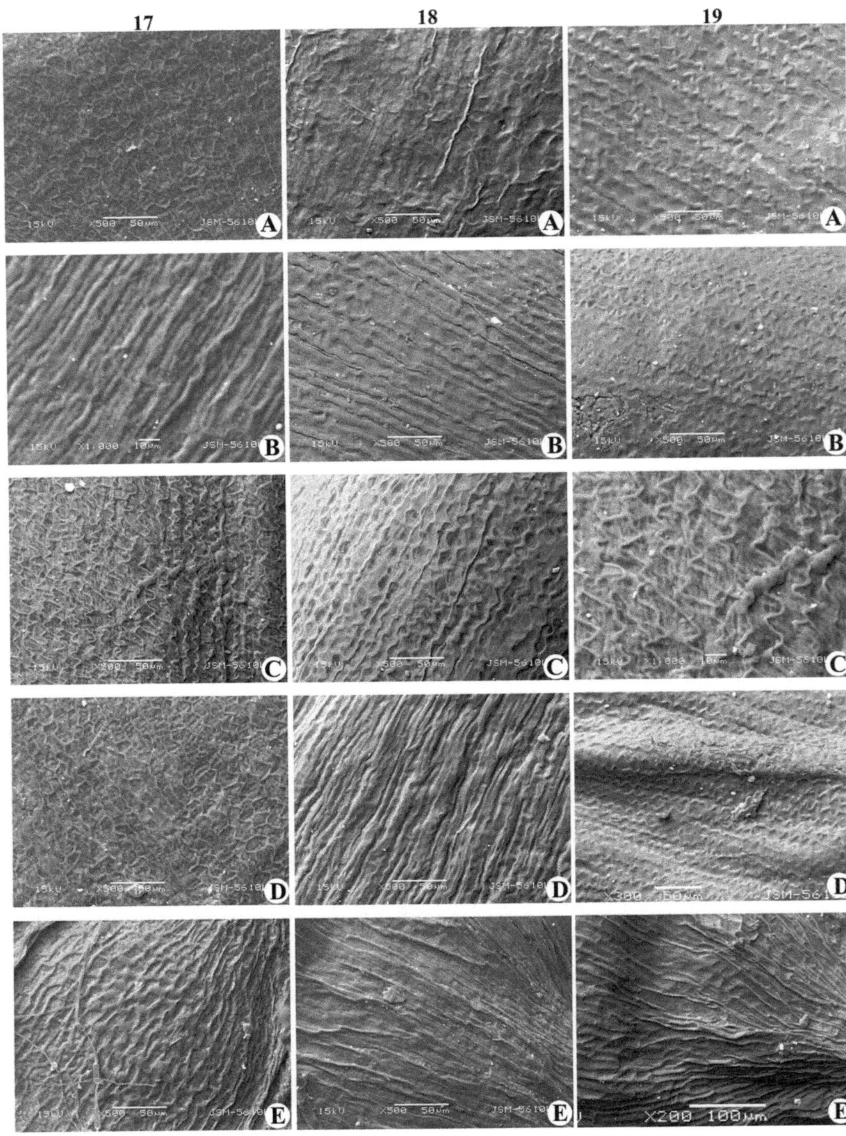

FIGURE 20 Scanning electron microscopic study of the grass caryopses (A. Dorsal surface; B. Ventral surface; C. Lateral surface; D. Embryo; E. Hilum; 17. *Ischaemum indicum*; 18. *Ischaemum molle*; 19. *Ischaemum pilosum*).

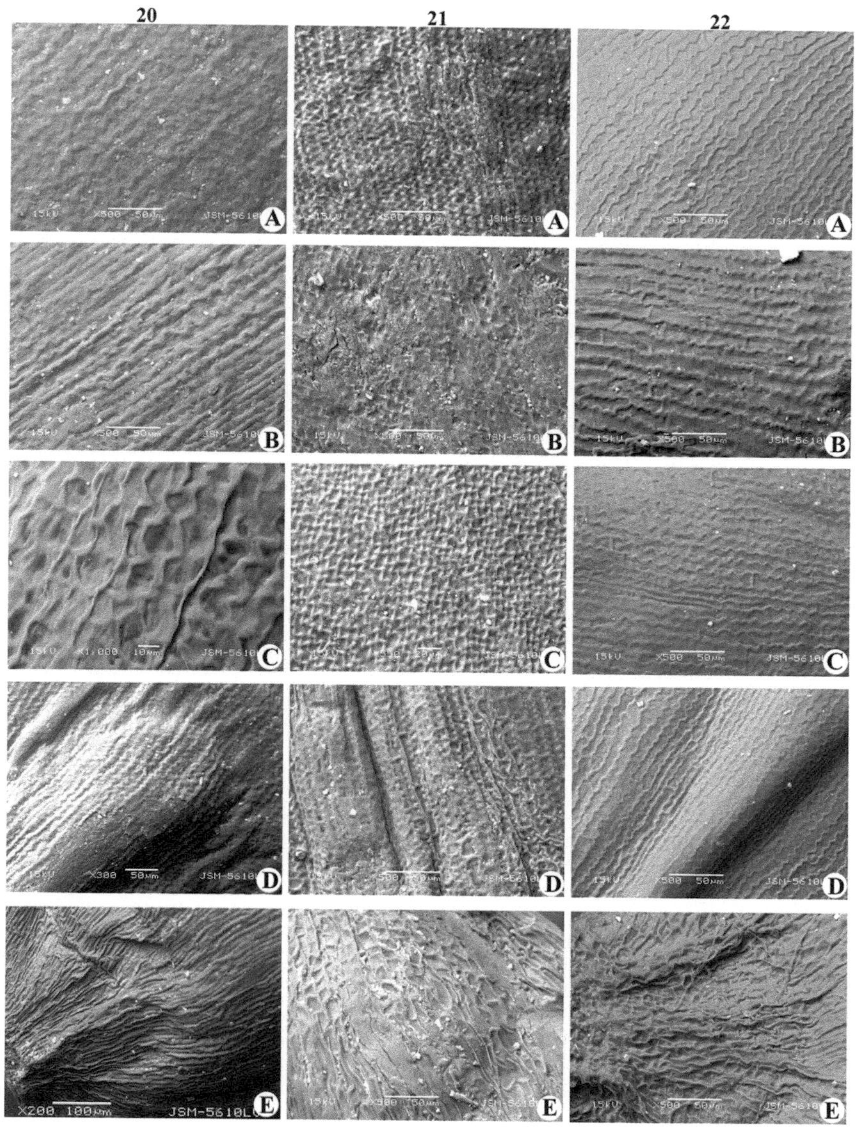

FIGURE 21 Scanning electron microscopic study of the grass caryopses (A. Dorsal surface; B. Ventral surface; C. Lateral surface; D. Embryo; E. Hilum; 20. *Ischaemum rugosum*; 21. *Iseilema laxum*; 22. *Ophiuros exaltatus*).

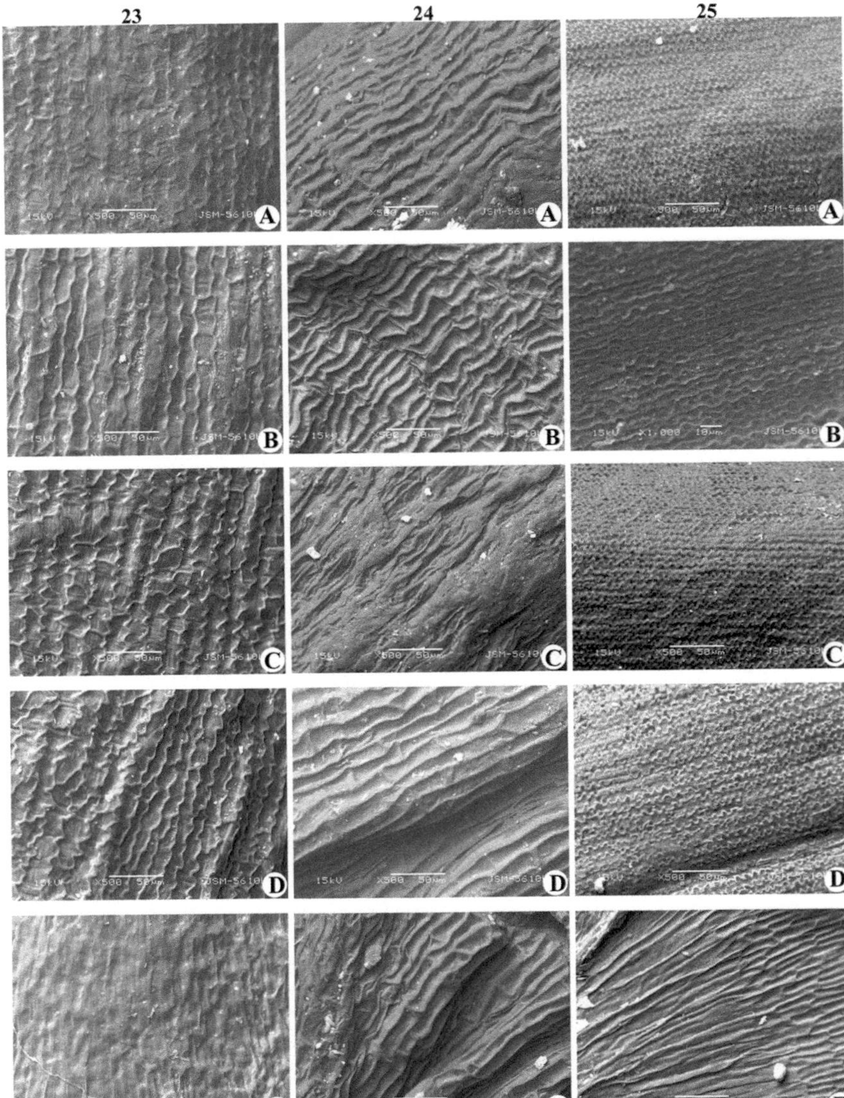

FIGURE 22 Scanning electron microscopic study of the grass caryopses (A. Dorsal surface; B. Ventral surface; C. Lateral surface; D. Embryo; E. Hilum; 23. *Rottboellia exaltata*; 24. *Saccharum spontaneum*; 25. *Sehima ischaemoides*).

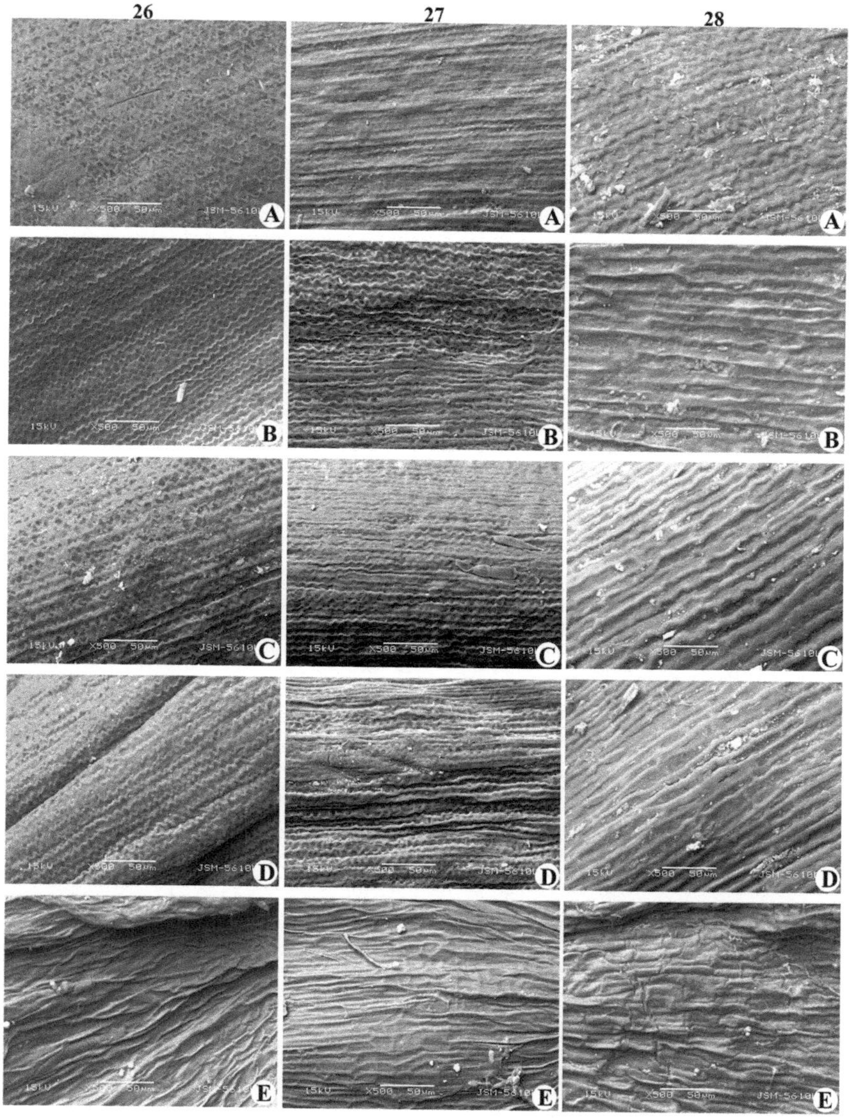

FIGURE 23 Scanning electron microscopic study of the grass caryopses (A. Dorsal surface; B. Ventral surface; C. Lateral surface; D. Embryo; E. Hilum; 26. *Sehima nervosum*; 27. *Sehima sulcatum*; 28. *Sorghum halepense*).

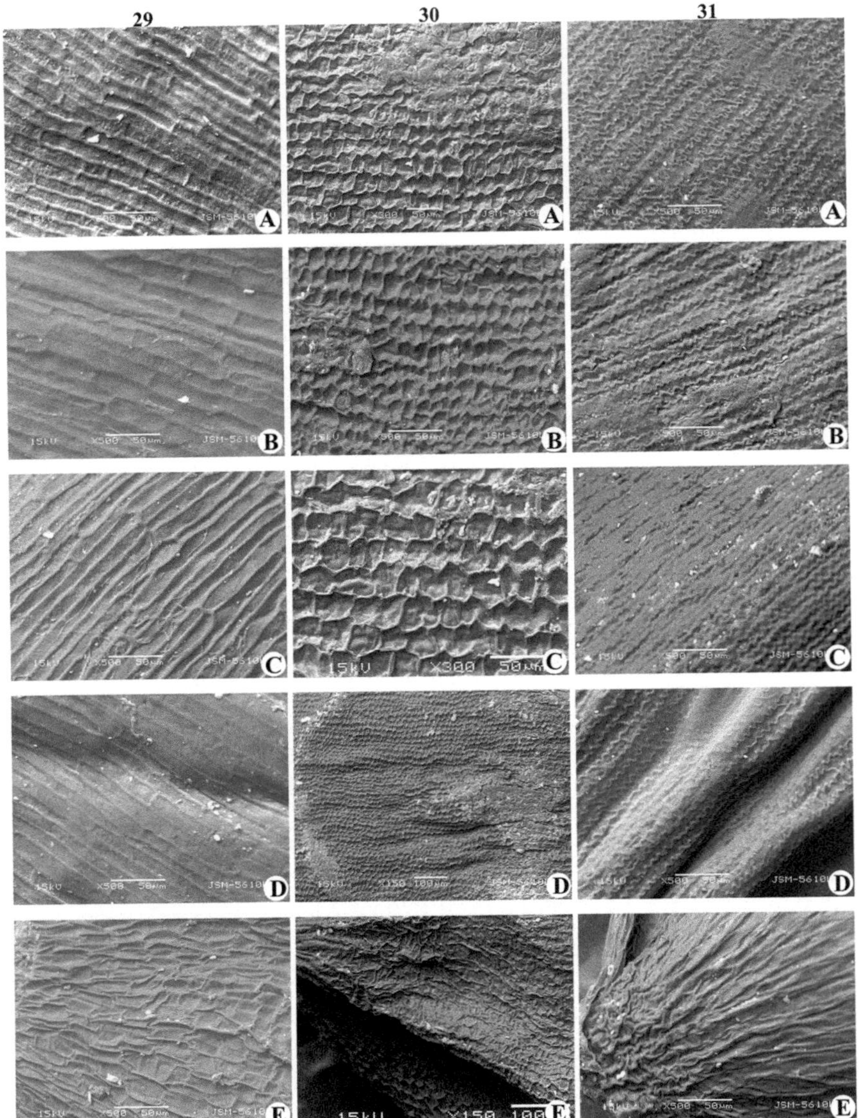

FIGURE 24 Scanning electron microscopic study of the grass caryopses (A. Dorsal surface; B. Ventral surface; C. Lateral surface; D. Embryo; E. Hilum; 29. *Sorghum purpureo-sericeum*; 30. *Thelepogn elegans*; 31. *Themeda cymbaria*).

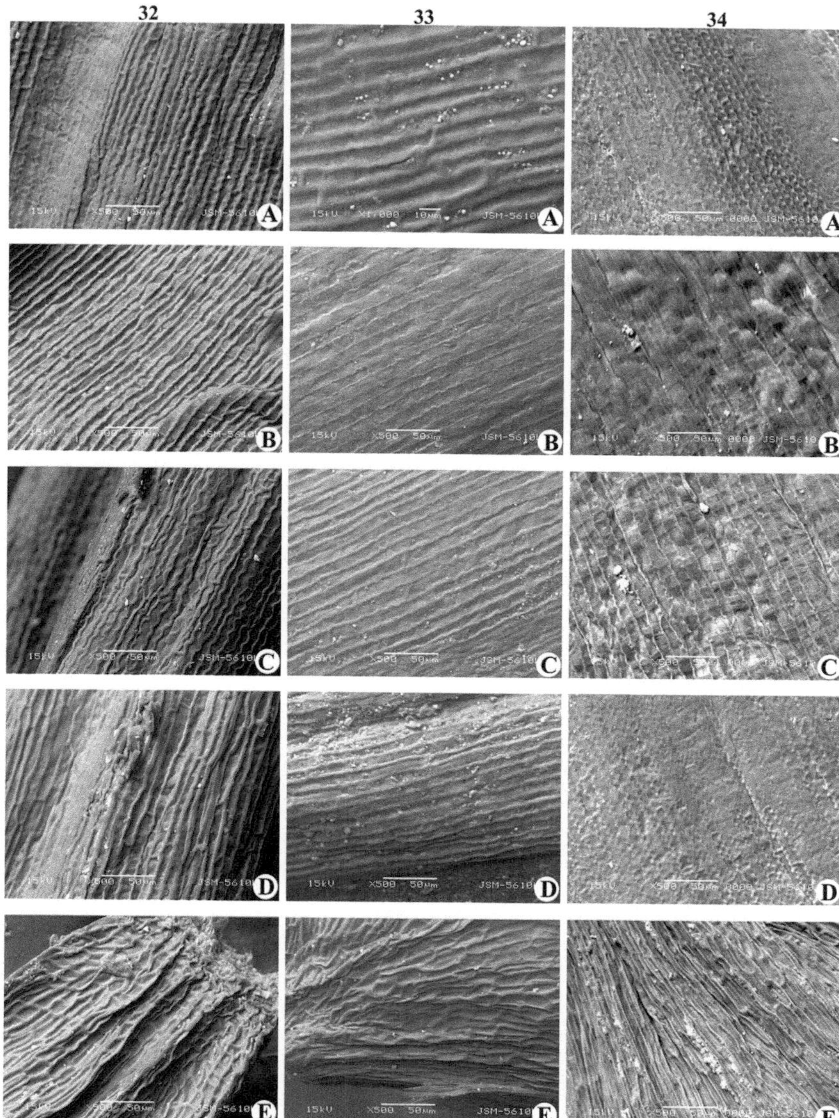

FIGURE 25 Scanning electron microscopic study of the grass caryopses (A. Dorsal surface; B. Ventral surface; C. Lateral surface; D. Embryo; E. Hilum; 32. *Themeda laxa*; 33. *Themeda triandra*; 34. *Themeda quadrivalvis*).

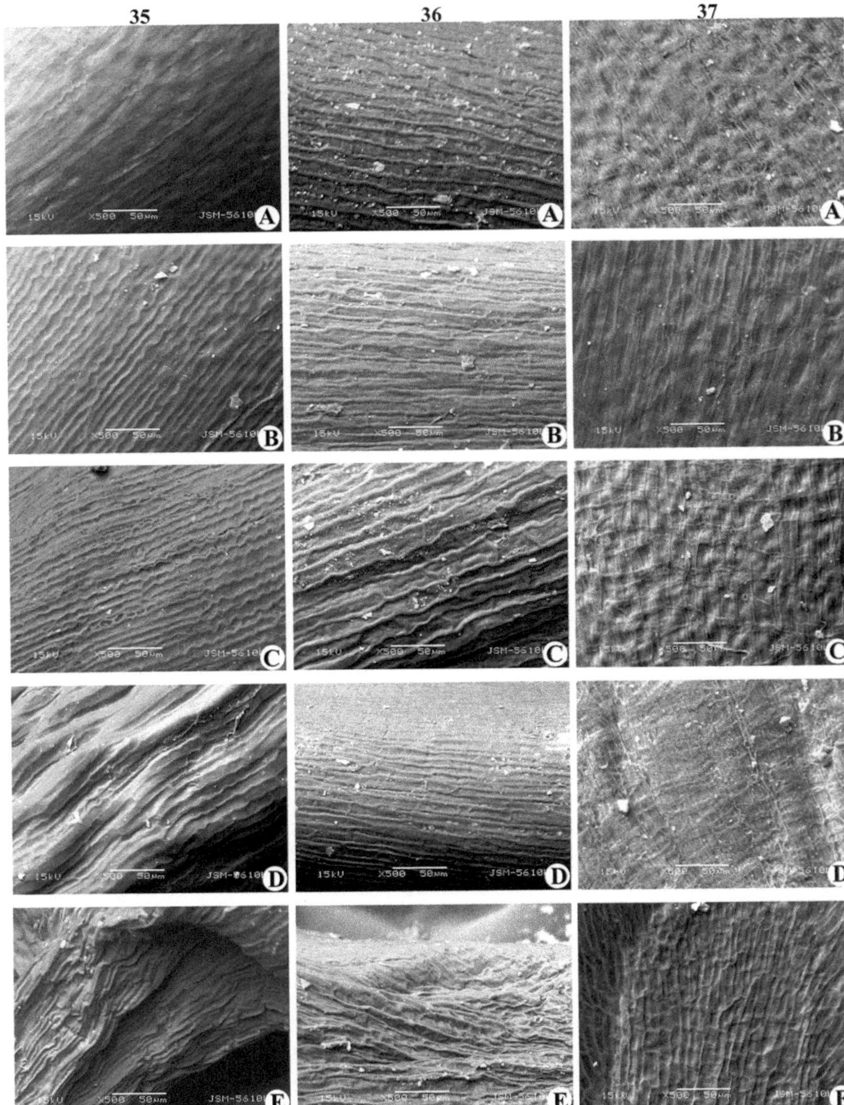

FIGURE 26 Scanning electron microscopic study of the grass caryopses (A. Dorsal surface; B. Ventral surface; C. Lateral surface; D. Embryo; E. Hilum; 35. *Triplopogon ramosissimus*; 36. *Vetivaria zizanioides*; 37. *Alloteropsis cimicina*).

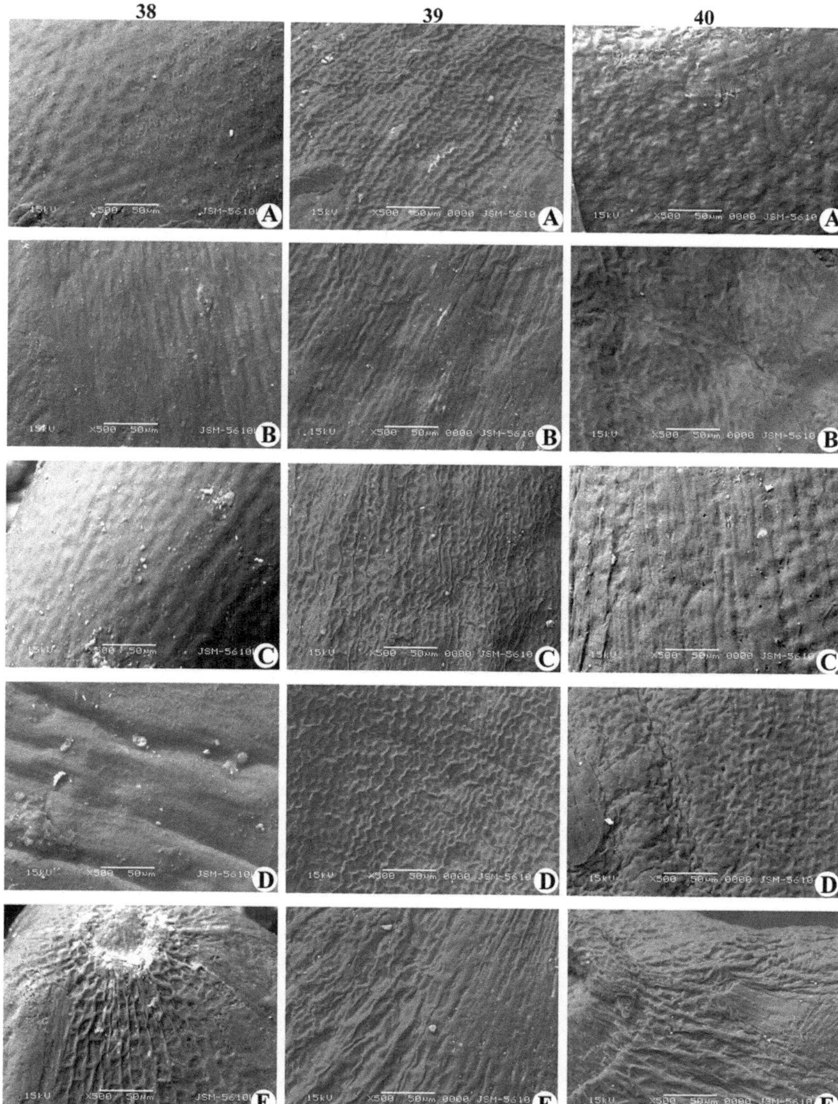

FIGURE 27 Scanning electron microscopic study of the grass caryopses (A. Dorsal surface; B. Ventral surface; C. Lateral surface; D. Embryo; E. Hilum; 38. *Bracharia cruciformis*; 39. *Bracharia distachya*; 40. *Bracharia ramosa*).

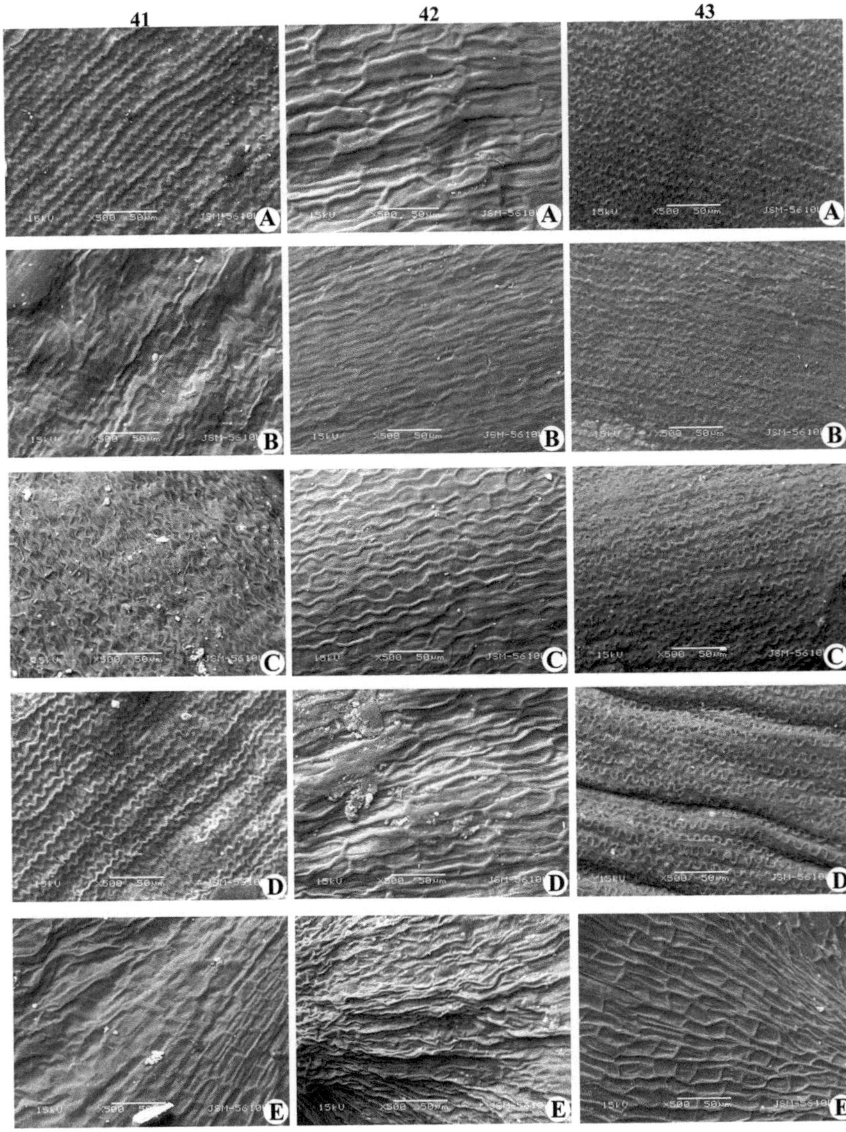

FIGURE 28 Scanning electron microscopic study of the grass caryopses (A. Dorsal surface; B. Ventral surface; C. Lateral surface; D. Embryo; E. Hilum; 41. *Bracharia reptans*; 42. *Cenchrus biflorus*; 43. *Cenchrus ciliaris*).

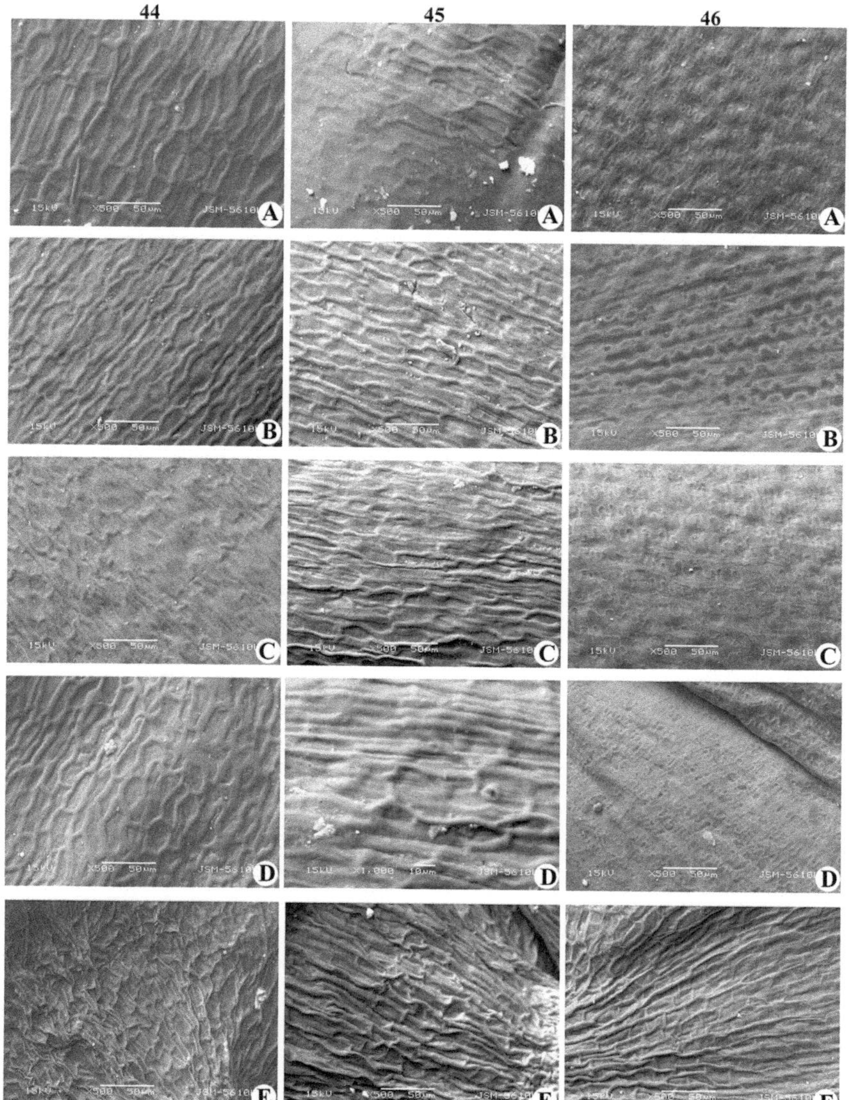

FIGURE 29 Scanning electron microscopic study of the grass caryopses (A. Dorsal surface; B. Ventral surface; C. Lateral surface; D. Embryo; E. Hilum; 44. *Cenchrus setigerus5.*; 45. *Cenchrus preurii*; 46. *Digitaria adscendens*).

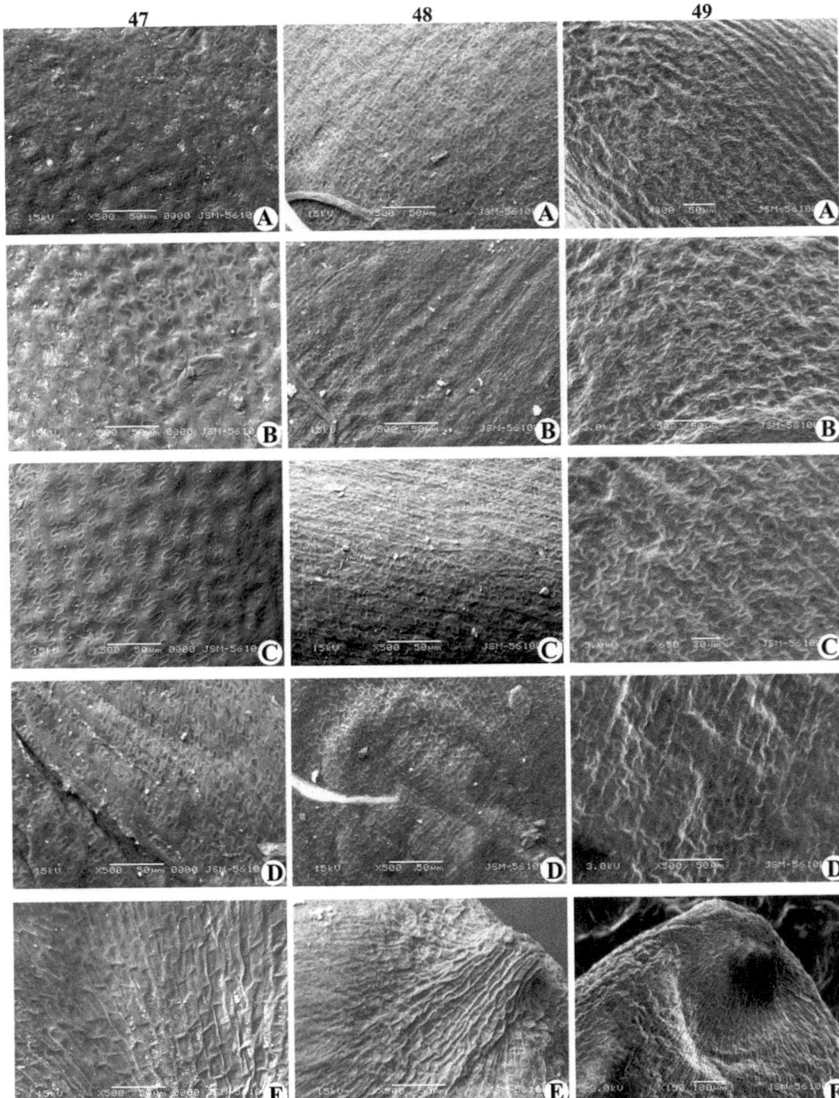

FIGURE 30 Scanning electron microscopic study of the grass caryopses (A. Dorsal surface; B. Ventral surface; C. Lateral surface; D. Embryo; E. Hilum; 47. *Digitaria ciliaris*; 48. *Digitaria granularis*; 49. *Echinochloa colonum*).

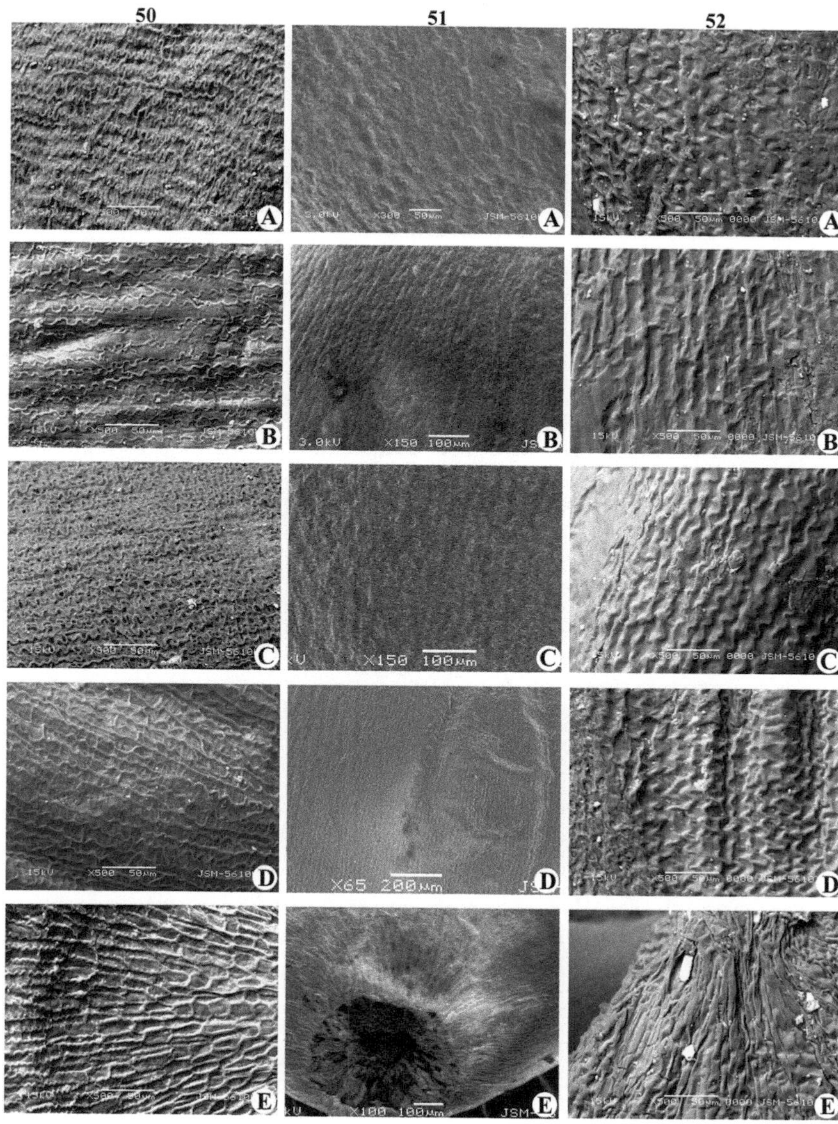

FIGURE 31 Scanning electron microscopic study of the grass caryopses (A. Dorsal surface; B. Ventral surface; C. Lateral surface; D. Embryo; E. Hilum; 50. *Echinochloa crusgalli*; 51. *Echinochloa stagnina*; 52. *Eriochloa procera*).

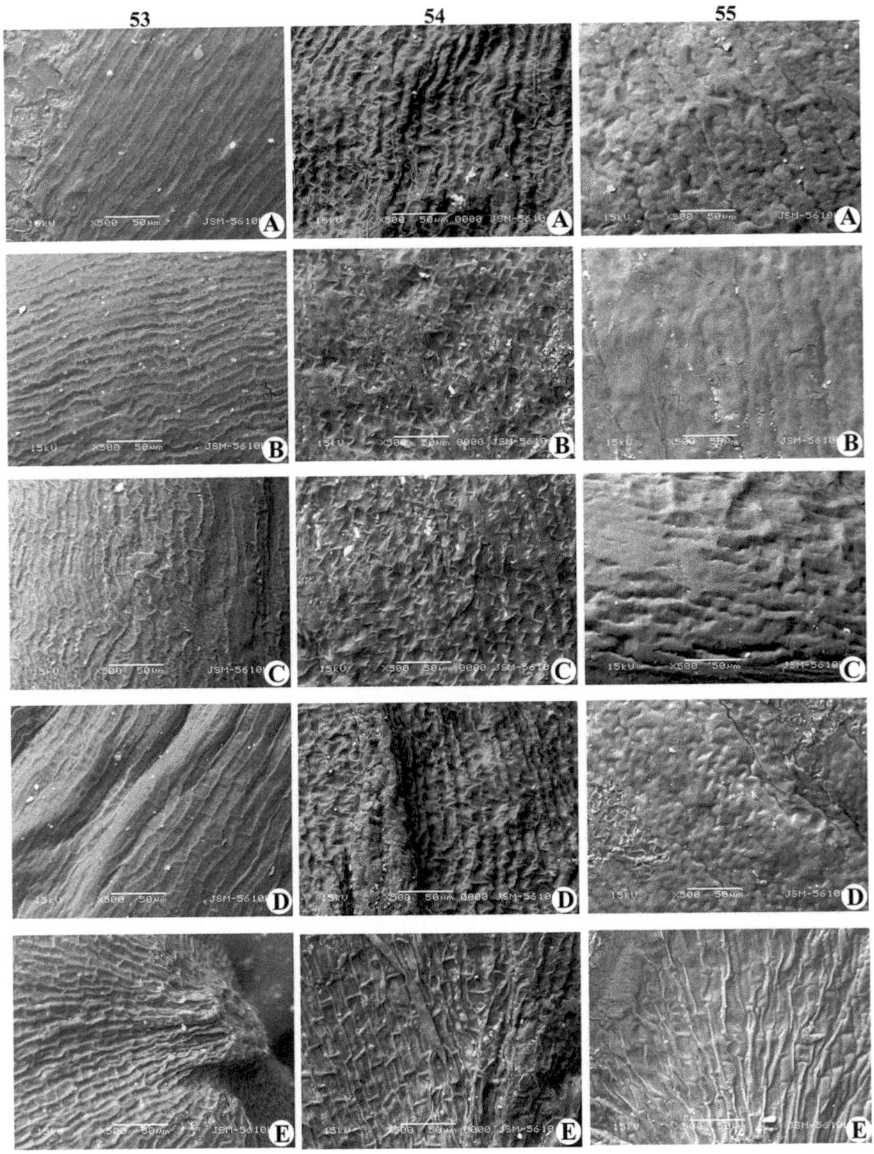

FIGURE 32 Scanning electron microscopic study of the grass caryopses (A. Dorsal surface; B. Ventral surface; C. Lateral surface; D. Embryo; E. Hilum; 53. *Oplismenus burmannii*; 54. *Oplismenus compositus*; 55. *Panicum antidotale*).

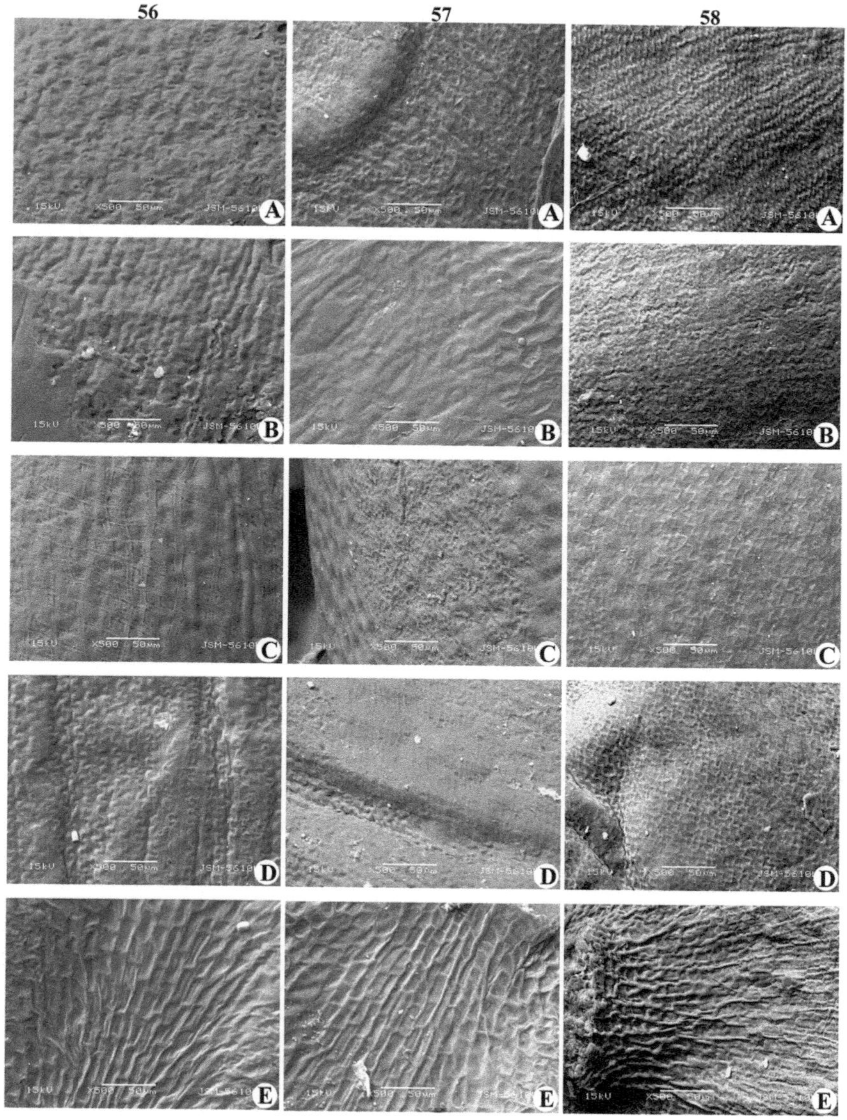

FIGURE 33 Scanning electron microscopic study of the grass caryopses (A. Dorsal surface; B. Ventral surface; C. Lateral surface; D. Embryo; E. Hilum; 56. *Panicum trypheron*; 57. *Panicum miliaceum*; 58. *Paspalidium flavidum*).

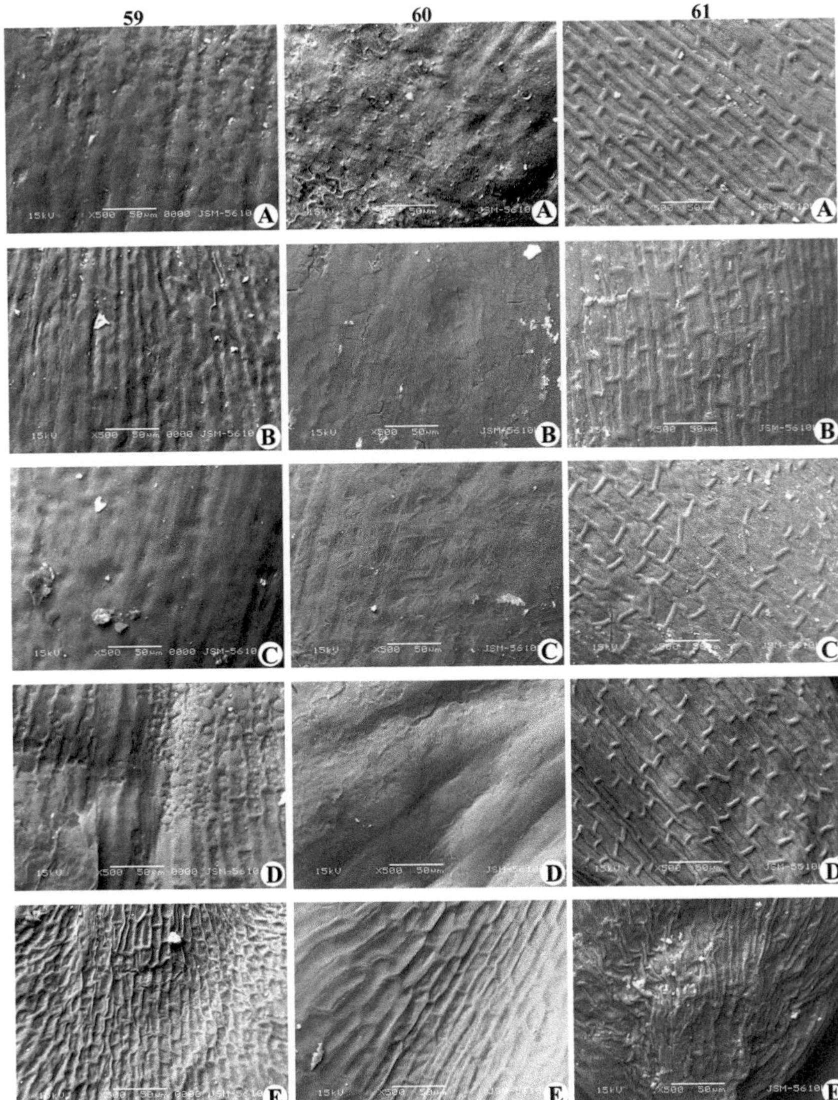

FIGURE 34 Scanning electron microscopic study of the grass caryopses (A. Dorsal surface; B. Ventral surface; C. Lateral surface; D. Embryo; E. Hilum; 59. *Paspalidium geminatum*; 60. *Paspalum scrobiculatum*; 61. *Pennisetum setosum*).

62 63 64

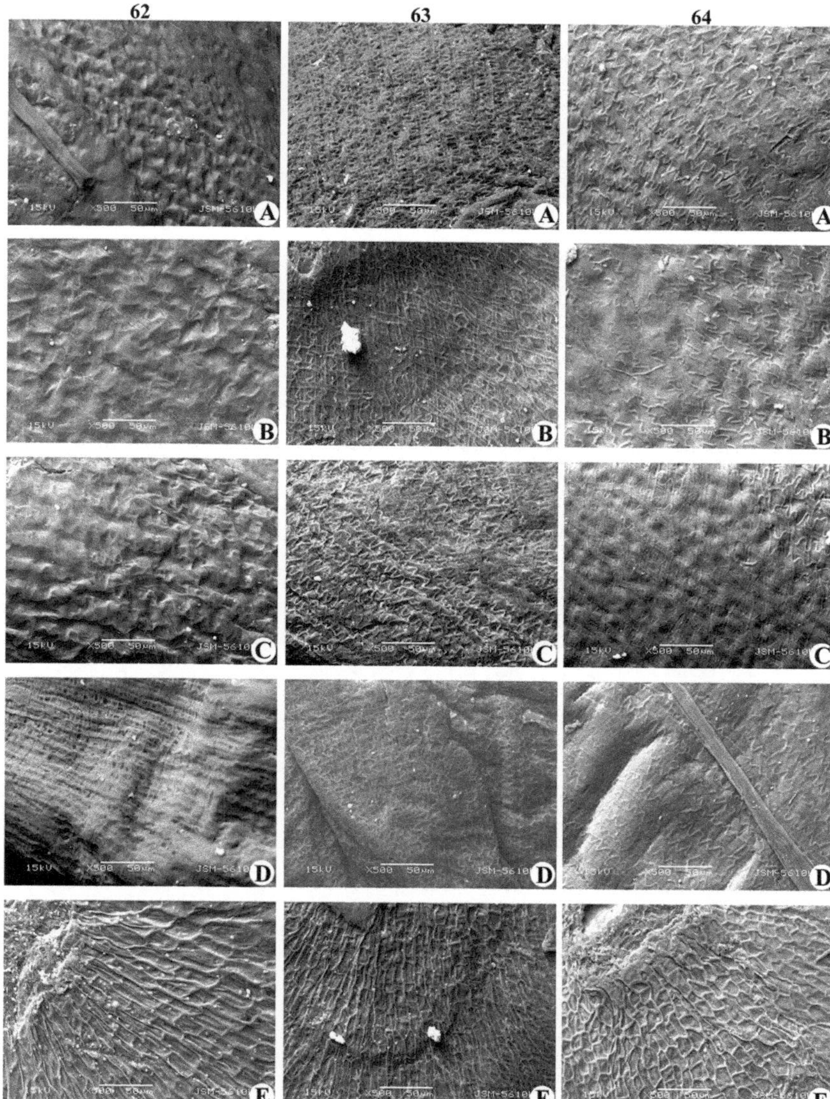

FIGURE 35 Scanning electron microscopic study of the grass caryopses (A. Dorsal surface; B. Ventral surface; C. Lateral surface; D. Embryo; E. Hilum; 62. *Setaria glauca*; 63. *Setaria tomentosa*; 64. *Setaria verticillata*).

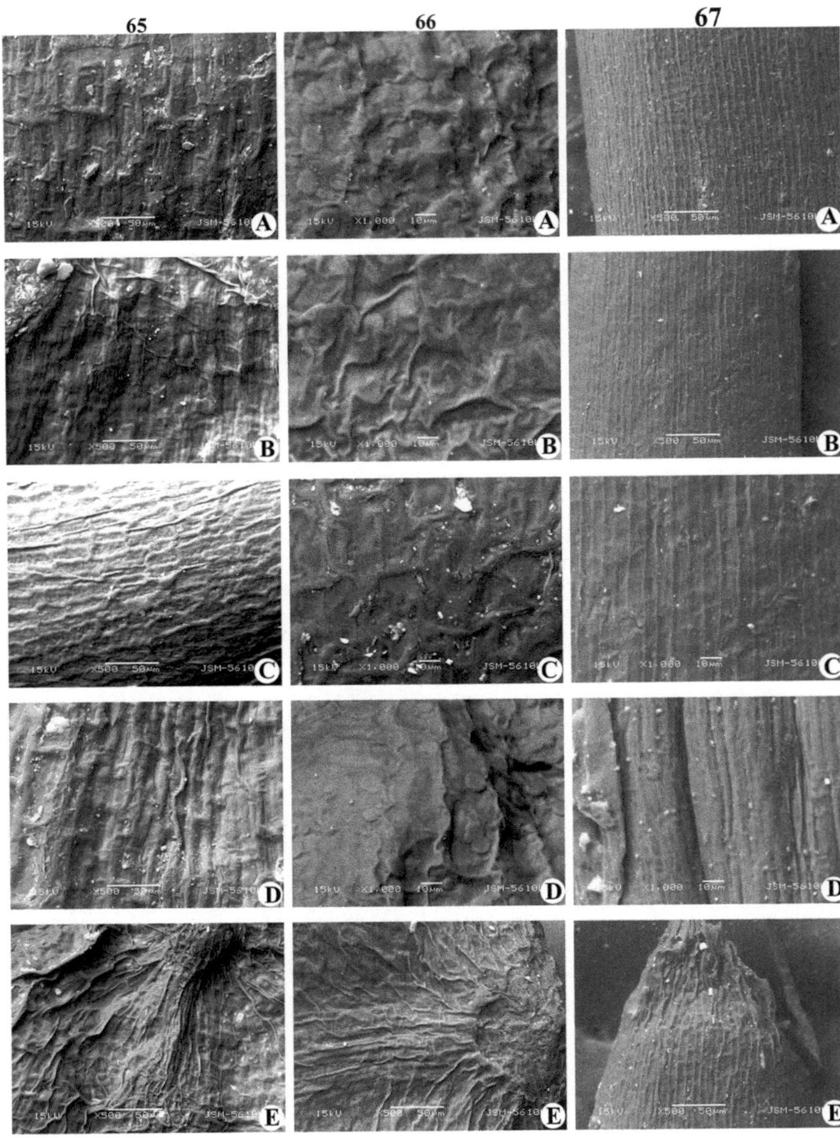

FIGURE 36 Scanning electron microscopic study of the grass caryopses (A. Dorsal surface; B. Ventral surface; C. Lateral surface; D. Embryo; E. Hilum; 65. *Aeluropus lagopoides*; 66. *Isachne globosa*; 67. *Aristida adscensionis*).

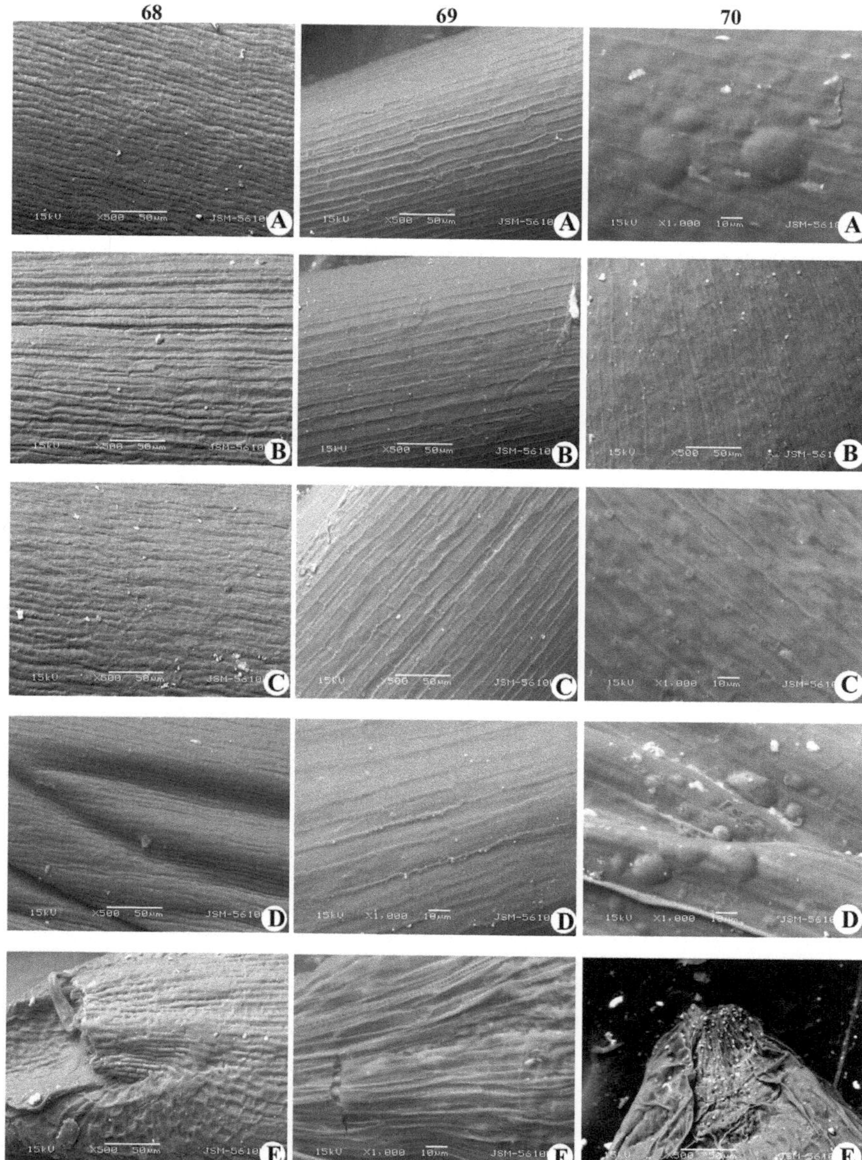

FIGURE 37 Scanning electron microscopic study of the grass caryopses (A. Dorsal surface; B. Ventral surface; C. Lateral surface; D. Embryo; E. Hilum; 68. *Aristida funiculata*; 69. *Perotis indica*; 70. *Chloris barbata*).

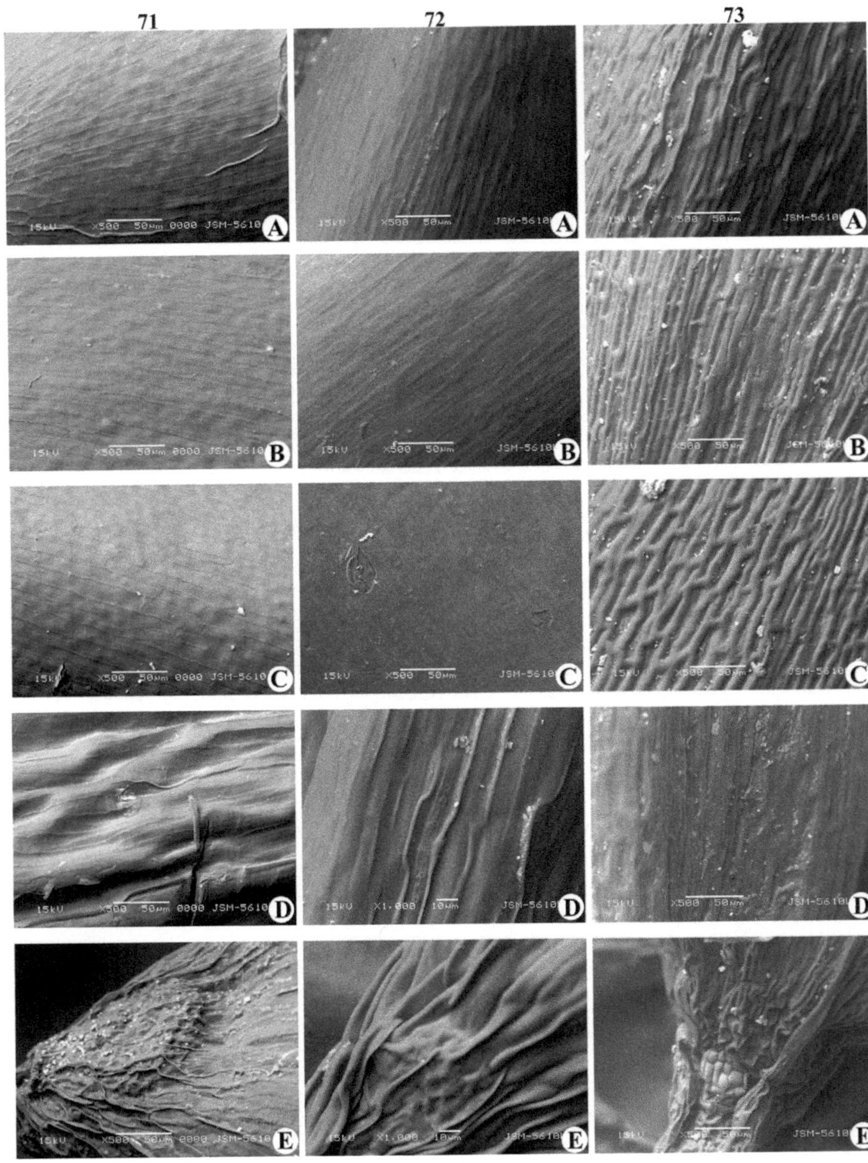

FIGURE 38 Scanning electron microscopic study of the grass caryopses (A. Dorsal surface; B. Ventral surface; C. Lateral surface; D. Embryo; E. Hilum; 71. *Choris montana*; 72. *Chloris virgata*; 73. *Cynadon dactylon*).

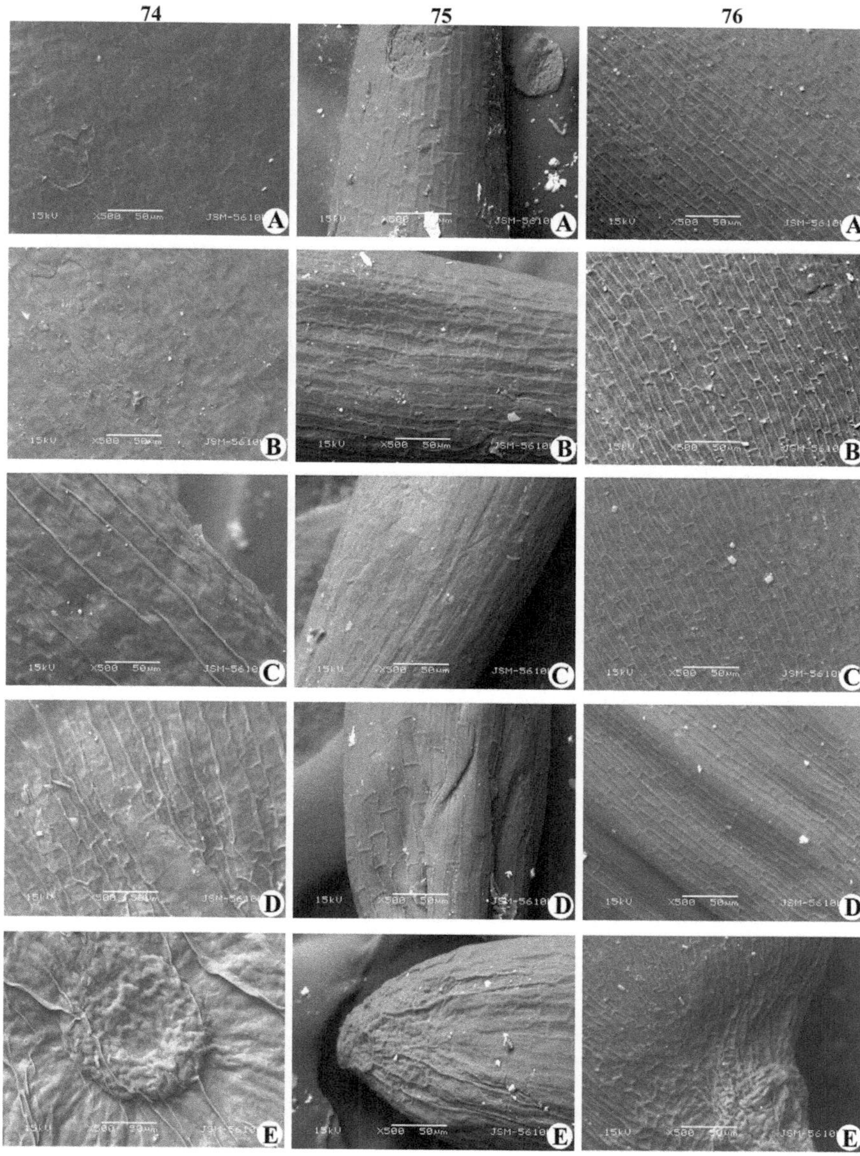

FIGURE 39 Scanning electron microscopic study of the grass caryopses (A. Dorsal surface; B. Ventral surface; C. Lateral surface; D. Embryo; E. Hilum; 74. *Melanocenchris jaequemontii*; 75. *Oropetium villosulum*; 76. *Sachoenefeldia gracilis*).

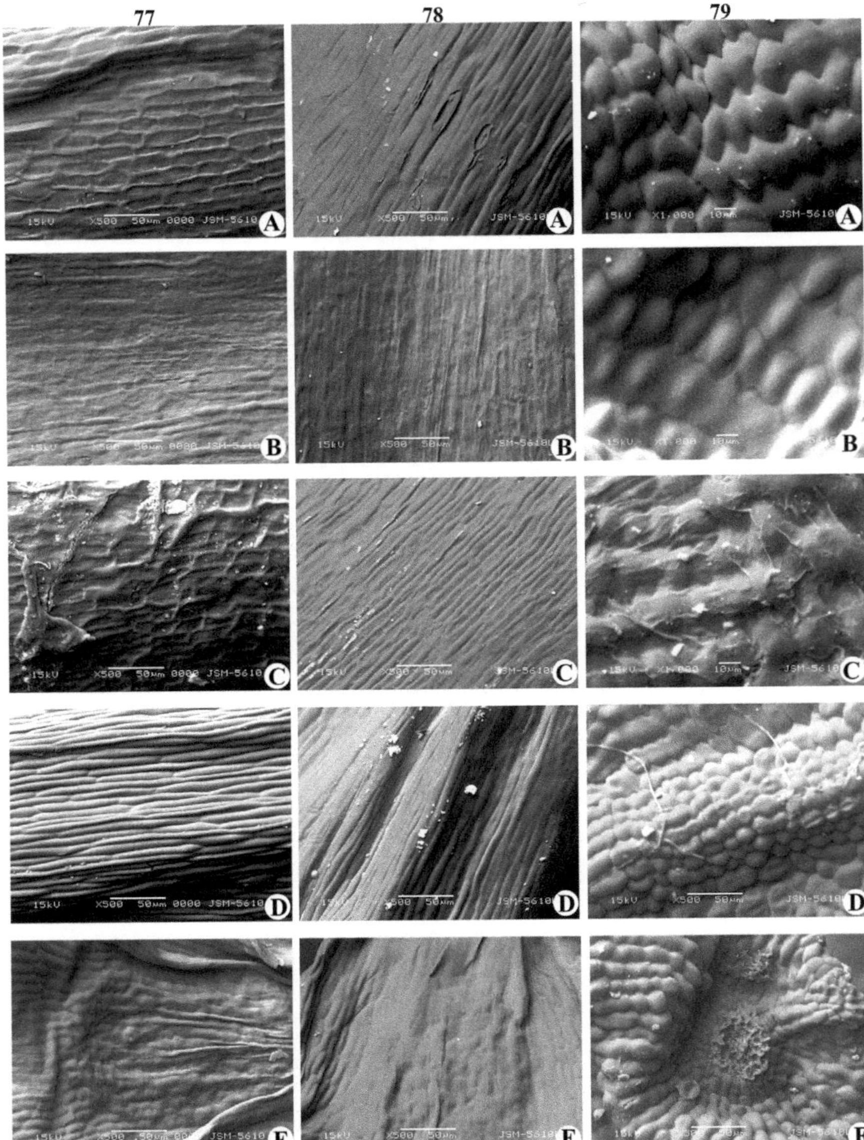

FIGURE 40 Scanning electron microscopic study of the grass caryopses (A. Dorsal surface; B. Ventral surface; C. Lateral surface; D. Embryo; E. Hilum; 77. *Tetrapogon tenellus*; 78. *Tetrapogon villosus*; 79. *Acrachne racemosa*).

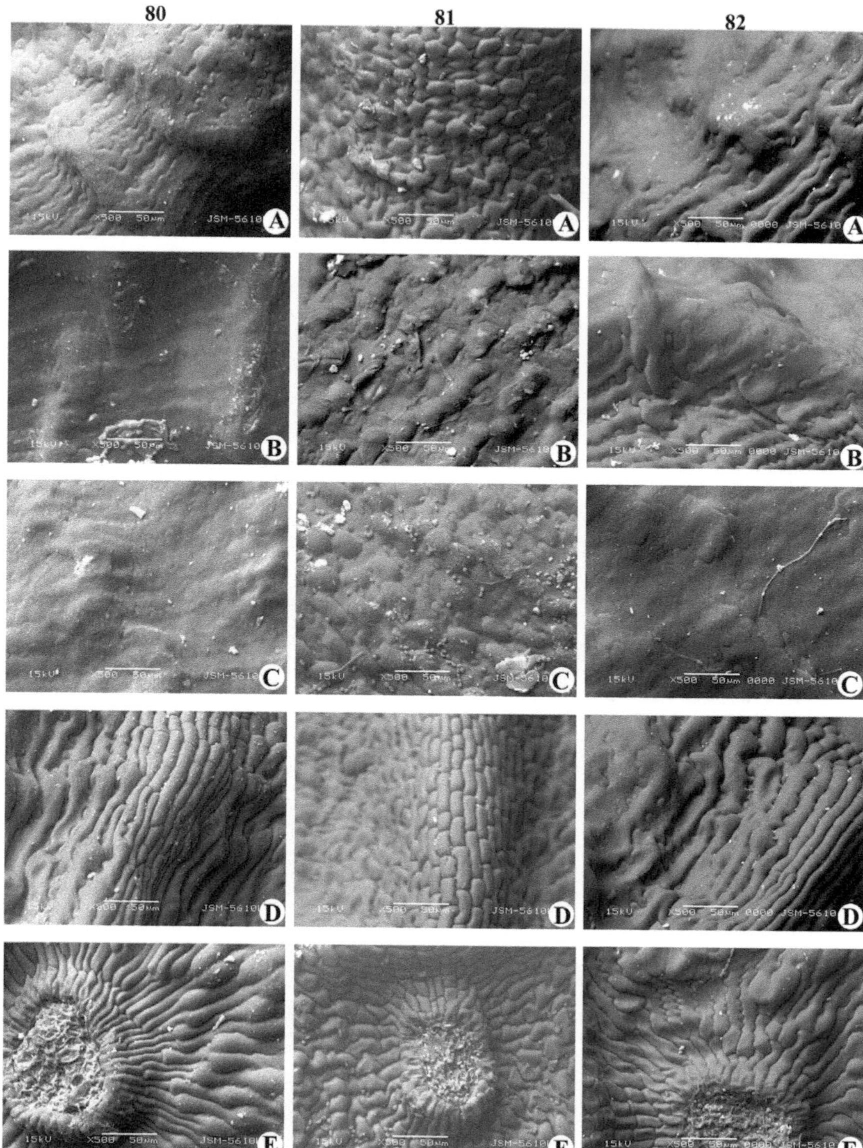

FIGURE 41 Scanning electron microscopic study of the grass caryopses (A. Dorsal surface; B. Ventral surface; C. Lateral surface; D. Embryo; E. Hilum; 80. *Dactyloctenium aegyptium*; 81. *Dactyloctenium sindicum*; 82. *Dactyloctenium giganteum*).

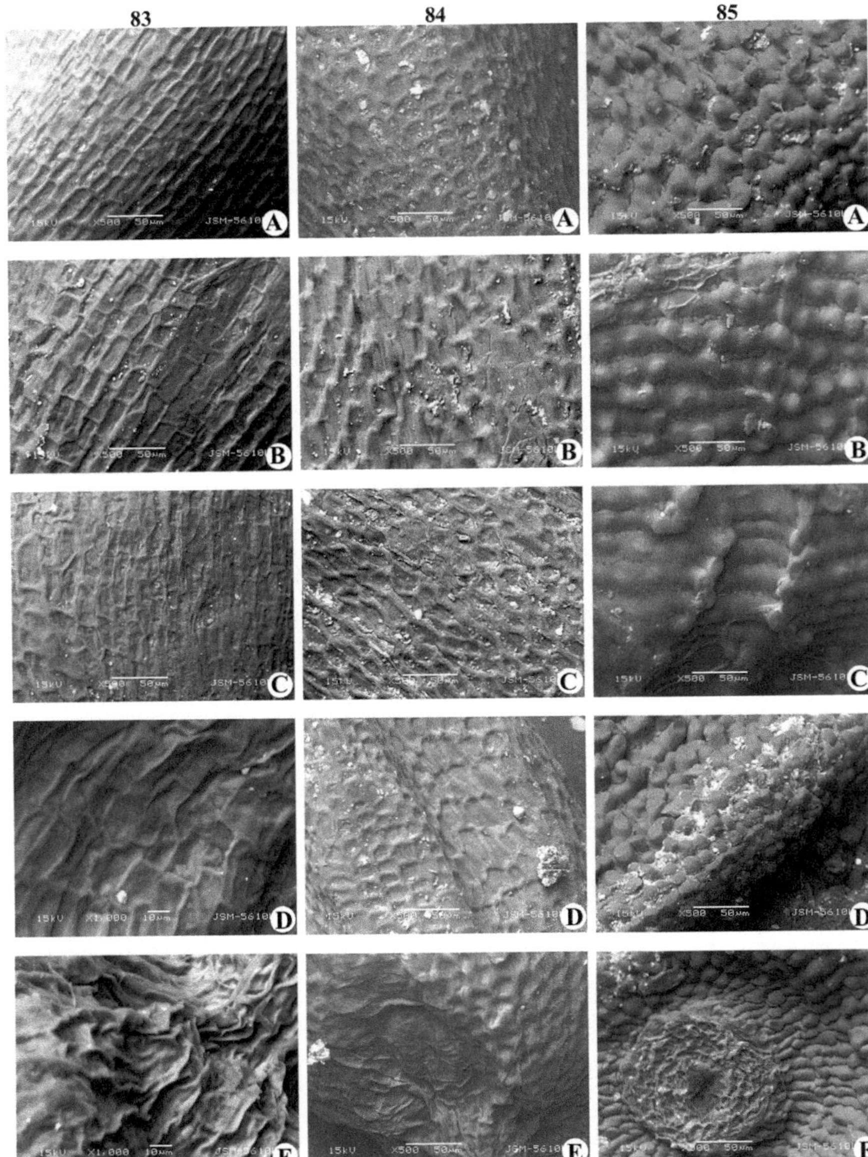

FIGURE 42 Scanning electron microscopic study of the grass caryopses (A. Dorsal surface; B. Ventral surface; C. Lateral surface; D. Embryo; E. Hilum; 83. *Desmostachya bipinnata*; 84. *Dinebra retroflexa*; 85. *Eleusine indica*).

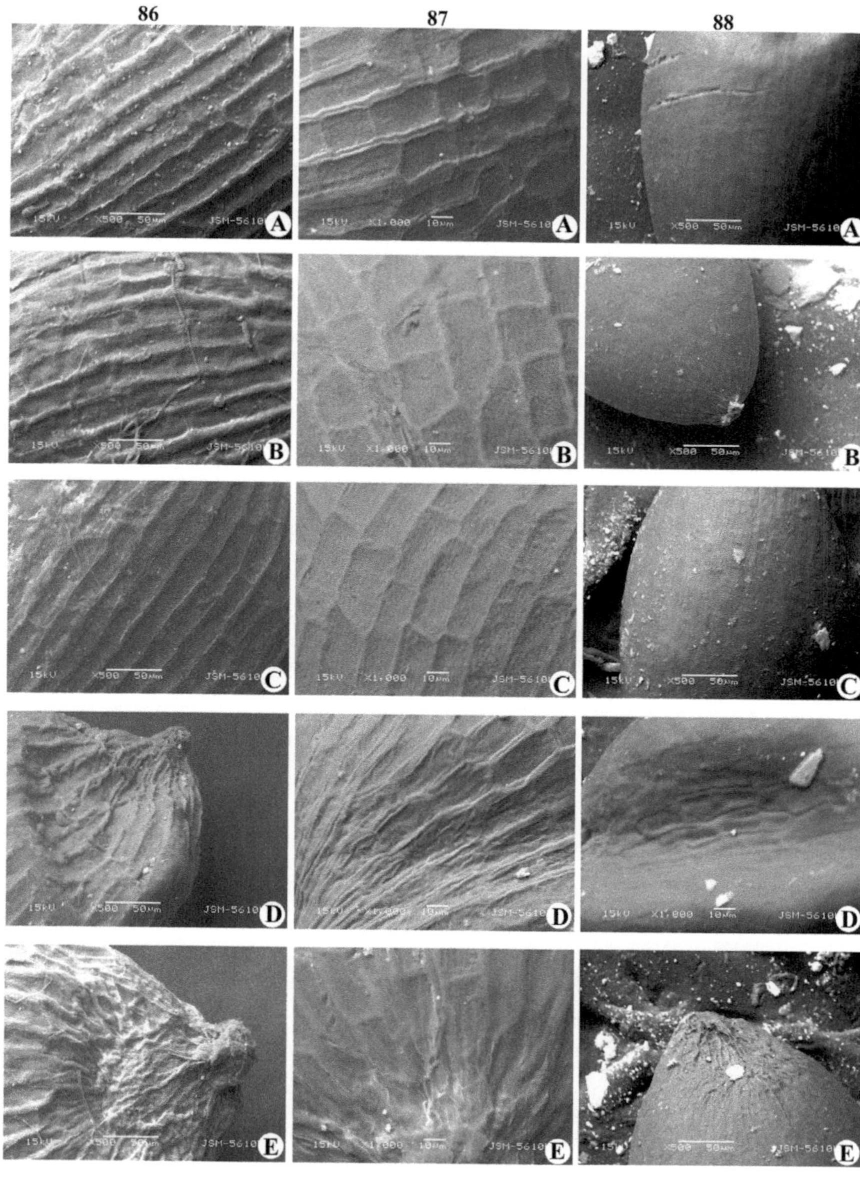

FIGURE 43 Scanning electron microscopic study of the grass caryopses (A. Dorsal surface; B. Ventral surface; C. Lateral surface; D. Embryo; E. Hilum; 86. *Eragrostiella bifaria*; 87. *Eragrostis cilianensis*; 88. *Eragrostis ciliaris*).

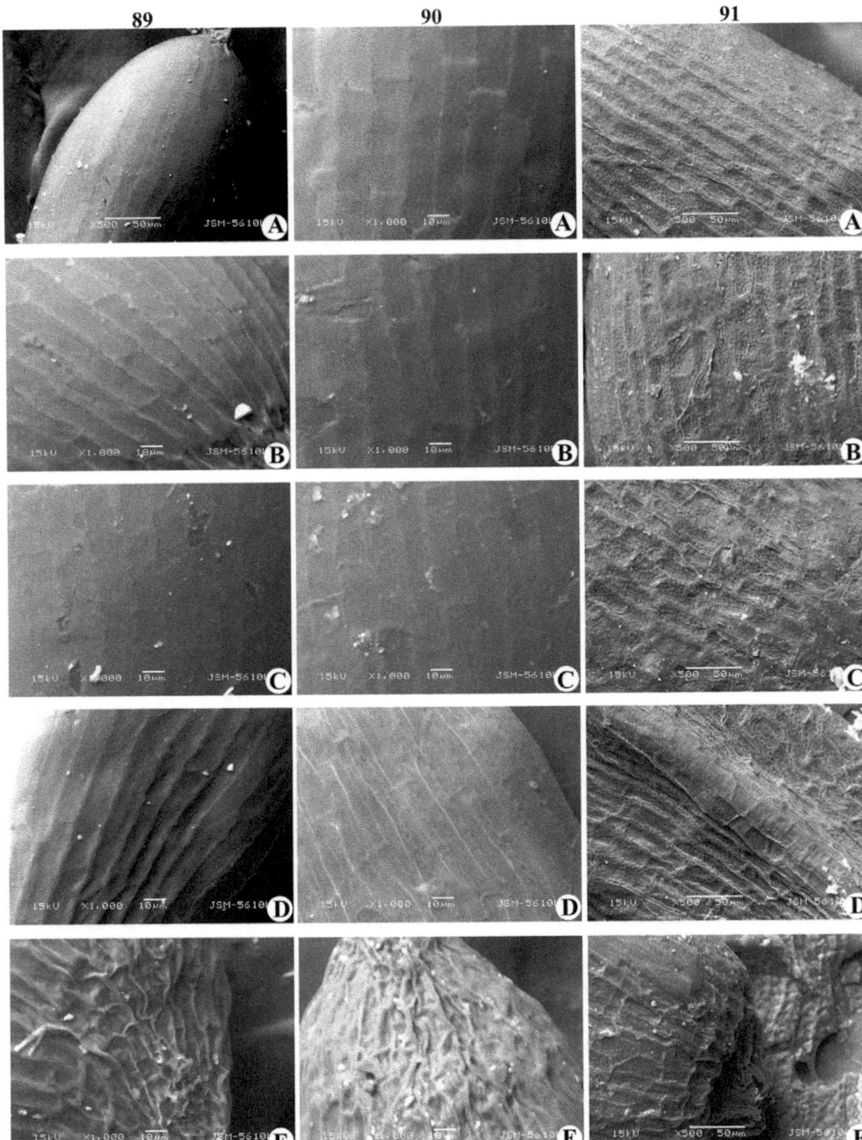

FIGURE 44 Scanning electron microscopic study of the grass caryopses (A. Dorsal surface; B. Ventral surface; C. Lateral surface; D. Embryo; E. Hilum; 89. *Eragrostis japonica*; 90. *Eragrostis nutans*; 91. *Eragrostis pilosa*).

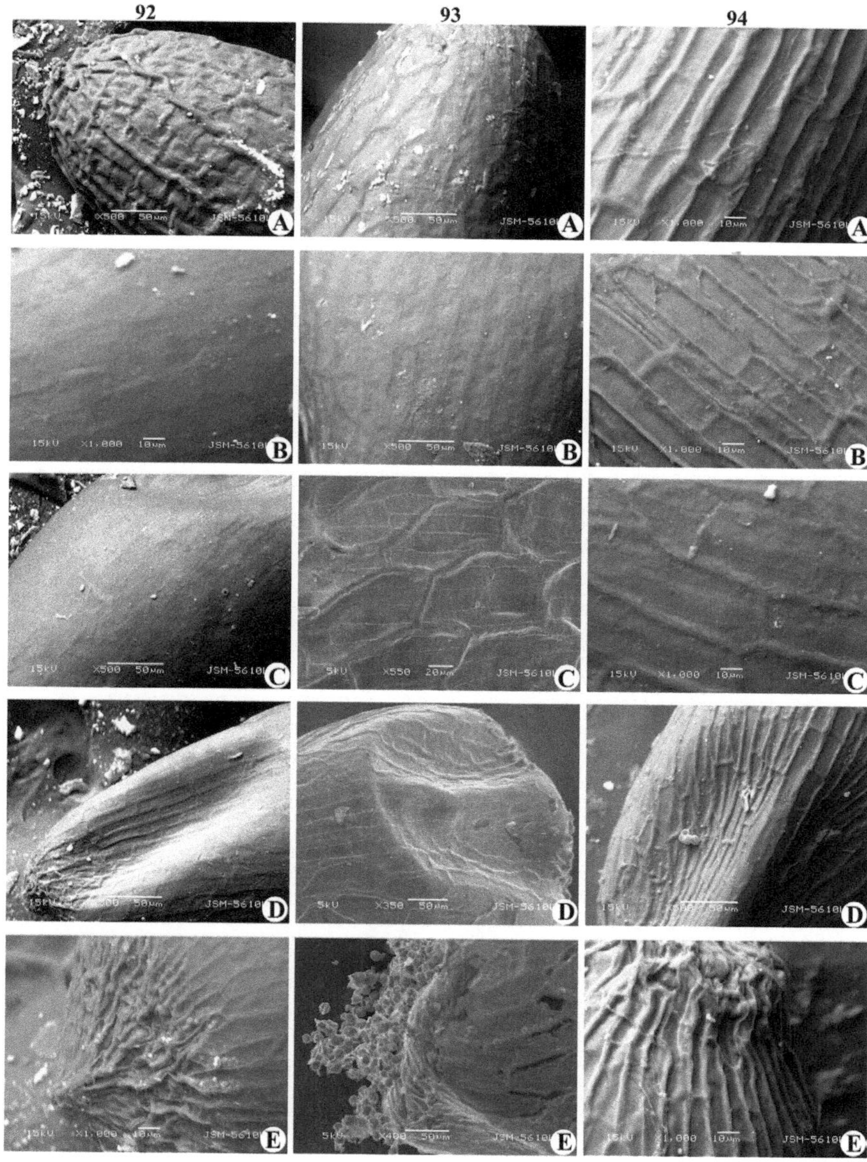

FIGURE 45 Scanning electron microscopic study of the grass caryopses (A. Dorsal surface; B. Ventral surface; C. Lateral surface; D. Embryo; E. Hilum; 92. *Eragrostis tenella*; 93. *Eragrostis tremula*; 94. *Eragrostis unioloides*).

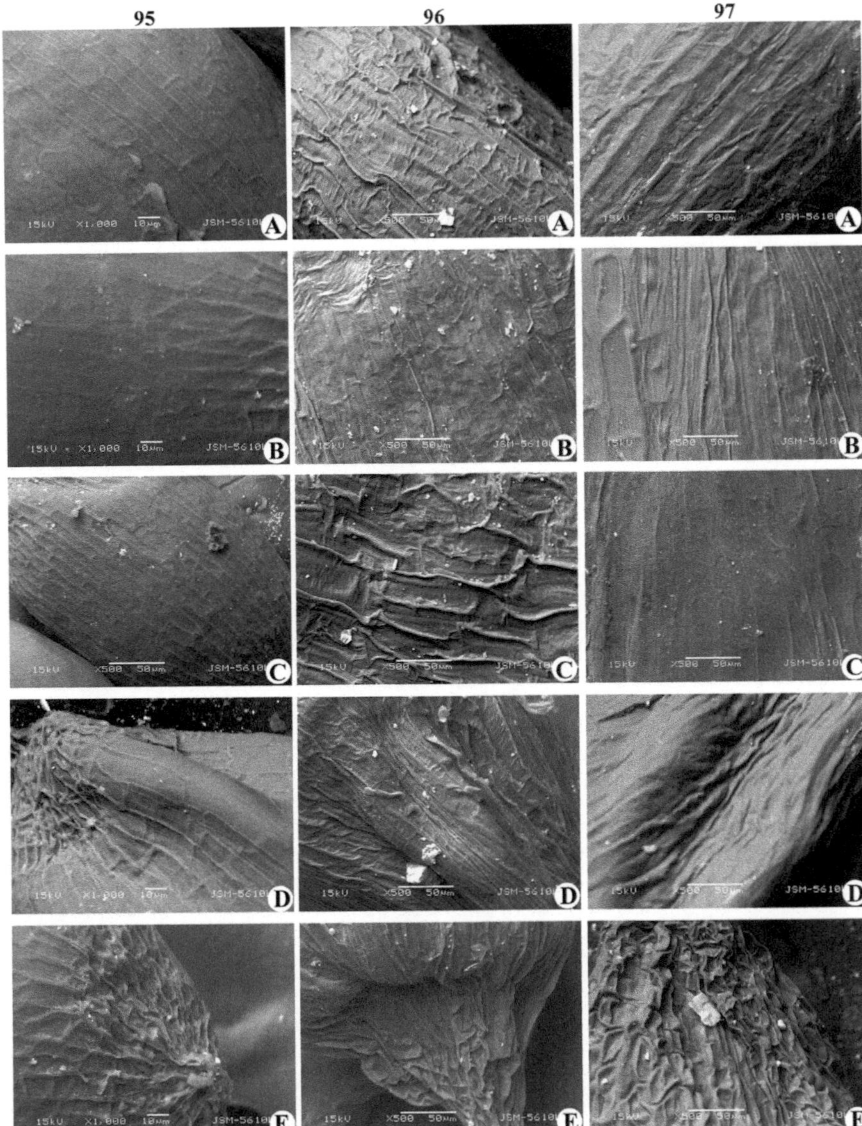

FIGURE 46 Scanning electron microscopic study of the grass caryopses (A. Dorsal surface; B. Ventral surface; C. Lateral surface; D. Embryo; E. Hilum; 95. *Eragrostis viscosa*; 96. *Sporobolus coromardelianus*; 97. *Sporobolus diander*).

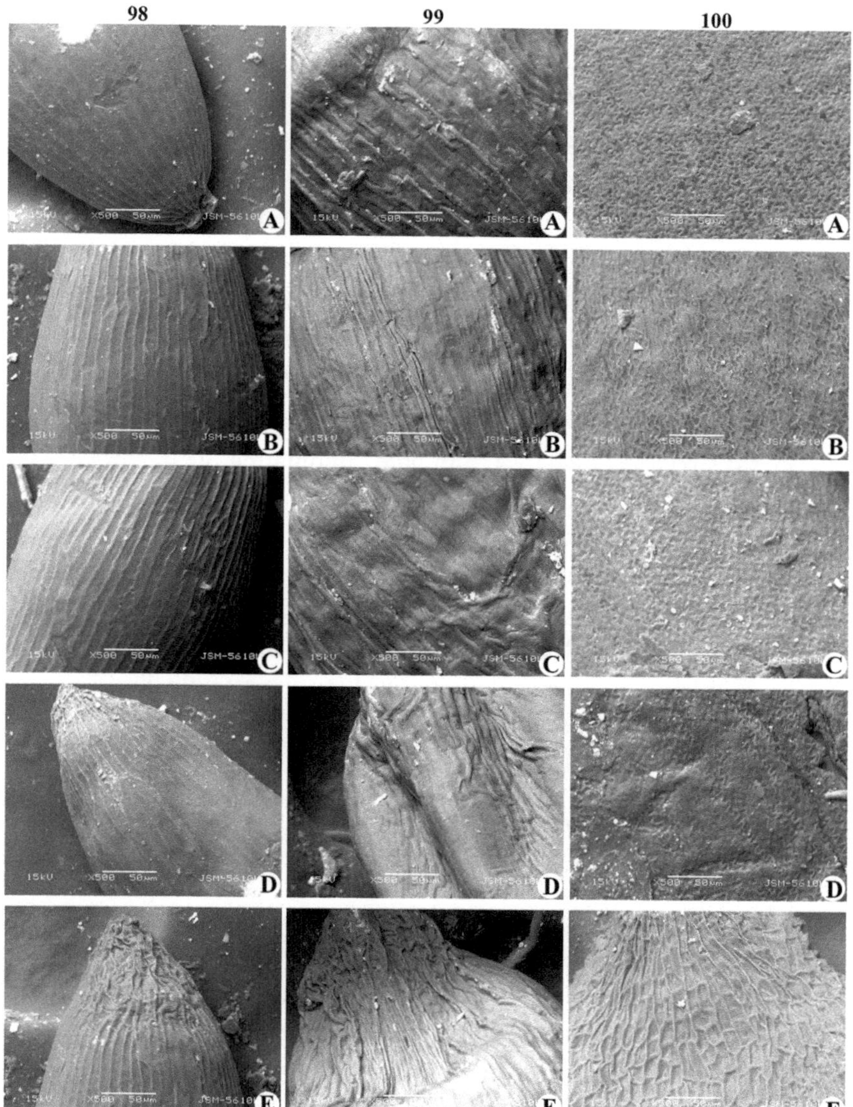

FIGURE 47 Scanning electron microscopic study of the grass caryopses (A. Dorsal surface; B. Ventral surface; C. Lateral surface; D. Embryo; E. Hilum; 98. *Sporobolus indicum*; 99. *Urochondra setulosa*; 100. *Tragrus biflorus*).

1.4 LIGHT AND SCANNING ELECTRON MICROSCOPIC FEATURES

1. *Chionachne koenigii*

Light Microscopy (Figure 4(1))

Color: White to creamish
Shape: Ovate to obovate
Texture: Smooth
Compression: Dorsally compressed
Dorsal/Lateral straitions: Absent
Ventral groove: Absent
Scutellum shape: Sickle
Embryo type: L
Embryo class: Large
Hilum visibility: Prominent
Hilum type: Basal
Hilum shape: Oval

Scanning Electron Microscopy (Figure 15(1))

Dorsal surface: Rugose ruminate, undulating wall, strong even undulation
Ventral surface: Rugose ruminate, undulating wall, strong even undulation
Lateral surface: Scalariform, both walls elevated, wide
Embryo surface: Reticulate, long rectangular reticulum, undulating walls, depressions between reticulum
Hilum surface: Rugose, both walls feebly undulating elevated and flat wall

2. *Coix lachrymal-jobi*

Light Microscopy (Figure 4(2))

Color: Dull yellowish to lightish brown
Shape: Suborbicular
Texture: Smooth, dull
Compression: Dorsally compressed
Dorsal/Lateral straitions: Absent
Ventral groove: Absent
Scutellum shape: Sickle
Embryo type: Large
Embryo class: L
Hilum visibility: Prominent
Hilum type: Basal
Hilum shape: Oval

Scanning Electron Microscopy (Figure 14)

Dorsal surface: Ruminate rugose, uneven elevated walls, big and irregular depressions
Ventral surface: Reticulate rugose-foveate, uneven elevated, thick and flat walls, smooth and irregular elevations, deep depressions between reticulum

Lateral surface: Ruminate rugose, undulate uneven elevated walls, big and irregular depressions

Embryo surface: Ruminate rugose, convolute walls, proximal end undulate and distal end reticulate blister, rugose and foveate, concave interspace

Hilum surface: Reticulate rugose, smooth undulate and narrow wall, irregular depressions towards the proximal end of caryopsis

3. *Andropogon pumilus*

Light Microscopy (Figure 4(3))

Color: Light yellowish to cream
Shape: Oblong, slender
Texture: Smooth
Compression: Dorsally compressed
Dorsal/Lateral straitions: Absent
Ventral groove: Absent
Scutellum shape: Sickle
Embryo type: Large
Embryo class: L
Hilum visibility: Prominent
Hilum type: Basal
Hilum shape: Fan shaped

Scanning Electron Microscopy (Figure 15(3))

Dorsal surface: Striate reticulate, elongated and thin reticulum, thick and feebly undulating walls

Ventral surface: Reticulate, rectangular reticulum, thin, narrow, slightly undulating wall

Lateral surface: Striate reticulate, elongated and thin reticulum, thick and feebly undulating walls

Embryo surface: Reticulate, squarish to rectangular reticulum, thick, flat wall, straight horizontal depressions present on the surface

Hilum surface: Striate ribbed

4. *Apluda mutica*

Light Microscopy (Figure 4(4))

Color: Light brown-to-brown
Shape: Ovate
Texture: Smooth, shiny
Compression: Laterally compressed
Dorsal/Lateral straitions: Absent
Ventral groove: Absent
Scutellum shape: Sickle
Embryo type: Large
Embryo class: L
Hilum visibility: Prominent
Hilum type: Basal
Hilum shape: Oval

Scanning Electron Microscopy (Figure 15(4))

Dorsal surface: Reticulate, undulating periclinal walls, '∩' and '∧' shaped undulation, flat, straight or slant anticlinal walls

Ventral surface: Reticulate, undulating periclinal walls, '∧' shaped undulation pattern, flat, straight anticlinal walls

Lateral surface: Reticulate, undulating periclinal walls, '∩' and '∧' shaped undulation, flat, straight or slant anticlinal walls

Embryo surface: Reticulate, undulating periclinal walls, '∩' shaped undulation, flat, straight anticlinal walls

Hilum surface: Reticulate, feebly undulating walls

5. *Arthraxon lanceolatus*

Light Microscopy (Figure 4(5))

Color: Whitish green to light brown **Shape:** Oblong

Texture: Smooth, shiny **Compression:** Dorsally compressed

Dorsal/Lateral straitions: Present **Ventral groove:** Absent

Scutellum shape: Sickle **Embryo type:** Large

Embryo class: L **Hilum visibility**: Prominent

Hilum type: Basal **Hilum shape:** V shaped

Scanning Electron Microscopy (Figure 16(5))

Dorsal surface: Reticulate, feebly undulating walls

Ventral surface: Reticulate, undulating periclinal walls, 'W' shaped undulation, straight anticlinal walls

Lateral surface: Reticulate, undulating periclinal walls, 'W' shaped undulation, straight anticlinal walls

Embryo surface: Reticulate, undulating periclinal walls, 'W' shaped undulation, straight anticlinal walls

Hilum surface: Ruminate rugose, convoluted walls

6. *Bothriochloa pertusa*

Light Microscopy (Figure 4(6))

Color: Light brown to brown **Shape:** Obovate

Texture: Smooth **Compression:** Dorsally compressed

Dorsal/Lateral straitions: Present
Ventral groove: Absent
Scutellum shape: Sickle
Embryo type: Large
Embryo class: N
Hilum visibility: Prominent
Hilum type: Basal
Hilum shape: Fan shaped

Scanning Electron Microscopy (Figure 16(6))

Dorsal surface: Reticulate, undulating periclinal walls, 'V' shaped undulation, straight anticlinal walls
Ventral surface: Reticulate, undulating periclinal walls, straight anticlinal walls
Lateral surface: Reticulate, undulating periclinal walls, straight anticlinal walls
Embryo surface: Reticulate, undulating periclinal walls, straight anticlinal walls
Hilum surface: Reticulate, elongated reticulams, straight to convoluted walls

7. *Capillipedium hugelii*

Light Microscopy (Figure 4(7))

Color: Light brown to yellow
Shape: Oblong to obovate
Texture: Smooth
Compression: Dorsally compressed
Dorsal/Lateral straitions: Present
Ventral groove: Absent
Scutellum shape: Sickle
Embryo type: Large
Embryo class: N
Hilum visibility: Prominent
Hilum type: Basal
Hilum shape: V shaped

Scanning Electron Microscopy (Figure 16(7))

Dorsal surface: Reticulums not very clear, undulating elevated walls periclinal walls, depressed anticlinal walls
Ventral surface: Reticulums not very clear, are undulating walls, '∩' shaped undulation
Lateral surface: Reticulums not very clear, undulating elevated walls periclinal walls, anticlinal walls not clear
Embryo surface: Reticulums not clear, undulating walls, anticlinal walls not clear
Hilum surface: Ruminate rugose, convoluted walls, depressed interspace

8. *Chrysopogon fulvus*

Light Microscopy (Figure 4(8))

Color: White to creamish
Shape: Oblong to elliptic, linear
Texture: Smooth
Compression: Laterally compressed
Dorsal/Lateral straitions: Present
Ventral groove: Absent
Scutellum shape: V
Embryo type: Large
Embryo class: N
Hilum visibility: Prominent
Hilum type: Basal
Hilum shape: Linear

Scanning Electron Microscopy (Figure 17(8))

Dorsal surface: Reticulate, elongated reticulum, straight walls, depressed interspace

Ventral surface: Striate pattern, no reticulum formation seen, feebly features

Lateral surface: Reticulate, elongated reticulum, straight walls

Embryo surface: Reticulate, straight and elevated walls, irregular elevations, depressed interspace

Hilum surface: Ruminate rugose, convoluted walls

9. *Cymbopogon martinii*

Light Microscopy (Figure 4(9))

Color: Pale yellow to light brown
Shape: Oblong
Texture: Smooth
Compression: Dorsally compressed
Dorsal/Lateral straitions: Absent
Ventral groove: Present
Scutellum shape: Sickle
Embryo type: Large
Embryo class: N
Hilum visibility: Prominent
Hilum type: Sub basal
Hilum shape: Circular

Scanning Electron Microscopy (Figure 17(9))

Dorsal surface: Reticulum not seen, undulating wall, 'W' shaped, sharp pointed undulation

Ventral surface: Reticulate, undulating walls, '∩' and '∧' shaped mixed undulation

Lateral surface: Reticulum not seen, undulating wall, undulation shape is not clear

Embryo surface: Reticulum not seen, undulating wall, 'W' shaped, sharp pointed undulation

Hilum surface: Ribbed striate, convolute walls

10. *Dichanthium annulatum*

Light Microscopy (Figure 4(10))

Color: Light brown
Shape: Oblong to ovate
Texture: Smooth
Compression: Dorsally compressed
Dorsal/Lateral straitions: Present
Ventral groove: Absent
Scutellum shape: Sickle
Embryo type: Large
Embryo class: N
Hilum visibility: Prominent
Hilum type: Basal
Hilum shape: Fan shaped

Scanning Electron Microscopy (Figure 17(10))

Dorsal surface: Reticulum feebly seen, feebly undulating wall
Ventral surface: Reticulate, feebly undulating periclinal wall, slanting and straight anticlinal wall
Lateral surface: Reticulate, undulating periclinal wall, straight anticlinal wall
Embryo surface: Reticulum feebly seen, thick undulating elevated wall
Hilum surface: Reticulate, straight walls, convex interspace

11. *Dichanthium caricosum*

Light Microscopy (Figure 5(11))

Color: Whitish brown to light brown
Shape: Ovate
Texture: Smooth
Compression: Dorsally compressed
Dorsal/Lateral straitions: Present
Ventral groove: Absent
Scutellum shape: Sickle
Embryo type: Large
Embryo class: N
Hilum visibility: Prominent
Hilum type: Basal
Hilum shape: Fan shaped

Scanning Electron Microscopy (Figure 18(11))

Dorsal surface: Reticulate, periclinal wall undulating and elevated, 'Λ' shaped undulation, straight or slant anticlinal wall
Ventral surface: Reticulate, undulating, thick, smooth periclinal wall, 'W' shaped broad undulation, straight anticlinal wall
Lateral surface: Reticulate, undulating periclinal wall, 'Λ' shaped undulation, straight anticlinal wall

Embryo surface: Reticulum feebly seen, undulating and thick wall, broad undulation

Hilum surface: Ruminate rugose, convoluted walls, depressed interspace

12. *Dimeria orinthopoda*

Light Microscopy (Figure 5(12))

Color: Light brown
Shape: Obovate
Texture: Smooth
Compression: Not compressed
Dorsal/Lateral straitions: Absent
Ventral groove: Absent
Scutellum shape: V
Embryo type: Short
Embryo class: L
Hilum visibility: Prominent
Hilum type: Basal
Hilum shape: Circular

Scanning Electron Microscopy (Figure 18(12))

Dorsal surface: Reticulate, elongated reticulum, straight or feebly undulating, thin periclinal wall, straight anticlinal wall

Ventral surface: Reticulate, elongated reticulum, are straight thin wall

Lateral surface: Reticulate, thin and elongated reticulum, feebly undulating or straight periclinal wall, straight anticlinal wall

Embryo surface: Reticulate, straight or feebly undulating periclinal wall, straight anticlinal wall, concave interspace

Hilum surface: Ruminate rugose, convoluted walls, depressed interspace

13. *Heteropogon contortus var. contortus sub var. typicus*

Light Microscopy (Figure 5(13))

Color: White to light brown
Shape: Linear
Texture: Smooth
Compression: Dorsally compressed
Dorsal/Lateral straitions: Absent
Ventral groove: Absent
Scutellum shape: Sickle
Embryo type: Large
Embryo class: N
Hilum visibility: Prominent
Hilum type: Basal
Hilum shape: Fan shaped

Scanning Electron Microscopy (Figure 18(13))

Dorsal surface: Reticulate, undulating periclinal wall, 'W' shaped undulation, smooth angle, straight anticlinal wall

Ventral surface: Reticulate, straight wall

Lateral surface: Reticulate, feebly undulating or straight walls

Embryo surface: Reticulate, periclinal wall of embryo axis feebly undulating while periclinal wall of scutellum straight, straight anticlinal wall

Hilum surface: Ruminate pattern, rugose surface, irregular and deep channels

14. *Heteropogon contortus var. contortus sub var. genuinus*

Light Microscopy (Figure 5(14))

Color: White to light brown

Shape: Linear

Texture: Smooth

Compression: Dorsally compressed

Dorsal/Lateral straitions: Absent

Ventral groove: Absent

Scutellum shape: Sickle

Embryo type: Large

Embryo class: N

Hilum visibility: Prominent

Hilum type: Basal

Hilum shape: Fan shaped

Scanning Electron Microscopy (Figure 19(14))

Dorsal surface: Reticulate, undulating periclinal wall, 'W' shaped undulation with sharp angle, straight anticlinal wall, concave interspace

Ventral surface: Reticulate, undulating periclinal wall, 'W' shaped undulation with sharp angle, straight anticlinal wall

Lateral surface: Reticulate, undulating periclinal wall, 'W' shaped undulation with sharp angle, straight anticlinal wall, concave interspace

Embryo surface: Reticulate, undulating periclinal wall, 'W' shaped undulation with sharp angle, straight anticlinal wall

Hilum surface: Ruminate, rugose, convoluted walls, irregular depressions, multidirectional crimpes

15. *Heteropogon ritchiei*

Light Microscopy (Figure 5(15))

Color: Light brown **Shape:** Linear
Texture: Smooth
Compression: Dorsally compressed
Dorsal/Lateral straitions: Absent
Ventral groove: Absent
Scutellum shape: Sickle
Embryo type: Large
Embryo class: N
Hilum visibility: Prominent
Hilum type: Basal
Hilum shape: Fan shaped

Scanning Electron Microscopy (Figure 19(15))

Dorsal surface: Reticulate, undulating periclinal wall, straight anticlinal wall
Ventral surface: Reticulate, undulating periclinal wall, sharp undulation, straight anticlinal wall, concave interspace
Lateral surface: Reticulate, undulating periclinal wall, straight anticlinal wall
Embryo surface: Reticulate, straight walls
Hilum surface: Reticulate, straight or slant, elevated wall

16. *Heteropogon triticeus*

Light Microscopy (Figure 5(16))

Color: Brown
Shape: Ovate
Texture: Smooth
Compression: Dorsally compressed
Dorsal/Lateral straitions: Present
Ventral groove: Absent
Scutellum shape: Sickle
Embryo type: Large
Embryo class: N
Hilum visibility: Prominent
Hilum type: Basal
Hilum shape: Fan shaped

Scanning Electron Microscopy (Figure 19(16))

Dorsal surface: Reticulate, straight and thick walls, depressed interspace
Ventral surface: Reticulate, straight or feebly undulating walls, depressed interspace
Lateral surface: Reticulate, straight and thick walls, depressed interspace
Embryo surface: Reticulate, straight walls, depressed and narrow interspace

Hilum surface: Ruminate rugose, depressed interspace, irregular depressions

17. *Ischaemum indicum*

Light Microscopy (Figure 5(17))

Color: Light brown to brown
Shape: Oblong
Texture: Smooth, dull
Compression: Dorsally compressed
Dorsal/Lateral straitions: Present
Ventral groove: Absent
Scutellum shape: Sickle
Embryo type: Large
Embryo class: N
Hilum visibility: Prominent
Hilum type: Sub basal
Hilum shape: Fan shaped

Scanning Electron Microscopy (Figure 20(17))

Dorsal surface: Reticulate, undulating periclinal wall, 'ʌ' shaped sharp angled undulation, straight anticlinal wall, slightly depressed interspace

Ventral surface: Reticulate, straight, thick and wide walls

Lateral surface: Reticulate, undulating periclinal wall, 'ʌ' shaped smooth angled undulation, straight anticlinal wall, slightly depressed interspace

Embryo surface: Reticulate, undulating periclinal wall, 'ʌ' shaped sharp angled undulation, straight anticlinal wall, slightly depressed interspace

Hilum surface: Reticulate, squarish reticulum, straight, thick and wide walls, depressed interspace

18. *Ischaemum molle*

Light Microscopy (Figure 5(18))

Color: Light brown
Shape: Oblong, ovate to ovoid
Texture: Smooth, dull
Compression: Dorsally compressed
Dorsal/Lateral straitions: Present
Ventral groove: Absent
Scutellum shape: Sickle
Embryo type: Large
Embryo class: N
Hilum visibility: Prominent
Hilum type: Basal
Hilum shape: Fan shaped

Scanning Electron Microscopy (Figure 20(18))

Dorsal surface: Reticulate, feebly reticulum, straight walls intermittent with undulating walls, slightly depressed interspace

Ventral surface: Reticulate, straight walls intermittent with undulating walls, slightly depressed interspace

Lateral surface: Reticulate, undulating periclinal wall, straight anticlinal wall

Embryo surface: Reticulate, straight walls intermittent with undulating walls, slightly depressed interspace

Hilum surface: Striate ribbed, elevated walls

19. *Ischaemum pilosum*

Light Microscopy (Figure 5(19))

Color: Brown

Shape: Oblong

Texture: Smooth

Compression: Dorsally compressed

Dorsal/Lateral straitions: Present

Ventral groove: Absent

Scutellum shape: Sickle

Embryo type: Large

Embryo class: N

Hilum visibility: Prominent

Hilum type: Basal

Hilum shape: Fan shaped

Scanning Electron Microscopy (Figure 20(19))

Dorsal surface: Reticulate foveate, undulating periclinal wall, broad 'ʌ' shaped undulation, striate anticlinal wall, pitted interspace

Ventral surface: Reticulate, undulating periclinal wall, narrow 'ʌ' shaped undulation, striate anticlinal wall, filled interspace

Lateral surface: Reticulate, undulating periclinal wall, 'ʌ' shaped undulation, striate or slant anticlinal wall, depressed interspace

Embryo surface: Reticulate, undulating periclinal wall, 'ʌ' shaped undulation, striate anticlinal wall

Hilum surface: Striate, narrow and convoluted wall, depressed interspace

20. *Ischaemum rugosum*

Light Microscopy (Figure 5(20))

Color: Light brown

Shape: Ovate

Texture: Smooth

Compression: Dorsally compressed

Dorsal/Lateral straitions: Present

Ventral groove: Absent

Scutellum shape: Sickle
Embryo type: Large
Embryo class: N
Hilum visibility: Prominent
Hilum type: Basal
Hilum shape: Fan shaped

Scanning Electron Microscopy (Figure 21(20))

Dorsal surface: Reticulate not clear, undulating and thick walls, smooth undulation

Ventral surface: Reticulate not clear, somewhat ribbed, undulating and thick walls, smooth undulation

Lateral surface: Reticulate not clear, undulating and thick walls, broad '∧' shaped blunt angle undulation

Embryo surface: Reticulate not clear, undulating and thick walls, smooth undulation

Hilum surface: Ruminate rugose, convoluted walls, irregular, narrow depressions

21. *Iseilema laxum*

Light Microscopy (Figure 6(21))

Color: Lightish green to creamish
Shape: Oblong
Texture: Smooth
Compression: Dorsally compressed
Dorsal/Lateral straitions: Present
Ventral groove: Absent
Scutellum shape: Sickle
Embryo type: Large
Embryo class: N
Hilum visibility: Prominent
Hilum type: Basal
Hilum shape: Fan shaped

Scanning Electron Microscopy (Figure 21(21))

Dorsal surface: Reticulate, undulating periclinal wall, '∧' shaped sharp angled undulation, straight anticlinal wall, slightly depressed interspace

Ventral surface: Reticulate, undulating periclinal wall, straight anticlinal wall

Lateral surface: Reticulate, undulating periclinal wall, '∧' shaped sharp angled undulation, straight anticlinal wall, slightly depressed interspace

Embryo surface: Reticulate pattern, anticlinal wall is undulating, periclinal wall is striate, undulating wall is mixed with striate wall

Hilum surface: Reticulate pattern, walls are striate, narrow, 4–5 sided reticulum, between two reticulum narrow channels are running, deeply sunken interspace

22. *Ophiuros exaltatus*

Light Microscopy (Figure 6(22))

Color: Light brown to brown
Shape: Oblong
Texture: Smooth
Compression: Not compressed
Dorsal/Lateral straitions: Present
Ventral groove: Absent
Scutellum shape: Sickle
Embryo type: Large
Embryo class: N
Hilum visibility: Prominent
Hilum type: Basal
Hilum shape: V shaped

Scanning Electron Microscopy (Figure 21(22))

Dorsal surface: Reticulate, undulating periclinal wall, sharp undulation, straight or undulating anticlinal wall

Ventral surface: Reticulate, undulating periclinal wall, sharp undulation, straight anticlinal wall

Lateral surface: Reticulate, undulating periclinal wall, sharp and uneven undulation, straight anticlinal wall

Embryo surface: Reticulate, undulating periclinal wall, sharp undulation, straight anticlinal wall

Hilum surface: Ruminate rugose, uneven depressions

23. *Rottboellia exaltata*

Light Microscopy (Figure 6(23))

Color: Greenish to light color
Shape: Oblong
Texture: Smooth
Compression: Dorsally compressed
Dorsal/Lateral straitions: Present
Ventral groove: Absent
Scutellum shape: Sickle
Embryo type: Large
Embryo class: N
Hilum visibility: Prominent
Hilum type: Sub basal
Hilum shape: Fan shaped

Scanning Electron Microscopy (Figure 22(23))

Dorsal surface: Reticulate, undulating periclinal wall, sharp angle undulation, straight anticlinal wall

Ventral surface: Reticulate, undulating periclinal wall, straight or slant anticlinal wall

Lateral surface: Reticulate, undulating periclinal wall, sharp angle 'V' shaped undulation, straight anticlinal wall

Embryo surface: Reticulate, undulating periclinal wall, uneven undulation, straight anticlinal wall

Hilum surface: Ruminate reticulate, narrow depressions between reticulum

24. *Saccharum spontaneum*

Light Microscopy (Figure 6(24))

Color: Light brown

Shape: Lanceolate

Texture: Rough

Compression: Dorsally compressed

Dorsal/Lateral straitions: Present

Ventral groove: Absent

Scutellum shape: Sickle

Embryo type: Large

Embryo class: N

Hilum visibility: Prominent

Hilum type: Basal

Hilum shape: Linear

Scanning Electron Microscopy (Figure 22(24))

Dorsal surface: Reticulate, uneven reticulum, wide and thick periclinal wall, thin, straight anticlinal wall, concave interspace

Ventral surface: Reticulate-scalariform rugose, is thick, elevated, undulating anticlinal wall, undulating periclinal wall, concave interspace

Lateral surface: Rugose and rough surface

Embryo surface: Reticulate, elevated periclinal wall, straight anticlinal wall, concave interspace

Hilum surface: Reticulate, uneven reticulum, elevated, thick wall, concave interspace

25. *Sehima ischaemoides*

Light Microscopy (Figure 6(25))

Color: Yellowish to light brown

Shape: Oblong

Texture: Smooth

Compression: Dorsally compressed

Dorsal/Lateral straitions: Absent

Ventral groove: Present

Scutellum shape: Sickle

Embryo type: Large

Embryo class: N

Hilum visibility: Prominent

Hilum type: Basal

Hilum shape: Fan shaped

Scanning Electron Microscopy (Figure 22(25))

Dorsal surface: Reticulate, undulating wall, 'V' or 'W' shaped, narrow and sharp angled undulation

Ventral surface: Reticulate, undulating wall, smooth angled, broad undulation

Lateral surface: Reticulate, undulating wall, 'V' or 'W' shaped, uneven, compact and sharp angled undulation

Embryo surface: Reticulate, undulating wall, 'V' or 'W' shaped, uneven and sharp angled undulation

Hilum surface: Reticulate ribbed, penta to hexagonal reticulum

26. *Sehima nervosum*

Light Microscopy (Figure 6(26))

Color: Pale yellow to light brown

Shape: Fusiform to lanceolate

Texture: Smooth, dull

Compression: Dorsally compressed

Dorsal/Lateral straitions: Absent

Ventral groove: Present

Scutellum shape: Sickle

Embryo type: Large

Embryo class: N

Hilum visibility: Prominent

Hilum type: Basal

Hilum shape: Fan shaped

Scanning Electron Microscopy (Figure 23(26))

Dorsal surface: Striate, undulating wall, 'V' or 'W' shaped, narrow and sharp angled undulation

Ventral surface: Striate, undulating wall, smooth angled, broad undulation, deep depression present between two walls of reticulum

Lateral surface: Striate, undulating wall, 'V' or 'W' shaped, uneven and sharp angled undulation

Embryo surface: Striate, undulating wall, 'V' or 'W' shaped, uneven and sharp angled undulation, filled interspace

Hilum surface: Rugose, irregular ribs seen

27. *Sehima sulcatum*

Light Microscopy (Figure 6(27))

Color: Yellowish to light brown

Shape: Oblong to fusiform

Texture: Smooth

Compression: Dorsally compressed

Dorsal/Lateral straitions: Absent
Ventral groove: Present
Scutellum shape: Sickle
Embryo type: Large
Embryo class: N
Hilum visibility: Prominent
Hilum type: Basal
Hilum shape: Fan shaped

Scanning Electron Microscopy (Figure 23(27))

Dorsal surface: Striate, feebly undulating wall
Ventral surface: Reticulate, undulating slightly elevated wall, Smooth angled undulation, concave interspace
Lateral surface: Reticulate, undulating periclinal wall, sharp angled undulation, straight anticlinal wall
Embryo surface: Reticulate, periclinal wall undulating, sharp angled undulation, straight anticlinal wall
Hilum surface: Reticulate rugose, uneven sized reticulum, straight and feebly seen wall

28. *Sorghum halepense*

Light Microscopy (Figure 6(28))

Color: Brownish black
Shape: Ovate
Texture: Smooth, dull
Compression: Dorsally compressed
Dorsal/Lateral straitions: Present
Ventral groove: Absent
Scutellum shape: Sickle
Embryo type: Large
Embryo class: N
Hilum visibility: Prominent
Hilum type: Basal
Hilum shape: V shaped

Scanning Electron Microscopy (Figure 23(28))

Dorsal surface: Striate, undulating, thick walls
Ventral surface: Striate reticulate, straight walls, slightly concave interspace
Lateral surface: Striate, feebly seen reticulum, undulating, thick and elevated walls
Embryo surface: Striate, slightly undulating and thick walls
Hilum surface: Ribbed rugose, horizontal elevations

29. *Sorghum purpureo-sericeum*

Light Microscopy (Figure 6(29))

Color: Brown to dark brown
Shape: Ovate
Texture: Smooth, dull
Compression: Dorsally compressed

Dorsal/Lateral straitions: Present **Ventral groove:** Absent
Scutellum shape: Sickle **Embryo type:** Large
Embryo class: N **Hilum visibility:** Prominent
Hilum type: Basal **Hilum shape:** V shaped

Scanning Electron Microscopy (Figure 24(29))

Dorsal surface: Reticulate, straight, thick and slightly elevated walls, pitted interspace

Ventral surface: Reticulate, feebly undulate walls, few reticulum shows pitted interspace

Lateral surface: Reticulate, straight, thick, sharp and slightly elevated walls, pitted interspace

Embryo surface: Reticulate, straight, thin walls, pitted interspace

Hilum surface: Rugose, uneven multidirectional crimps, concave interspace

30. *Thelepogn elegans*

Light Microscopy (Figure 6(30))

Color: Light brown to black **Shape:** Oblong
Texture: Rough **Compression:** Dorsally compressed
Dorsal/Lateral straitions: Present **Ventral groove:** Absent
Scutellum shape: Sickle **Embryo type:** Large
Embryo class: L **Hilum visibility:** Prominent
Hilum type: Basal **Hilum shape:** V shaped

Scanning Electrwwon Microscopy (Figure 24(30))

Dorsal surface: Reticulate, uneven sized reticulum, elevated undulating periclinal wall, sharp angled and broad undulation, straight anticlinal wall, concave interspace

Ventral surface: Reticulate, uneven sized reticulum, elevated undulating periclinal wall, sharp angled and broad undulation, straight anticlinal wall, concave interspace

Lateral surface: Reticulate, uneven sized reticulum, elevated undulating periclinal wall, sharp angled and broad undulation, straight anticlinal wall, concave interspace

Embryo surface: Reticulate, uneven sized reticulum, elevated undulating periclinal wall, sharp angled and broad undulation, straight anticlinal wall, concave interspace

Hilum surface: Rugose, uneven multidirectional crimpes

31. *Themeda cymbaria*

Light Microscopy (Figure 7(31))

Color: Yellowish to light brown
Shape: Ellipse to fusiform
Texture: Smooth, dull
Compression: Dorsally compressed
Dorsal/Lateral straitions: Absent
Ventral groove: Absent
Scutellum shape: Sickle
Embryo type: Large
Embryo class: N
Hilum visibility: Prominent
Hilum type: Basal
Hilum shape: V shaped

Scanning Electron Microscopy (Figure 24(31))

Dorsal surface: Reticulate, undulating elevated periclinal wall, slightly smooth angled and narrow reticulum, straight or slant anticlinal wall

Ventral surface: Reticulate, undulating periclinal wall, slightly smooth angled and broad reticulum, straight anticlinal wall

Lateral surface: Reticulate, undulating periclinal wall, slightly smooth angled and broad reticulum, straight anticlinal wall

Embryo surface: Reticulate, undulating periclinal wall, smooth angled undulation, straight anticlinal wall

Hilum surface: Rugose-foveate, uneven multidirectional crimpes

32. *Themeda laxa*

Light Microscopy (Figure 7(32))

Color: Light brown
Shape: Ovate
Texture: Smooth
Compression: Dorsally compressed
Dorsal/Lateral straitions: Absent
Ventral groove: Absent
Scutellum shape: Sickle
Embryo type: Large

Embryo class: N
Hilum visibility: Prominent
Hilum type: Basal
Hilum shape: V shaped

Scanning Electron Microscopy (Figure 25(32))

Dorsal surface: Reticulate, feebly undulating, thick wall

Ventral surface: Reticulate, feebly undulating, thin and elevated periclinal wall, straight thin elevated anticlinal wall, concave interspace

Lateral surface: Reticulate, undulating periclinal wall, smooth angled and broad undulation, straight anticlinal wall

Embryo surface: Reticulate, straight to feebly undulating periclinal wall, straight anticlinal wall, thick wide and elevated walls

Hilum surface: Reticulate, straight, thick and elevated walls, depressed interspace

33. *Themeda triandra*

Light Microscopy (Figure 7(33))

Color: Whitish brown to brown
Shape: Lanceolate to oblong
Texture: Smooth
Compression: Dorsally compressed
Dorsal/Lateral straitions: Absent
Ventral groove: Absent
Scutellum shape: Sickle
Embryo type: Large
Embryo class: N
Hilum visibility: Prominent
Hilum type: Basal
Hilum shape: V shaped

Scanning Electron Microscopy (Figure 25(33))

Dorsal surface: Reticulate, straight, thick, elevated walls, depressed interspace

Ventral surface: Reticulate, feebly undulating periclinal wall, straight or slant anticlinal wall

Lateral surface: Reticulate, feebly undulating periclinal wall, straight or slant anticlinal wall, walls are slightly elevated, concave interspace

Embryo surface: Reticulate pattern, walls are straight, periclinal wall is straight or slant

Hilum surface: Reticulate-ruminate rugose, multidirectional crimpes, depressed interspace

34. *Themeda quadrivalvis*

Light Microscopy (Figure 7(34))

Color: Whitish brown to brown
Shape: Lanceolate
Texture: Smooth
Compression: Dorsally compressed
Dorsal/Lateral straitions: Absent
Ventral groove: Absent
Scutellum shape: Sickle
Embryo type: Large
Embryo class: N
Hilum visibility: Faint
Hilum type: Basal
Hilum shape: V shaped

Scanning Electron Microscopy (Figure 25(34))

Dorsal surface: Reticulate, undulating periclinal wall, sharp angled and broad undulation, straight and slant anticlinal wall, concave interspace
Ventral surface: Reticulate pitted, straight, thin walls
Lateral surface: Reticulate-ruminate, slightly wavy and thin periclinal wall, straight anticlinal wall
Embryo surface: Reticulate, undulating periclinal wall, sharp angled undulation, straight anticlinal wall
Hilum surface: Striate, unidirectional crimpes, narrow depressed interspace

35. *Triplopogon ramosissimus*

Light Microscopy (Figure 7(35))

Color: Brown to dark brown
Shape: Ellipsoid
Texture: Smooth, shiny
Compression: Laterally compressed
Dorsal/Lateral straitions: Absent
Ventral groove: Absent
Scutellum shape: V
Embryo type: Large
Embryo class: L
Hilum visibility: Prominent
Hilum type: Basal
Hilum shape: V shaped

Scanning Electron Microscopy (Figure 26(35))

Dorsal surface: Reticulate, undulating periclinal wall, smooth angled undulation, straight anticlinal wall
Ventral surface: Reticulate, undulating periclinal wall intermittent with straight, smooth angled undulation, straight anticlinal wall, concave interspace

Lateral surface: Reticulate, undulating periclinal wall intermittent with straight, smooth angled undulation, straight anticlinal wall, concave interspace

Embryo surface: Reticulate, elevated and thick walls, straight intermittent with undulating wall periclinal wall, straight or slant anticlinal wall, concave interspace

Hilum surface: Rugose pattern, anticlinal wall is smooth, elevated and deeply depressed periclinal walls

36. ***Vetivaria zinzanoides***

Light Microscopy (Figure 7(36))

Color: Light green to whitish brown | **Shape:** Lanceolate

Texture: Smooth | **Compression:** Laterally compressed

Dorsal/Lateral straitions: Present | **Ventral groove:** Absent

Scutellum shape: V | **Embryo type:** Large

Embryo class: N | **Hilum visibility:** Prominent

Hilum type: Basal | **Hilum shape:** V shaped

Scanning Electron Microscopy (Figure 26(36))

Dorsal surface: Reticulate, straight, thick and elevated walls, feebly undulating anticlinal wall

Ventral surface: Reticulate, straight, thick and elevated walls, feebly undulating anticlinal wall, foveate interspace

Lateral surface: Reticulate, straight, thick and elevated walls, feebly undulating anticlinal wall, foveate and concave interspace

Embryo surface: Reticulate, elevated, thick, flat and straight periclinal wall, straight or slant anticlinal wall, depressed interspace

Hilum surface: Reticulate, narrow, smooth and elevated walls, depressed interspace

37. ***Alloteropsis cimicina***

Light Microscopy (Figure 7(37))

Color: Light brown to brown | **Shape:** Oblong

Texture: Smooth | **Compression:** Dorsally compressed

Dorsal/Lateral straitions: Absent | **Ventral groove:** Absent

Scutellum shape: Sickle | **Embryo type:** Large

Embryo class: N
Hilum visibility: Prominent
Hilum type: Basal
Hilum shape: Oval

Scanning Electron Microscopy (Figure 26(37))

Dorsal surface: Rugose ruminate feature, flat walls
Ventral surface: Reticulate Rugose ruminate, squarish reticulum, flat walls
Lateral surface: Reticulate Rugose ruminate, squarish reticulum, flat and slightly elevated walls
Embryo surface: Rugose ruminate, flat and running irregularly walls, periclinal wall is more thick than the anti-clinal wall
Hilum surface: Reticulate, straight slightly elevated walls, concave interspace

38. ***Brachiaria eruciformis***

Light Microscopy (Figure 7(38))

Color: Yellowish to light brown
Shape: Ovoid to oblong
Texture: Smooth
Compression: Dorsally compressed
Dorsal/Lateral straitions: Present
Ventral groove: Absent
Scutellum shape: V
Embryo type: Large
Embryo class: N
Hilum visibility: Prominent
Hilum type: Sub basal
Hilum shape: Sub basal

Scanning Electron Microscopy (Figure 27(38))

Dorsal surface: Reticulate, undulating and thin periclinal wall, '∩' shaped with smooth angled undulation, straight or slant anticlinal wall
Ventral surface: feebly striate
Lateral surface: Striate, horizontal elevation on the surface
Embryo surface: Reticulate, undulating and thin periclinal wall, uneven undulation, straight anticlinal wall
Hilum surface: Reticulate, straight, thick and elevated walls, depressed interspace

39. ***Brachiaria distachya***

Light Microscopy (Figure 7(39))

Color: Greenish to whitish
Shape: Ovoid to oblong

Texture: Smooth, dull
Dorsal/Lateral straitions: Present
Scutellum shape: V
Embryo class: N
Hilum type: Sub basal
Compression: Dorsally compressed
Ventral groove: Absent
Embryo type: Large
Hilum visibility: Prominent
Hilum shape: Oval

Scanning Electron Microscopy (Figure 27(39))

Dorsal surface: Reticulate, undulating periclinal wall, sharp angled undulation, straight anticlinal wall

Ventral surface: Reticulate, undulating periclinal wall, wide and smooth angled undulation, straight anticlinal wall, depressed interspace

Lateral surface: Reticulate, undulating elevated periclinal wall, uneven undulation, straight anticlinal wall, depressed interspace

Embryo surface: Reticulate, undulating elevated periclinal wall, uneven smooth and sharp angled undulation, straight anticlinal wall, depressed interspace

Hilum surface: Reticulate rugose, straight flat walls, slightly concave interspace

40. *Brachiaria ramosa*

Light Microscopy (Figure 7(40))

Color: Greenish to whitish
Texture: Smooth
Dorsal/Lateral straitions: Present
Scutellum shape: V
Embryo class: N
Hilum type: Sub basal
Shape: Ovoid to oblong
Compression: Dorsally compressed
Ventral groove: Absent
Embryo type: Large
Hilum visibility: Faint
Hilum shape: Oval

Scanning Electron Microscopy (Figure 27(40))

Dorsal surface: Reticulate, undulating periclinal wall, '∩' shaped with smooth angled undulation, straight anticlinal wall

Ventral surface: Reticulate pitted, straight, elevated walls, depressed interspace

Lateral surface: Striate, feebly reticulum, undulating wall at some place, horizontally uneven elevations

Embryo surface: Reticulate, undulating slightly elevated periclinal wall, wide angled undulation, straight anticlinal wall

Hilum surface: Reticulate rugose, straight, elevated, thick and crimped walls, depressed interspace

41. *Brachiaria reptans*

Light Microscopy (Figure 8(41))

Color: White to greenish
Shape: Ovoid to orbicular
Texture: Smooth
Compression: Dorsally compressed
Dorsal/Lateral straitions: Present
Ventral groove: Absent
Scutellum shape: V
Embryo type: Large
Embryo class: N
Hilum visibility: Prominent
Hilum type: Sub basal
Hilum shape: Oval

Scanning Electron Microscopy (Figure 28(41))

Dorsal surface: Reticulate elevated, undulating periclinal wall, smooth angled, ʌ shape with wall undulation, straight and elevated anticlinal wall, depressed interspace

Ventral surface: Reticulate, elevated and uneven walls, depressed interspace

Lateral surface: Reticulate foveolate, undulating elevated periclinal wall, smooth angled and walled with broad undulation, straight anticlinal walls, concave interspace

Embryo surface: Reticulate, undulating elevated periclinal wall, smooth angled, ʌ shape undulation, straight and elevated anticlinal wall, depressed interspace

Hilum surface: Reticulate, straight, slightly elevated and flat walls, superficial interspace with broad pit

42. *Cenchrus biflorus*

Light Microscopy (Figure 8(42))

Color: Creamish yellow
Shape: Ovate to suborbicular
Texture: Smooth, shiny
Compression: Dorsally compressed
Dorsal/Lateral straitions: Absent
Ventral groove: Absent

Scutellum shape: Sickle
Embryo type: Large
Embryo class: N
Hilum visibility: Prominent
Hilum type: Basal
Hilum shape: Oval

Scanning Electron Microscopy (Figure 28(42))

Dorsal surface: Reticulate, elevated wide walls, concave interspace, thin

Ventral surface: Reticulate, flat, wavy and thick wall, superficial interspace

Lateral surface: Reticulate, feebly undulating, elevated, thick and flat periclinal wall, slant and straight anticlinal wall, shallow interspace

Embryo surface: Reticulate pattern, walls are elevated, thick and wide, concave interspace, thin, straight lines are running in the interspace

Hilum surface: Ruminate pattern, rugose surface, irregular and narrow channels, running in different direction

43. ***Cenchrus ciliaris***

Light Microscopy (Figure 8(43))

Color: Greenish brown
Shape: Oblong
Texture: Smooth, shiny
Compression: Dorsally compressed
Dorsal/Lateral straitions: Present
Ventral groove: Absent
Scutellum shape: Sickle
Embryo type: Large
Embryo class: N
Hilum visibility: Prominent
Hilum type: Basal
Hilum shape: Oval

Scanning Electron Microscopy (Figure 28(43))

Dorsal surface: Reticulate pattern, undulating and elevated periclinal wall, '∩' shaped with smooth angled undulation, straight anticlinal wall, concave interspace

Ventral surface: Reticulate, strongly undulating periclinal wall, smooth angled and broad undulation, straight anticlinal wall

Lateral surface: Reticulate pattern, undulating periclinal wall, '∩' shaped with smooth angled undulation, straight anticlinal wall

Embryo surface: Reticulate, strongly undulating and elevated periclinal wall, '∩' shaped with smooth angled undulation, straight anticlinal wall

Hilum surface: Reticulate, straight, elevated and thick wall, depressed interspace

44. *Cenchrus setigerus*

Light Microscopy (Figure 8(44))

Color: Dull white to greenish
Shape: Ovoid to oblong
Texture: Smooth
Compression: Dorsally compressed
Dorsal/Lateral straitions: Present
Ventral groove: Absent
Scutellum shape: Sickle
Embryo type: Large
Embryo class: N
Hilum visibility: Prominent
Hilum type: Basal
Hilum shape: Oval

Scanning Electron Microscopy (Figure 29(44))

Dorsal surface: Reticulate, straight, thick, wide and elevated walls, slightly concave interspace

Ventral surface: Reticulate, wavy, elevated, smooth angled undulation periclinal wall, slant and straight at some place anticlinal wall, shallow interspace

Lateral surface: Reticulate foveolate, thin and elongated reticulum, straight thick walls

Embryo surface: Reticulate, straight, thick, wide, flat and elevated walls, slightly concave interspace

Hilum surface: Reticulate ruminate, thick, crimped and elevated walls, multidirectional crimpes, depressed interspace

45. *Cenchrus preurii*

Light Microscopy (Figure 8(45))

Color: Dull white to greenish
Shape: Ovoid to oblong
Texture: Smooth
Compression: Dorsally compressed
Dorsal/Lateral straitions: Present
Ventral groove: Absent
Scutellum shape: Sickle
Embryo type: Large
Embryo class: N
Hilum visibility: Prominent
Hilum type: Basal
Hilum shape: Oval

Scanning Electron Microscopy (Figure 29(45))

Dorsal surface: Reticulate, penta to hexagonal reticulum, straight and thick walls

Ventral surface: Reticulum, penta to hexagonal reticulum, straight, elevated and thick walls, depressed interspace

Lateral surface: Reticulum, penta to hexagonal reticulum, straight but intermittent feebly undulating, elevated and thick periclinal wall, straight anticlinal wall, depressed interspace

Embryo surface: Reticulum, penta to hexagonal reticulum, straight to feebly undulating, elevated thick periclinal wall, straight anticlinal wall, depressed interspace

Hilum surface: Reticulate ruminate, thick, crimped and elevated walls, multidirectional crimpes, depressed interspace

46. *Digitaria ciliaris*

Light Microscopy (Figure 8(45))

Color: White creamish to light pale yellow

Shape: Oblong

Texture: Smooth

Compression: Dorsally compressed

Dorsal/Lateral straitions: Present

Ventral groove: Absent

Scutellum shape: Sickle

Embryo type: Large

Embryo class: N

Hilum visibility: Prominent

Hilum type: Sub basal

Hilum shape: Oval

Scanning Electron Microscopy (Figure 29(45))

Dorsal surface: Reticulate, undulating and thin walls, '∩' shaped, smooth angled undulation, horizontally elevations on surfaces

Ventral surface: Reticulate, undulating periclinal wall, thin and elevated, smooth angled broad undulation, straight anticlinal wall, depressed interspace

Lateral surface: Reticulate, undulating and thin walls, 'Ω' shaped, smooth angled undulation, horizontally elevations on surface

Embryo surface: Reticulate, undulating and thin walls, 'Ω' shaped, smooth angled undulation, horizontally elevations on surface

Hilum surface: Reticulate, straight, slightly elevated walls, depressed interspace

47. *Digitaria microbachne*

Light Microscopy (Figure 8(47))

Color: White to creamish brown
Shape: Ellipsoid
Texture: Smooth
Compression: Dorsally compressed
Dorsal/Lateral straitions: Present
Ventral groove: Absent
Scutellum shape: Sickle
Embryo type: Large
Embryo class: N
Hilum visibility: Prominent
Hilum type: Sub basal
Hilum shape: Oval

Scanning Electron Microscopy (Figure 30(47))

Dorsal surface: Reticulate pitted, feebly features, undulating and thin walls, '∩' shaped, smooth angled undulation

Ventral surface: Reticulate, horizontally elevations on surface, undulating and thin walls, smooth angled, and broad undulation, shallow interspace

Lateral surface: Reticulate, horizontally elevations on surface, undulating and thin walls, '∩' shaped, smooth angled undulation

Embryo surface: Reticulate, undulating and thin walls, '∩' shaped, smooth angled undulation, horizontally elevations on surface, depressed interspace

Hilum surface: Reticulate, straight, slightly elevated walls, depressed interspace

48. *Digitaria granularis*

Light Microscopy (Figure 8(48))

Color: White to creamish
Shape: Ovoid
Texture: Smooth, shiny
Compression: Dorsally compressed
Dorsal/Lateral straitions: Present
Ventral groove: Absent
Scutellum shape: Sickle
Embryo type: Large
Embryo class: L
Hilum visibility: Prominent
Hilum type: Sub basal
Hilum shape: Oval

Scanning Electron Microscopy (Figure 30(48))

Dorsal surface: Reticulate, horizontally elevations on surface, undulating and thin walls, '∩' and 'Ω' shaped, smooth angled undulation

Ventral surface: Reticulate, horizontally elevations on surface, undulating and thin walls, smooth angled and broad undulation

Lateral surface: Reticulate, horizontally elevations on surface, undulating and thin walls, smooth angled and broad undulation

Embryo surface: Reticulate, uneven surface, undulating and thin walls, '∩' and 'Ω' shaped, smooth angled undulation

Hilum surface: Reticulate, elevated walls, depressed interspace

49. *Echinochloa colona*

Light Microscopy (Figure 8(49))

Color: White to creamish

Shape: Ovoid to ellipsoid

Texture: Smooth, slightly shiny

Compression: Dorsally compressed

Dorsal/Lateral straitions: Present

Ventral groove: Absent

Scutellum shape: V

Embryo type: Large

Embryo class: N

Hilum visibility: Faint

Hilum type: Basal

Hilum shape: Fan shaped

Scanning Electron Microscopy (Figure 30(49))

Dorsal surface: Reticulate rugose, undulating and thin walls, '∩' and '∧' shaped undulation, depressed interspace

Ventral surface: Reticulate rugose surface, undulating and thin walls, '∩' and '∧' shaped undulation, depressed interspace

Lateral surface: Reticulate rugose, undulating and thin walls, '∩' and '∧' shaped undulation, depressed interspace

Embryo surface: Reticulate rugose, undulating and elevated walls, depressed interspace

Hilum surface: Reticulate, straight and thin walls, concave interspace

50. *Echinochloa crusgalli*

Light Microscopy (Figure 8(50))

Color: Dull white
Shape: Ellipsoid
Texture: Smooth
Compression: Dorsally compressed
Dorsal/Lateral straitions: Present
Ventral groove: Absent
Scutellum shape: V
Embryo type: Large
Embryo class: N
Hilum visibility: Prominent
Hilum type: Basal
Hilum shape: Fan shaped

Scanning Electron Microscopy (Figure 31(50))

Dorsal surface: Reticulate, undulating, elevated thin periclinal wall, '∩' and '∧' shaped, smooth narrow undulation, straight anticlinal wall, depressed interspace

Ventral surface: Reticulate, undulating, thin and slightly elevated walls, 'Ω' shaped and broad undulation, depressed interspace

Lateral surface: Reticulate, undulating, thin and slightly elevated walls, '∩' shaped and broad, smooth angled undulation, depressed interspace, thick and flat anticlinal bands in interspace

Embryo surface: Reticulate, undulating periclinal wall, '∧' shaped, sharp angled and broad undulation, straight anticlinal wall, concave interspace, anticlinal elevations in interspace

Hilum surface: Reticulate, thick and elevated walls, concave interspace

51. *Echinochloa stagnina*

Light Microscopy (Figure 9(51))

Color: Whitish to dull white
Shape: Ellipsoid to obovate
Texture: Smooth
Compression: Dorsally compressed
Dorsal/Lateral straitions: Present
Ventral groove: Absent
Scutellum shape: V
Embryo type: Large
Embryo class: N
Hilum visibility: Prominent
Hilum type: Basal
Hilum shape: Fan shaped

Scanning Electron Microscopy (Figure 31(51))

Dorsal surface: Reticulate rugose pitted, undulating elevated walls, depressed interspace

Ventral surface: Reticulate striate to rugose, straight walls, narrow depressions between the reticulum

Lateral surface: Reticulate rugose, undulating and straight walls

Embryo surface: Reticulate, undulating, elevated walls, '∩' shaped, broad, smooth angled undulation, depressed interspace

Hilum surface: Reticulate, thick and elevated walls, concave interspace

52. *Eriochloa procera*

Light Microscopy (Figure 9(52))

Color: Creamish to light brown
Shape: Ellipsoid
Texture: Smooth
Compression: Dorsally compressed
Dorsal/Lateral straitions: Present
Ventral groove: Absent
Scutellum shape: Sickle
Embryo type: Large
Embryo class: N
Hilum visibility: Faint
Hilum type: Sub basal
Hilum shape: Linear

Scanning Electron Microscopy (Figure 31(52))

Dorsal surface: Reticulate, undulating, thick and flat walls, '∩' shaped, smooth angled and broad undulation, concave interspace

Ventral surface: Reticulate, elevated, thick and feebly undulating walls intermittent, concave interspace

Lateral surface: Reticulate, undulating periclinal wall, '∧' shaped smooth angled and broad undulation, straight anticlinal wall, concave interspace

Embryo surface: Reticulate, undulating periclinal wall, '∧' shaped, sharp angled and narrow undulation, straight anticlinal wall, concave interspace

Hilum surface: Ruminate rugose surface, smooth and uneven crimpes, uneven deep depressions present

53. *Oplismenus burmannii*

Light Microscopy (Figure 9(53))

Color: Yellowish to light brown
Texture: Smooth
Dorsal/Lateral straitions: Absent
Scutellum shape: Sickle
Embryo class: L
Hilum type: Basal
Shape: Oblong to ovate
Compression: Dorsally compressed
Ventral groove: Absent
Embryo type: Large
Hilum visibility: Prominent
Hilum shape: Linear

Scanning Electron Microscopy (Figure 32(53))

Dorsal surface: Reticulate, feebly undulating periclinal wall, straight anticlinal wall, depressed interspace, anti-cinally elevations in interspace

Ventral surface: Reticulate, undulating periclinal wall, sharp angled, uneven and broad undulation, straight anticlinal wall, depressed interspace

Lateral surface: Reticulate, undulating periclinal wall, sharp angled, uneven broad undulation, straight anticlinal wall, depressed interspace

Embryo surface: Reticulate, slightly wavy periclinal wall, straight anticlinal wall, depressed interspace, anticinally elevations in interspace

Hilum surface: Reticulate, straight, elevated, thick slightly crimped walls, unidirectional crimpes, concave interspace

54. *Oplismenus compositus*

Light Microscopy (Figure 9(54))

Color: Light brown
Texture: Smooth
Dorsal/Lateral straitions: Present
Scutellum shape: Sickle
Embryo class: N
Hilum type: Basal
Shape: Oblong to ovate
Compression: Dorsally compressed
Ventral groove: Absent
Embryo type: Large
Hilum visibility: Prominent
Hilum shape: Linear

Scanning Electron Microscopy (Figure 32(54))

Dorsal surface: Reticulate, undulating slightly thick periclinal wall, '∧' shaped sharp angled, narrow undulation, straight anticlinal wall, deeply pitted concave interspace

Ventral surface: Reticulate, undulating and slightly thick periclinal wall, '∧' shaped, sharp angled, broad and undulation, straight anticlinal wall, concave interspace

Lateral surface: Reticulate, undulating slightly thick periclinal wall, '∧' shaped, sharp broad uneven sized undulation, straight anticlinal wall, concave interspace

Embryo surface: Reticulate, undulating and thin periclinal wall, '∧' shaped, sharp narrow uneven sized undulation, straight anticlinal wall, pitted concave interspace

Hilum surface: Reticulate, straight, slightly elevated and thin walls, pitted concave interspace

55. *Panicum antidotale*

Light Microscopy (Figure 9(55))

Color: Yellowish
Shape: Ovoid
Texture: Smooth
Compression: Dorsally compressed
Dorsal/Lateral straitions: Absent
Ventral groove: Absent
Scutellum shape: V
Embryo type: Large
Embryo class: N
Hilum visibility: Prominent
Hilum type: Basal
Hilum shape: Oval

Scanning Electron Microscopy (Figure 32(55))

Dorsal surface: Reticulate, undulating, elevated, thick and flat periclinal wall, '∩' shaped, smooth narrow undulation, straight/slant anticlinal wall, concave interspace

Ventral surface: Reticulate, undulating elevated, and thick periclinal wall, '∧' shaped, sharp broad undulation, straight/slant anticlinal wall, narrow concave interspace

Lateral surface: Reticulate, undulating elevated, and thick periclinal wall, '∧' shaped, sharp broad undulation, straight/slant anticlinal wall, concave interspace

Embryo surface: Reticulate, undulating, elevated, thick flat periclinal walls, '∧' shaped, smooth broad undulation, straight/slant anticlinal wall, concave interspace

Hilum surface: Reticulate blister pattern, crimped interspace, unidirectional crimpes

56. *Panicum trypheron*

Light Microscopy (Figure 9(56))

Color: White to cream
Shape: Ovoid to orbicular
Texture: Smooth, shiny
Compression: Dorsally compressed
Dorsal/Lateral straitions: Absent
Ventral groove: Absent
Scutellum shape: V
Embryo type: Large
Embryo class: N
Hilum visibility: Faint
Hilum type: Basal
Hilum shape: Oval

Scanning Electron Microscopy (Figure 33(56))

Dorsal surface: Reticulate, undulating, thick, flat and slightly elevated walls, 'Ω' shaped, smooth angled and broad undulation, depressed interspace

Ventral surface: Reticulate, undulating, thick, flat and slightly elevated walls, '∧' shaped, smooth angled and broad undulation, concave interspace

Lateral surface: Reticulate, squarish reticulum, thick, flat wall, undulating, thin and flat walls intermittent, '∩' shaped, feebly, smooth angled and broad undulation

Embryo surface: Reticulate, undulating, thick, flat slightly elevated walls, 'Ω' shaped, smooth angled and broad undulation, depressed interspace

Hilum surface: Reticulate, straight, elevated thick walls, concave interspace, very narrow depression between reticulum

57. *Panicum miliaceum*

Light Microscopy (Figure 9(57))

Color: Whitish to light brown
Shape: Ovoid to orbicular
Texture: Smooth
Compression: Dorsally compressed
Dorsal/Lateral straitions: Absent
Ventral groove: Absent
Scutellum shape: V
Embryo type: Large
Embryo class: N
Hilum visibility: Prominent
Hilum type: Basal
Hilum shape: Linear

Scanning Electron Microscopy (Figure 33(57))

Dorsal surface: Reticulate, straight, thick and elevated walls, depressed and pitted interspace

Ventral surface: Reticulate, thick and undulating walls intermittent with straight wall, uneven undulation, concave interspace

Lateral surface: Reticulate, straight, thick and elevated walls, concave and pitted interspace

Embryo surface: Reticulate, squarish reticulum, thin walls, concave interspace

Hilum surface: Reticulate, straight, elevated and thick walls, concave interspace

58. *Paspalidium flavidum*

Light Microscopy (Figure 9(58))

Color: Creamish, yellowish to light greenish
Shape: Ovoid to orbicular
Texture: Smooth
Compression: Dorsally compressed
Dorsal/Lateral straitions: Absent
Ventral groove: Absent
Scutellum shape: V
Embryo type: Large
Embryo class: N
Hilum visibility: Prominent
Hilum type: Basal
Hilum shape: Oval

Scanning Electron Microscopy (Figure 33(58))

Dorsal surface: Reticulate, undulating and slightly elevated periclinal wall, '∧' shaped, sharp angled and narrow undulation, straight anticlinal wall, concave interspace

Ventral surface: Reticulate, undulating and slightly elevated periclinal wall, '٨' shaped, sharp angled and broad undulation, straight anticlinal wall

Lateral surface: Reticulate, undulating and slightly elevated periclinal wall, '٨' shaped, sharp angled and broad undulation, straight anticlinal wall, depressed interspace

Embryo surface: Reticulate, undulating and slightly elevated periclinal wall, '٨' shaped, sharp angled and broad undulation, straight anticlinal wall, depressed interspace

Hilum surface: Reticulate, straight, elevated, thick walls, concave interspace

59. *Paspalidium geminatum*

Light Microscopy (Figure 9(59))

Color: Light brown to brown
Shape: Ovoid to oblong
Texture: Smooth
Compression: Dorsally compressed
Dorsal/Lateral straitions: Absent
Ventral groove: Absent
Scutellum shape: V
Embryo type: Large
Embryo class: N
Hilum visibility: Prominent
Hilum type: Basal
Hilum shape: Oval

Scanning Electron Microscopy (Figure 34(59))

Dorsal surface: Reticulate, feebly feature, pitted granular interspace

Ventral surface: Reticulate, walls straight or slightly wavy, elevated, depressed interspace

Lateral surface: Reticulate, feebly feature, thick, flat and straight walls, depressed pitted interspace

Embryo surface: Reticulate, straight, thick and slightly elevated walls, pimple-foveate and concave interspace

Hilum surface: Reticulate, straight, elevated, thick and crimped walls, unidirectional crimpes, concave interspace

60. *Paspalum scrobiculatum*

Light Microscopy (Figure 9(60))

Color: Light brown to brown
Shape: Ovoid
Texture: Smooth
Compression: Dorsally compressed

Dorsal/Lateral straitions: Absent **Ventral groove:** Absent
Scutellum shape: V **Embryo type:** Large
Embryo class: N **Hilum visibility:** Prominent
Hilum type: Basal **Hilum shape:** Oval

Scanning Electron Microscopy (Figure 34(60))

Dorsal surface: Rugose, very feebly undulating, '∩' shaped, smooth angled and broad undulation
Ventral surface: Reticulate, thin flat walls, flat interspace
Lateral surface: Rugose, undulating wall, '∩ shaped, uneven, smooth angled and broad undulation
Embryo surface: Rugose-foveate, undulating thin wall, '∩' shaped, smooth angled and broad undulation
Hilum surface: Reticulate, straight, elevated and thick walls, concave interspace

61. *Pennisetum setosum*

Light Microscopy (Figure 10(61))

Color: Greenish to whitish **Shape:** Obovate
Texture: Smooth **Compression:** Dorsally compressed
Dorsal/Lateral straitions: Absent **Ventral groove:** Absent
Scutellum shape: Sickle **Embryo type:** Large
Embryo class: N **Hilum visibility:** Prominent
Hilum type: Basal **Hilum shape:** Oval

Scanning Electron Microscopy (Figure 34(61))

Dorsal surface: Reticulate, straight, thick and elevated walls, concave interspace
Ventral surface: Reticulate, straight, thick, flat and slightly elevated walls, little depressed interspace
Lateral surface: Reticulate, straight or slightly wavy, thick and slightly elevated walls, concave interspace
Embryo surface: Reticulate, straight, thick and elevated walls, concave interspace
Hilum surface: Reticulate rugose surface, elevated and crimped walls, multidirectional crimpes, depressed interspace

62. *Setaria glauca*

Light Microscopy (Figure 10(62))

Color: Whitish cream to greenish brown
Shape: Ovoid to ovate
Texture: Smooth
Compression: Dorsally compressed
Dorsal/Lateral straitions: Absent
Ventral groove: Absent
Scutellum shape: V
Embryo type: Large
Embryo class: N
Hilum visibility: Prominent
Hilum type: Sub basal
Hilum shape: Oval

Scanning Electron Microscopy (Figure 35(62))

Dorsal surface: Reticulate rugose, undulating and elevated walls, '$\wedge$' shaped, sharp angled and broad undulation, depressed interspace

Ventral surface: Reticulate rugose pitted, undulating and elevated walls, '$\wedge$' shaped, sharp angled and broad undulation, concave interspace

Lateral surface: Reticulate rugose and pitted, undulating and elevated walls, '$\wedge$' shaped, sharp angled and broad undulation, concave interspace

Embryo surface: Reticulate, undulating walls, '$\wedge$' shaped, sharp and smooth angled, short and broad undulation, depressed interspace

Hilum surface: Reticulate, walls are straight, thin and elevated, concave interspace

63. *Setaria tomentosa*

Light Microscopy (Figure 10(63))

Color: Whitish to creamish
Shape: Oblong to ovoid
Texture: Smooth
Compression: Dorsally compressed
Dorsal/Lateral straitions: Absent
Ventral groove: Absent
Scutellum shape: V
Embryo type: Large
Embryo class: N
Hilum visibility: Prominent
Hilum type: Sub basal
Hilum shape: Oval

Scanning Electron Microscopy (Figure 35(63))

Dorsal surface: Reticulate rugose-ruminate, undulating slightly elevated and thin walls, '$\cap$' shaped, smooth angled and broad undulation, depressed interspace

Ventral surface: Reticulate, undulating and thin walls, '∧' shaped, smooth angled and broad undulation

Lateral surface: Reticulate, undulating, thin and slightly elevated walls, '∧' shaped, smooth angled and narrow undulation, depressed interspace

Embryo surface: Reticulate, undulating, slightly elevated and thin walls, '∧' shaped, smooth angled and broad undulation, depressed interspace

Hilum surface: Reticulate slightly rugose, straight, thin and elevated walls, concave interspace

64. *Setaria verticillata*

Light Microscopy (Figure 10(64))

Color: Whitish cream to very light brown

Shape: Ovate to obovate

Texture: Slightly rough surface

Compression: Dorsally compressed

Dorsal/Lateral straitions: Absent

Ventral groove: Absent

Scutellum shape: V

Embryo type: Large

Embryo class: N

Hilum visibility: Prominent

Hilum type: Sub basal

Hilum shape: Oval

Scanning Electron Microscopy (Figure 35(64))

Dorsal surface: Reticulate pitted, undulating and thin periclinal wall, '∧' shaped, sharp to smooth angled and broad undulation, straight anticlinal wall, very shallow interspace

Ventral surface: Reticulate pitted, undulating and thin periclinal wall, '∧' shaped, uneven smooth angled and broad undulation, straight anticlinal

Lateral surface: Reticulate pitted, undulating and thin walls, '∧' and '∩' shaped, smooth angled and broad undulation, uneven undulations

Embryo surface: Reticulate, undulating and thin periclinal walls, '∧' shaped, smooth angled and broad undulation, straight anticlinal wall, depressed interspace

Hilum surface: Reticulate, straight, thick and elevated walls, concave interspace

65. *Aeluropus lagopoides*

Light Microscopy (Figure 10(65))

Color: Light brown
Texture: Rough
Dorsal/Lateral straitions: Absent
Scutellum shape: Sickle
Embryo class: N
Hilum type: Basal
Shape: Ovate
Compression: Dorsally compressed
Ventral groove: Absent
Embryo type: Large
Hilum visibility: Prominent
Hilum shape: Circular

Scanning Electron Microscopy (Figure 36(65))

Dorsal surface: Reticulate, straight and elevated walls, highly elevated thicker periclinal wall than the anticlinal wall, concave interspace

Ventral surface: Reticulate rugose-pitted, straight and thin walls, very depressed interspace

Lateral surface: Reticulate foveate, thick feebly undulating elevated walls, foveate interspace

Embryo surface: Reticulate rugose, elevated walls, highly elevated thicker periclinal wall than the anticlinal wall, concave interspace

Hilum surface: Reticulate rugose-ruminate, thin, elevated walls, multidirectional elevation

66. *Isachne globosa*

Light Microscopy (Figure 10(66))

Color: Brown to dark brown
Texture: Smooth
Dorsal/Lateral straitions: Present
Scutellum shape: V
Embryo class: L
Hilum type: Linear
Shape: Orbicular to subglobose
Compression: Not compressed
Ventral groove: Present
Embryo type: Short
Hilum visibility: Prominent
Hilum shape: Short

Scanning Electron Microscopy (Figure 36(66))

Dorsal surface: Reticulate rugose, undulating, thick and elevated walls, '∩' shaped, smooth angled broad undulation, depressed interspace, pitted circles in interspace

Ventral surface: Reticulate ruminate, feebly undulating, elevated and thin walls, '∩' shaped, smooth angled broad undulation, uneven elevation

Lateral surface: Reticulate, straight, thick and slightly elevated walls, depressed interspace

Embryo surface: Reticulate rugose, undulating elevated walls, smooth angled broad undulation, depressed interspace, pitted circles in interspace

Hilum surface: Reticulate ruminate-rugose, elevated, thin and crimped walls, multidirectional crimpes, depressed interspace

67. *Aristida adscensionis*

Light Microscopy (Figure 10(67))

Color: Light brown to brown

Shape: Acicular

Texture: Smooth

Compression: Laterally compressed

Dorsal/Lateral straitions: Present

Ventral groove: Absent

Scutellum shape: V

Embryo type: Short

Embryo class: L

Hilum visibility: Prominent

Hilum type: Linear

Hilum shape: Long

Scanning Electron Microscopy (Figure 36(67))

Dorsal surface: Reticulate, thin and elevated walls, anticlinal bands in interspace

Ventral surface: Reticulate, thin and elevated periclinal wall, straight or slant anticlinal wall, depressed interspace

Lateral surface: Reticulate, thin and elevated walls periclinal wall, straight or slant anticlinal wall, depressed interspace

Embryo surface: Reticulate, thin walls, anticlinal bands in interspace, narrow depressions between the reticulum

Hilum surface: Reticulate, straight and elevated walls, depressed interspace

68. *Aristida funiculata*

Light Microscopy (Figure 10(68))

Color: Whitish to light brown

Shape: Acicular

Texture: Smooth, shiny

Compression: Not compressed

Dorsal/Lateral straitions: Present

Ventral groove: Absent

Scutellum shape: V

Embryo type: Large

Embryo class: L
Hilum visibility: Prominent
Hilum type: Linear
Hilum shape: Long

Scanning Electron Microscopy (Figure 37(68))

Dorsal surface: Reticulate-striate, thick and elevated walls, narrow depressions between the walls

Ventral surface: Reticulate rugose, elevated and pimple walls, depressed interspace

Lateral surface: Reticulate, thin, undulating and elevated periclinal wall, '^' shaped smooth angled and broad undulation, straight anticlinal wall, ribbed interspace

Embryo surface: Reticulate, elongated reticulum, straight and thin walls, very depressed interspace

Hilum surface: Reticulate, squarish reticulum, straight, elevated and thick walls, very depressed interspace

69. *Perotis indica*

Light Microscopy (Figure 10(69))

Color: Brown
Shape: Oblong, thin
Texture: Smooth, glossy
Compression: Laterally compressed
Dorsal/Lateral straitions: Absent
Ventral groove: Absent
Scutellum shape: Sickle
Embryo type: Short
Embryo class: L
Hilum visibility: Prominent
Hilum type: Basal
Hilum shape: Fan shaped

Scanning Electron Microscopy (Figure 37(69))

Dorsal surface: Reticulate, straight, thin and slightly elevated walls, shallow interspace

Ventral surface: Reticulate, slightly wavy, thin and elevated periclinal wall, straight or slant anticlinal wall, depressed interspace

Lateral surface: Reticulate, slightly wavy, thin and elevated periclinal wall, straight or slant anticlinal wall, depressed interspace

Embryo surface: Reticulate, slightly wavy, thin and elevated wall, depressed interspace

Hilum surface: Striate foveate, irregularly elevated wall, multidirectional elevations

70. *Chloris barbata*

Light Microscopy (Figure 10(70))

Color: Creamish to light yellow
Texture: Smooth, shiny
Dorsal/Lateral straitions: Absent
Scutellum shape: Sickle
Embryo class: N
Hilum type: Basal
Shape: Fusiform
Compression: Dorsally compressed
Ventral groove: Absent
Embryo type: Large
Hilum visibility: Prominent
Hilum shape: Oval

Scanning Electron Microscopy (Figure 37(70))

Dorsal surface: Reticulate, thin and straight walls
Ventral surface: Reticulate, thin and straight walls, narrow depressions intermittent between the reticulum
Lateral surface: Reticulate little foveate appearance, thin and straight walls, border pits arranged horizontally at some place
Embryo surface: Reticulate, thin and straight walls
Hilum surface: Reticulate, walls are folded, irregular elevated walls, multidirectional elevation

71. *Chloris montana*

Light Microscopy (Figure 11(71))

Color: Brown
Texture: Smooth, shiny
Dorsal/Lateral straitions: Absent
Scutellum shape: V
Embryo class: N
Hilum type: Basal
Shape: Fusiform
Compression: Dorsally compressed
Ventral groove: Absent
Embryo type: Large
Hilum visibility: Faint
Hilum shape: Circular

Scanning Electron Microscopy (Figure 38(71))

Dorsal surface: Reticulate pitted, thin and straight walls, very depressed interspace
Ventral surface: Reticulate pitted, thin and straight walls, very depressed interspace
Lateral surface: Reticulate pitted, thin and straight walls, very depressed interspace
Embryo surface: Reticulate, folded surface, thin and straight walls, very depressed interspace

Hilum surface: Reticulate rugose, irregular elevated walls, multi-directional elevation

72. *Chloris virgata*

Light Microscopy (Figure 11(72))

Color: Creamish to light yellow
Shape: Ellipsoid
Texture: Smooth, shiny
Compression: Laterally compressed
Dorsal/Lateral straitions: Absent
Ventral groove: Absent
Scutellum shape: V
Embryo type: Large
Embryo class: N
Hilum visibility: Prominent
Hilum type: Basal
Hilum shape: Oval

Scanning Electron Microscopy (Figure 38(72))

Dorsal surface: Reticulate, straight, thin and slightly elevated walls, depressed interspace

Ventral surface: Reticulate, smooth surface, straight and thin walls, feebly features

Lateral surface: Reticulate, smooth surface, straight and thin walls, feebly features

Embryo surface: Reticulate, straight, thin and elevated walls, depressed interspace

Hilum surface: Ruminate rugose, big and irregular depressions with uneven elevations

73. *Cynodon dactylon*

Light Microscopy (Figure 11(73))

Color: Brownish black
Shape: Oblong
Texture: Smooth
Compression: Laterally compressed
Dorsal/Lateral straitions: Present
Ventral groove: Absent
Scutellum shape: Sickle
Embryo type: Large
Embryo class: N
Hilum visibility: Faint
Hilum type: Basal
Hilum shape: Circular

Scanning Electron Microscopy (Figure 38(73))

Dorsal surface: Reticulum ribbed, straight, highly elevated and thick walls, narrow depressed interspace

Ventral surface: Reticulum, straight, elevated, flat and thick walls, narrow concave interspace

Lateral surface: Reticulum, straight, elevated, flat and thick walls, narrow depressed interspace

Embryo surface: Reticulum ribbed, straight, elevated and thin walls, depressed interspace

Hilum surface: Blister in the center, around that ruminate, multi-directional elevations

74. *Melanocenchris jaequemontii*

Light Microscopy (Figure 11(74))

Color: Dark brown
Shape: Ovate
Texture: Rough
Compression: Dorsally compressed
Dorsal/Lateral straitions: Present
Ventral groove: Absent
Scutellum shape: Sickle
Embryo type: Large
Embryo class: L
Hilum visibility: Prominent
Hilum type: Sub basal
Hilum shape: Circular

Scanning Electron Microscopy (Figure 39(74))

Dorsal surface: Reticulate, feebly undulating and thin periclinal wall, straight anticlinal wall

Ventral surface: Reticulate and feebly pitted surface, feebly undulating and thin periclinal wall, straight anticlinal wall

Lateral surface: Reticulate and feebly pitted surface, thin elevation at some place, feebly undulating and thin periclinal wall, straight anticlinal wall

Embryo surface: Reticulate, elevated thin folded walls, feebly depressed interspace

Hilum surface: Ruminate rugose

75. *Oropetium villosulum*

Light Microscopy (Figure 11(75))

Color: Brown
Shape: Oblanceolate
Texture: Rough
Compression: Laterally compressed
Dorsal/Lateral straitions: Absent
Ventral groove: Absent
Scutellum shape: Sickle
Embryo type: Short
Embryo class: N
Hilum visibility: Faint
Hilum type: Basal
Hilum shape: Fan shaped

Scanning Electron Microscopy (Figure 39(75))

Dorsal surface: Reticulate, thin elevated walls, smooth depressed interspace

Ventral surface: Reticulate, thin elevated walls, depressed interspace

Lateral surface: Reticulate, thin flat walls, feebly features

Embryo surface: Reticulate, thin, feebly undulating and elevated walls, smooth depressed interspace

Hilum surface: Reticulate rugose, thick elevated walls, multidirectionally crimped, concave interspace

76. *Schoenefeldia gracilis*

Light Microscopy (Figure 11(76))

Color: Dark brown

Shape: Fusiform to obovate

Texture: Smooth, shiny

Compression: Laterally compressed

Dorsal/Lateral straitions: Absent

Ventral groove: Absent

Scutellum shape: Sickle

Embryo type: Large

Embryo class: L

Hilum visibility: Prominent

Hilum type: Basal

Hilum shape: V shaped

Scanning Electron Microscopy (Figure 39(76))

Dorsal surface: Reticulate, in the middle reticulum become broader, straight, thin and elevated walls, concave interspace

Ventral surface: Reticulate, straight, thin and elevated walls, concave interspace

Lateral surface: Reticulate, in the middle reticulum become broader, straight, thin and elevated walls, shallow interspace

Embryo surface: Reticulate, folded surface, straight, thin and elevated walls, concave interspace

Hilum surface: Blister at proximal end, below that reticulate with elevated, crimped and thin walls, concave interspace

77. *Tetrapogon tenellus*

Light Microscopy (Figure 11(77))

Color: Brown

Shape: Ovate

Texture: Smooth

Compression: Laterally compressed

Dorsal/Lateral straitions: Absent

Ventral groove: Absent

Scutellum shape: Sickle
Embryo type: Large
Embryo class: N
Hilum visibility: Prominent
Hilum type: Basal
Hilum shape: Oval

Scanning Electron Microscopy (Figure 40(77))

Dorsal surface: Reticulate, thick elevated walls, concave interspace
Ventral surface: Reticulate, feebly reticulum, thin and straight walls
Lateral surface: Reticulate, thick and elevated walls, concave interspace
Embryo surface: Blister-reticulate, narrow depressions between the reticulum
Hilum surface: Reticulate pitted-foveate, thin and straight walls, feebly reticulums

78. *Tetrapogon villosus*

Light Microscopy (Figure 11(78))

Color: Brown
Shape: Ellipsoid to oblong
Texture: Smooth
Compression: Dorsally compressed
Dorsal/Lateral straitions: Absent
Ventral groove: Absent
Scutellum shape: Sickle
Embryo type: Large
Embryo class: L
Hilum visibility: Prominent
Hilum type: Basal
Hilum shape: Basal

Scanning Electron Microscopy (Figure 40(78))

Dorsal surface: Blister-ribbed, multidirections depressions between the blisters
Ventral surface: Reticulate, feebly reticulum, thin and straight walls
Lateral surface: Reticulate, thick and elevated walls, depressed interspace
Embryo surface: Blister, narrow depressions between the reticulum
Hilum surface: Reticulate, pitted-foveate surface, straight and thin walls, feebly reticulum

79. *Acrachne racemosa*

Light Microscopy (Figure 11(79))

Color: Dark brown to blackish brown
Shape: Ovate, ellipsoid to subglobose
Texture: Rough
Compression: Dorsally compressed

Dorsal/Lateral straitions: Absent
Ventral groove: Present
Scutellum shape: Sickle
Embryo type: Large
Embryo class: N
Hilum visibility: Prominent
Hilum type: Sub basal
Hilum shape: Raised circular

Scanning Electron Microscopy (Figure 40(79))

Dorsal surface: Compound reticulate with blister appearance, strongly undulating walls, '∧' shaped, sharp to smooth angled broad undulation, foveate interspace
Ventral surface: Compound reticulate with blister appearance, strongly undulating walls, '∧' shaped, sharp to smooth angled broad undulation, foveate interspace
Lateral surface: Blister, pitted surface, depressions between the blisters
Embryo surface: Compound reticulate – blister, depressed blister wall
Hilum surface: Ruminate in the middle, around that blister with curve undulating blister wall

80. *Dactyloctenium aegyptium*

Light Microscopy (Figure 11(80))

Color: light brown
Shape: Ovoid to squarish
Texture: Rough
Compression: Laterally compressed
Dorsal/Lateral straitions: Absent
Ventral groove: Absent
Scutellum shape: Sickle
Embryo type: Large
Embryo class: N
Hilum visibility: Prominent
Hilum type: Sub basal
Hilum shape: Circular

Scanning Electron Microscopy (Figure 41(80))

Dorsal surface: Reticulate, deep undulating walls, 'Ω' shaped broad undulation, anticinally arranged ridges
Ventral surface: Reticulate, deep undulating walls, 'Ω' shaped broad undulation, anticinally arranged ridges
Lateral surface: Reticulate, undulating walls, 'Ω' shaped broad undulation, anticinally arranged ridges
Embryo surface: Blister, depressed periclinal and elevated anticlinal blister wall
Hilum surface: Ruminate in the middle, around that blister with curve undulating blister wall, intermittently at some places blister become bulgy

81. *Dactyloctenium scindicum*

Light Microscopy (Figure 12(81))

Color: Brown
Texture: Rough
Dorsal/Lateral straitions: Absent
Scutellum shape: Sickle
Embryo class: N
Hilum type: Sub basal
Shape: Ovoid to squarish
Compression: Laterally compressed
Ventral groove: Absent
Embryo type: Short
Hilum visibility: Prominent
Hilum shape: Circular

Scanning Electron Microscopy (Figure 41(81))

Dorsal surface: Compound reticulate with blister appearance, undulating walls, smooth broad undulation, bulgy blister between two reticulum, foveate interspace

Ventral surface: Reticulate, deep undulating walls, 'Ω' shaped broad feebly undulation, ridges present on the surface

Lateral surface: Reticulate, deep undulating walls, 'Ω' shaped broad undulation, ridges present on the surface

Embryo surface: Blister, depressed periclinal and elevated anticlinal blister wall

Hilum surface: Ruminate in the middle, around that blister with curve, undulating blister wall

82. *Dactyloctenium giganteum*

Light Microscopy (Figure 12(82))

Color: light brown
Texture: Rough
Dorsal/Lateral straitions: Absent
Scutellum shape: Sickle
Embryo class: N
Hilum type: Sub basal
Shape: Ovoid to squarish
Compression: Laterally compressed
Ventral groove: Absent
Embryo type: Short
Hilum visibility: Prominent
Hilum shape: Circular

Scanning Electron Microscopy (Figure 41(82))

Dorsal surface: Reticulate, deep strongly undulating walls, 'Ω' shaped broad undulation, irregular ridges on the surface, elevated interspace, which gives appearance of blister

Ventral surface: Reticulate, deep strongly undulating walls, 'Ω' shaped broad undulation, irregular ridges on the surface, elevated interspace, which gives appearance of blister

Lateral surface: Reticulate, strongly undulating walls, 'Ω' shaped broad undulation, irregular ridges on the surface, flat interspace

Embryo surface: Blister pattern, tetragonal curve blister wall

Hilum surface: Ruminate in the middle, around that blister with curve, undulating blister wall, at one side bulgy blister present

83. *Desmostachya bipinnata*

Light Microscopy (Figure 12(83))

Color: Brown
Shape: Ovate
Texture: Smooth
Compression: Laterally compressed
Dorsal/Lateral straitions: Absent
Ventral groove: Absent
Scutellum shape: Sickle
Embryo type: Short
Embryo class: N
Hilum visibility: Prominent
Hilum type: Basal
Hilum shape: Oval

Scanning Electron Microscopy (Figure 42(83))

Dorsal surface: Reticulate, straight, thick and elevated walls, concave interspace

Ventral surface: Reticulate, straight, thick and flat walls, depressed interspace

Lateral surface: Reticulate, straight, thin and flat walls, very depressed interspace

Embryo surface: Reticulate, straight, thick and elevated walls, small pitted concave interspace

Hilum surface: Ruminate rugose, convoluted walls, multidirectional elevations, narrow depressed interspace

84. *Dinebra retroflexa*

Light Microscopy (Figure 12(84))

Color: Brown
Shape: Oblong

Texture: Smooth, glossy
Compression: Laterally compressed
Dorsal/Lateral straitions: Absent
Ventral groove: Present
Scutellum shape: V
Embryo type: Short
Embryo class: N
Hilum visibility: Prominent
Hilum type: Basal
Hilum shape: Oval

Scanning Electron Microscopy (Figure 42(84))

Dorsal surface: Reticulate, straight, thick and elevated walls, penta to hexagonal reticulum, concave interspace

Ventral surface: Reticulate, straight, thick and elevated walls, penta to hexagonal reticulum, concave interspace

Lateral surface: Reticulate, straight, thick and flat walls, penta to hexagonal reticulum, shallow interspace

Embryo surface: Reticulate, straight, thick and elevated walls, penta to hexagonal reticulum, concave interspace

Hilum surface: Ruminate rugose, convoluted and thick walls, multidirectional elevations

85. *Eleusine indica*

Light Microscopy (Figure 12(85))

Color: Dark brown to black
Shape: Ovate, ellipsoid to subglobose
Texture: Rough
Compression: Dorsally compressed
Dorsal/Lateral straitions: Absent
Ventral groove: Present
Scutellum shape: Sickle
Embryo type: Short
Embryo class: N
Hilum visibility: Prominent
Hilum type: Sub basal
Hilum shape: Raised circular

Scanning Electron Microscopy (Figure 42(85))

Dorsal surface: Reticulate, deep strongly undulating walls, 'Ω' and '∩' shaped broad undulation, blister interspace, for example, small hills like structure present within interspace, horizontally arranged blisters

Ventral surface: Reticulate, walls deep undulating, deep strongly undulating walls, 'Ω' and '∩' shaped broad undulation, horizontally arranged blister present within interspace

Lateral surface: Reticulate, walls deep undulating, deep strongly undulating walls, 'Ω' and '∩' shaped broad undulation, horizontally arranged blister present within interspace, ridges arranged in semicircular manner

Embryo surface: Blister pattern, depressed blister wall, round and hill shaped blister

Hilum surface: Ruminate in the middle, around that blister with curve, undulating blister wall

86. *Eragrostiella bifaria*

Light Microscopy (Figure 12(86))

Color: Brown to dark brown
Shape: Ovoid to subglobose
Texture: Smooth, shiny
Compression: Not compressed
Dorsal/Lateral straitions: Present
Ventral groove: Absent
Scutellum shape: Sickle
Embryo type: Short
Embryo class: N
Hilum visibility: Prominent
Hilum type: Basal
Hilum shape: Circular

Scanning Electron Microscopy (Figure 43(86))

Dorsal surface: Reticulate, Straight, thick and elevated walls, concave interspace

Ventral surface: Reticulate, straight, thick and elevated walls, feebly undulating periclinal wall intermittent with elevated wall, concave interspace

Lateral surface: Reticulate pattern, thick and elevated walls, feebly undulating periclinal wall intermittent with elevated wall, concave interspace

Embryo surface: Reticulate, straight, thick and elevated walls, concave interspace

Hilum surface: Ruminate, multidirectional irregular elevations

87. *Eragrostis cilianensis*

Light Microscopy (Figure 12(87))

Color: Dark brown to black
Shape: Orbicular
Texture: Smooth, shiny
Compression: Not compressed
Dorsal/Lateral straitions: Present
Ventral groove: Absent

Scutellum shape: Sickle
Embryo type: Short
Embryo class: N
Hilum visibility: Prominent
Hilum type: Basal
Hilum shape: Circular

Scanning Electron Microscopy (Figure 43(87))

Dorsal surface: Reticulate, thick, flat and elevated wall, concave foveate interspace
Ventral surface: Reticulate, thin, flat and elevated wall, concave foveate interspace
Lateral surface: Reticulate, thin, flat and elevated wall, concave foveate interspace
Embryo surface: Reticulate, distinct elongated reticulum, thin, flat and elevated wall, concave interspace
Hilum surface: Reticulate pattern, walls are elevated, indistinct irregular reticulum, shallow interspace

88. ***Eragrostis ciliaris***

Light Microscopy (Figure 12(88))

Color: Dark brown
Shape: Oblong to obovate
Texture: Smooth, shiny
Compression: Not compressed
Dorsal/Lateral straitions: Absent
Ventral groove: Absent
Scutellum shape: V
Embryo type: Short
Embryo class: N
Hilum visibility: Prominent
Hilum type: Basal
Hilum shape: Circular

Scanning Electron Microscopy (Figure 43(88))

Dorsal surface: Smooth but superficial reticulate features present
Ventral surface: Smooth, only distal end showing converging elevations
Lateral surface: Reticulate, shallow features, indistinct straight elevations
Embryo surface: No prominent feature on embryo axis but scutellum shows elongated square reticulum, depressed walls
Hilum surface: Reticulate, elevated crimped walls, irregular wavy convolutions

89. *Eragrostis japonica*

Light Microscopy (Figure 12(89))

Color: Dark brown
Shape: Oblong
Texture: Smooth, shiny
Compression: Laterally compressed
Dorsal/Lateral straitions: Absent
Ventral groove: Absent
Scutellum shape: V
Embryo type: Short
Embryo class: N
Hilum visibility: Prominent
Hilum type: Basal
Hilum shape: Circular

Scanning Electron Microscopy (Figure 44(89))

Dorsal surface: Smooth but superficial reticulate features present
Ventral surface: Reticulate, elongated reticulum, slightly undulating walls, shallow interspace
Lateral surface: Feebly reticulate features, elongated reticulum, slightly undulating walls, shallow interspace
Embryo surface: Reticulated, elongated reticulum, slightly undulating and elevated walls, concave interspace
Hilum surface: Reticulate, irregular distinct reticulum, elevated and crimped walls

90. *Eragrostis nutans*

Light Microscopy (Figure 12(90))

Color: Dark brown
Shape: Oblong to ovate
Texture: Smooth, shiny
Compression: Not compressed
Dorsal/Lateral straitions: Absent
Ventral groove: Absent
Scutellum shape: V
Embryo type: Short
Embryo class: N
Hilum visibility: Prominent
Hilum type: Basal
Hilum shape: Circular

Scanning Electron Microscopy (Figure 44(90))

Dorsal surface: Reticulate, slightly undulating walls
Ventral surface: Reticulate, slightly undulating walls
Lateral surface: Reticulate, slightly undulating and elevated walls, very shallow interspace
Embryo surface: Reticulate, slightly undulating and elevated walls, horizontal thin elevations within reticulum

Hilum surface: Reticulate, irregular distinct reticulum elevated and crimped walls, multidirectional uneven crimpes

91. *Eragrostis pilosa*

Light Microscopy (Figure 13(91))

Color: Dark brown
Shape: Ovoid to oblong
Texture: Smooth
Compression: Laterally compressed
Dorsal/Lateral straitions: Present
Ventral groove: Absent
Scutellum shape: V
Embryo type: Large
Embryo class: N
Hilum visibility: Faint
Hilum type: Basal
Hilum shape: Circular

Scanning Electron Microscopy (Figure 44(91))

Dorsal surface: Reticulate-foveate, elevated and thin walls, concave interspace, punctate sculpturing within reticulum

Ventral surface: Reticulate-foveate, elevated and thin walls, concave interspace, punctate sculpturing within reticulum

Lateral surface: Reticulate-foveate, elevated and thin walls, concave interspace, punctate sculpturing within reticulum

Embryo surface: Embryo axis reticulated, reticulate-foveate scutellum, elevated and thin walls, concave interspace, punctate sculpturing within reticulum.

Hilum surface: Reticulate, irregular crimped wall

92. *Eragrostis tenella*

Light Microscopy (Figure 13(92))

Color: Dark brown
Shape: Oblong to obovate
Texture: Smooth, shiny
Compression: Not compressed
Dorsal/Lateral straitions: Absent
Ventral groove: Absent
Scutellum shape: V
Embryo type: Large
Embryo class: N
Hilum visibility: Prominent
Hilum type: Basal
Hilum shape: Circular

Scanning Electron Microscopy (Figure 45(93))

Dorsal surface: Reticulated, undulating, thick and elevated periclinal wall, undulating anticlinal wall, intermittent thick undulating tangential elevations

Ventral surface: Smooth

Lateral surface: Smooth

Embryo surface: Reticulate, straight, thick and elevated walls, concave interspace

Hilum surface: Ruminate-reticulate, elevated and crimped walls, multidirectional irregular crimpes converging towards the proximal end of carypsis

93. ***Eragrostis tremula***

Light Microscopy (Figure 13(93))

Color: Creamish to light brown
Shape: Round to oblong
Texture: Smooth, shiny
Compression: Laterally compressed
Dorsal/Lateral straitions: Present
Ventral groove: Absent
Scutellum shape: Sickle
Embryo type: Large
Embryo class: N
Hilum visibility: Prominent
Hilum type: Basal
Hilum shape: Circular

Scanning Electron Microscopy (Figure 45(93))

Dorsal surface: Reticulate, penta to hexagonal reticulum, straight, thick and elevated walls, shallow interspace

Ventral surface: Reticulate, straight, thin and slightly elevated walls, very shallow interspace

Lateral surface: Reticulate, straight, thick and elevated walls, penta to hexagonal reticulum, concave interspace, elongated reticulate reticulum superimposed on penta to hexagonal reticulum

Embryo surface: Reticulate, straight, thick and elevated walls, irregular sized penta to hexagonal reticulum

Hilum surface: Globular slimy glands at proximal end ventral surface

94. *Eragrostis unioloides*

Light Microscopy (Figure 13(94))

Color: Brown to dark brown
Texture: Smooth, shiny
Dorsal/Lateral straitions: Present
Scutellum shape: V
Embryo class: N
Hilum type: Basal
Shape: Ovoid
Compression: Laterally compressed
Ventral groove: Absent
Embryo type: Large
Hilum visibility: Prominent
Hilum shape: Circular

Scanning Electron Microscopy (Figure 45(94))

Dorsal surface: Reticulate, thick, straight and elevated walls, concave interspace
Ventral surface: Reticulate, thick, straight and elevated walls, concave interspace
Lateral surface: Reticulate, thick, straight and elevated walls, shallow interspace
Embryo surface: Reticulate, elongated reticulum, thick, straight and elevated walls, concave interspace
Hilum surface: Reticulate, thick, straight, elevated and undulating walls, concave interspace

95. *Eragrostis viscosa*

Light Microscopy (Figure 13(95))

Color: Dark brown
Texture: Smooth, shiny
Dorsal/Lateral straitions: Absent
Scutellum shape: V
Embryo class: N
Hilum type: Basal
Shape: Oblong to ovate
Compression: Laterally compressed
Ventral groove: Absent
Embryo type: Large
Hilum visibility: Prominent
Hilum shape: Circular

Scanning Electron Microscopy (Figure 46(95))

Dorsal surface: Reticulate, long rectangular reticulum intermittent with undulating margin, straight, smooth and thin walls, very shallow interspace
Ventral surface: Smooth

Lateral surface: Reticulate, long rectangular reticulum straight, thin and slightly elevated walls, shallow interspace

Embryo surface: Reticulate, thin, straight and elevated walls, concave interspace, reticulums converging towards the proximal end of caryopsis

Hilum surface: Reticulate, elevated crimped walls, highly convoluted crimpes

96. *Sporobolus coromandelianus*

Light Microscopy (Figure 13(96))

Color: Brown to dark brown

Shape: Oblong to ovoid

Texture: Smooth

Compression: Laterally compressed

Dorsal/Lateral straitions: Absent

Ventral groove: Absent

Scutellum shape: Sickle

Embryo type: Large

Embryo class: N

Hilum visibility: Faint

Hilum type: Basal

Hilum shape: V shaped

Scanning Electron Microscopy (Figure 46(96))

Dorsal surface: Reticulate, thick and slightly elevated walls, very shallow interspace

Ventral surface: Reticulate-striate, thick and elevated walls, shallow interspace, horizontal elevations within interspace

Lateral surface: Reticulate, thick elevated walls, irregular feebly undulation within interspace

Embryo surface: Reticulate, irregular shaped of reticulum, elevated crimped wall, multidirectional crimples, narrow depressed channels between reticulums

Hilum surface: Reticulate rugose, elevated crimped wall, irregular crimpes converging towards the proximal end of caryopes

97. *Sporobolus diander*

Light Microscopy (Figure 13(97))

Color: Creamy yellowish to light brown

Shape: Oblong to ovoid

Texture: Smooth, shiny

Compression: Laterally compressed

Dorsal/Lateral straitions: Absent **Ventral groove:** Absent
Scutellum shape: Sickle **Embryo type:** Large
Embryo class: N **Hilum visibility:** Prominent
Hilum type: Basal **Hilum shape:** V shaped

Scanning Electron Microscopy (Figure 46(97))

Dorsal surface: Reticulate, thick and slightly elevated walls, very shallow interspace

Ventral surface: Reticulate-striate, thick, elevated and wrinkled walls, unidirectional wrinkles, shallow interspace, horizontal elevations within interspace

Lateral surface: Reticulate, shallow features, thin and flat walls

Embryo surface: Reticulate, elevated crimped wall, multidirectional crimples, narrow depressed channels between reticulums

Hilum surface: Reticulate, elevated crimped wall, irregular crimples, depressed interspace, broad depressed channels between the reticulums

98. *Sporobolus indicus*

Light Microscopy (Figure 13(98))

Color: Dark brown to black **Shape:** Ovoid
Texture: Smooth, shiny **Compression:** Not compressed
Dorsal/Lateral straitions: Absent **Ventral groove:** Absent
Scutellum shape: Sickle **Embryo type:** Large
Embryo class: N **Hilum visibility:** Prominent
Hilum type: Basal **Hilum shape:** V shaped

Scanning Electron Microscopy (Figure 47(98))

Dorsal surface: Reticulate, flat surface, thin and very feebly undulating wall

Ventral surface: Reticulate, thin and very feebly undulating wall, anticlinal wall straight or slanting, shallow interspace

Lateral surface: Reticulate, thin and very feebly undulating wall, anticlinal wall straight or slanting, shallow interspace

Embryo surface: Reticulate, flat surface, thin and very feebly undulating wall

Hilum surface: Reticulate rugose, elevated and folded walls, irregular crimp in multidirectional, depressed interspace

99. *Urochondra setulosa*

Light Microscopy (Figure 13(99))

Color: Brown
Shape: Ellipsoid
Texture: Rough
Compression: Laterally compressed
Dorsal/Lateral straitions: Present
Ventral groove: Absent
Scutellum shape: Sickle
Embryo type: Large
Embryo class: N
Hilum visibility: Prominent
Hilum type: Basal
Hilum shape: V shaped

Scanning Electron Microscopy (Figure 47(99))

Dorsal surface: Reticulate-striate, uneven sized reticulum, thick prominently elevated walls, very shallow interspace

Ventral surface: Reticulate-striate, uneven sized reticulum, thick, folded and elevated walls, very shallow interspace, intermittent horizontal elevations within reticulum

Lateral surface: Reticulate-striate, pitted surface, uneven sized reticulum, thin indistinct walls, very shallow interspace, horizontal elevations within reticulum

Embryo surface: Reticulate-blister, curve blister periclinal wall, thin depressed anticlinal wall, narrows horizontal elevations between the blisters

Hilum surface: Reticulate rugose, elevated and folded walls, irregular crimped arranged horizontally in multidirectional, depressed interspace

100. *Tragrus biflorus*

Light Microscopy (Figure 13(100))

Color: Light brown
Shape: Ellipsoid to ovate
Texture: Smooth
Compression: Dorsally compressed
Dorsal/Lateral straitions: Absent
Ventral groove: Absent

Scutellum shape: Sickle
Embryo type: Large
Embryo class: N
Hilum visibility: Prominent
Hilum type: Basal
Hilum shape: Oval

Scanning Electron Microscopy (Figure 47(100))

Dorsal surface: Reticulate, strongly undulating and elevated walls, '∩' shaped, smooth angled and narrow to broad undulations, shallow to concave interspace
Ventral surface: Reticulate, strongly undulating elevated walls, '∩' shaped, smooth very broad undulations, shallow interspace, small circles within reticulum
Lateral surface: Reticulate, feebly undulating walls, ∩ shaped, smooth angled and broad undulations, shallow interspace
Embryo surface: Reticulate, feebly undulating and elevated walls, '∩' shaped, smooth angled and broad undulations, shallow interspace
Hilum surface: Reticulate, straight, elevated and thick walls, concave interspace

1.5 IDENTIFICATION KEY

Based on the characteristic features of identification, a diagnostic key for the studied species has been prepared.

1. Laterally compressed caryopses with rough surface.......................... **2**
1. Dorsally compressed or not compressed caryopses, if laterally compressed than with smooth surface **4**
2. Ventral groove present ....***Isachne globosa***
2. Ventral groove absent... **3**
3. Short embryo class .... ***Aristida adscensionis***
3. Large embryo class ....***Aristida funiculata***
4. Not compressed caryopses **5**
4. Dorsally or laterally compressed caryopses, if not compressed than with L-embryo type and V-shaped hilum.. **29**
5. V-shaped scutellum, if sickle shaped than with L-embryo type with short embryo class.. **6**

5. Sickle shaped scutellum, if V-shaped than with linear hilum **14**
6. L-embryo type... **7**
6. N-embryo type **8**
7. Circular shaped hilum .... ***Dimeria orinthopoda***
7. Fan shaped hilum .... ***Perotis indica***
8. Large embryo class **9**
8. Short embryo class..**11**
9. Oval shaped hilum .... ***Chloris virgata***
9. Circular shaped hilum **10**
10. Caryopses length is 0.55 mm .... ***Eragrostis tenella***
10. Caryopses length is o.25 mm .... ***Eragrostis viscosa***
11. Ventral groove present .... ***Dinebra retroflexa***
11. Ventral groove absent.. **12**
12. Dorsal surface have reticulate pattern (under SEM)... ***Eragrostis nutans***
12. Smooth dorsal surface, superficial features seen (under SEM).. **13**
13. Ventral surface smooth (under SEM)...................... ***Eragrostis ciliaris***
13. Ventral surface have reticulate pattern (under SEM).. ***Eragrostis japonica***
14. V-shaped scutellum **15**
14. Sickle shaped scutellum **17**
15. Large embryo class .... ***Dactyloctenium aegyptium***
15. Short embryo class.. **16**
16. Dorsal surface have compound reticulate pattern with blister (under SEM) .... ***Dactyloctenium sindicum***
16. Dorsal surface have c reticulate pattern (under SEM)... ***Dactyloctenium giganteum***
17. Ventral groove absent... **18**
17. Ventral groove present **19**
18. Large embryo class .... ***Acrachne racemosa***
18. Short embryo class.. ***Eleusine indica***
19. Short embryo class... **20**
19. Large embryo class **23**
20. Rough surface of caryopses .... ***Oropetium villosulum***
20. Smooth surface of caryopses.. **21**

21. Oval shaped hilum ....***Desmostachya bipinnata***
21. Circular shaped hilum **22**
22. Caryopses length is 0.66 mm ....***Eragrostiella bifaria***
22. Caryopses length is 0.49 mm ....***Eragrostis cilianensis***
23. Dorsal/lateral striations present.. **24**
23. Dorsal/lateral striations absent **25**
24. Sub basal type of hilum....................................... ***Rottboellia exaltata***
24. Basal type of hilum **26**
25. Smooth surface of caryopses................................ ***Ophiuros exaltatus***
25. Rough surface of caryopses .... ***Urochondra setulosa***
26. L-embryo type.. **27**
26. N-embryo type **28**
27. Linear shaped hilum..................................... ***Oplismenus burmannii***
27. Oval shaped hilum .... ***Tetrapogon villosus***
28. Oval shaped hilum ....***Tetrapogon tenellus***
28. V-shaped hilum ....***Sporobolus indicus***
29. Caryopses surface either smooth or smooth shiny.......................... **30**
29. Caryopses surface either smooth dull or rough............................... **41**
30. Dorsal/lateral striations absent **31**
30. Dorsal/lateral striations present... **34**
31. L-embryo type.. ***Coix lachryma-jobi***
31. N-embryo type **32**
32. Ventral groove present ....***Sehima nervosum***
32. Ventral groove absent... **33**
33. V-shaped hilum ....***Themeda cymbaria***
33. Circular shaped hilum ....***Aeluropus lagopoides***
34. L-embryo type.. **35**
34. N-embryo type **36**
35. V-shaped hilum .... ***Thelepogn elegans***
35. Circular shaped hilum ....***Melanocenchris jaequemontii***
36. Sub basal type of hilum... **37**
36. Basal type of hilum **39**
37. V-shaped scutellum .... ***Bracharia distachya***
37. Sickle shaped scutellum **38**
38. Caryopses thickness is 0.40 mm .... ***Ischaemum indicum***
38. Caryopses thickness is 0.70 mm ....***Ischaemum molle***

39. Linear shaped hilum .... ***Saccharum spontaneum***
39. V-shaped hilum **40**
40. Caryopses length is 2.95 mm .... ***Sorghum halepense***
40. Caryopses length is 4.14 mm .... ***Sorghum purpureo-sericeum***
41. Dorsally compressed caryopses have striations
with L-embryo type .... ***Digitaria granularis***
41. Dorsally/laterally/not compressed, if dorsally compressed
have presence of striations with N-embryo type **42**
42. Laterally compressed caryopses or not compressed **43**
42. Dorsally compressed caryopses **54**
43. Not compressed caryopses .... ***Arthraxon lanceolatus***
43. Laterally compressed caryopses **44**
44. Dorsal/lateral striations absent **45**
44. Dorsal/lateral striations present **49**
45. N-embryo type **46**
45. L-embryo type **47**
46. T:B ratio 217.12 .... ***Sporobolus coromardelianus***
46. T:B ratio 132.18 .... ***Sporobolus diander***
47. V-shaped scutellm .... ***Triplopogon ramosissimus***
47. Sickle shaped scutellum **48**
48. Oval shaped hilum .... ***Apluda mutica***
48. V-shaped hilum .... ***Sachoenefeldia gracilis***
49. Smooth shiny surface of caryopses **50**
49. Smooth normal or dull surface of caryopses **51**
50. Sickle shaped scutellum .... ***Eragrostis tremula***
50. V-shaped scutellum .... ***Eragrostis unioloides***
51. Sickle shaped scutellum .... ***Cynadon dactylon***
51. V-shaped scutellum **52**
52. Caryopses shape is lanceolate .... ***Vetivaria zizanioides***
52. Caryopses shape is ovoid to oblong to elliptic **53**
53. Linear shaped hilum .... ***Chrysopogon fulvus***
53. Circular shaped hilum .... ***Eragrostis pilosa***
54. Sub basal circular shaped hilum .... ***Cymbopogon martini***
54. Linear or basal shaped hilum, if sub basal than either
oval or fan shaped hilum **55**
55. V-shaped scutellum with smooth dull surface **56**

55. Sickle shaped scutellum, if v shaped than have smooth shiny surface **68**
56. Dorsal/lateral striations present.......... **57**
56. Dorsal/lateral striations absent.......... **61**
57. Fan shaped hilum **58**
57. Oval shaped hilum **59**
58. Caryopses thickness is 0.70 mm ....***Echinochloa crusgalli***
58. Caryopses thickness is 1.54 mm ....***Echinochloa stagnina***
59. Caryopses thickness is 0.38 mm .... ***Bracharia cruciformis***
59. Caryopses thickness in between 0.61 and 0.62 mm.......... **60**
60. Dorsal surface have ∩ shaped undulation (under SEM).......... ***Bracharia ramosa***
60. Dorsal surface have ∧ shaped undulation (under SEM).......... ***Bracharia reptans***
61. Sub basal type of hilum.......... **62**
61. Basal type of hilum **64**
62. Dorsal surface have ∩ shaped undulation (under SEM)..........***Setaria tomentosa***
62. Dorsal surface have ∧ shaped undulation (under SEM) **63**
63. T:B ratio 51.10 .... ***Setaria glauca***
63. T:B ratio 60.71 ....***Setaria verticillata***
64. Linear shaped hilum..........***Panicum miliare***
64. Oval shaped hilum **65**
65. Caryopses breadth is 2.06 mm .... ***Paspalum scrobiculatum***
65. Caryopses breadth in between 0.95 and 1.19 mm.......... **66**
66. Embryo % is 83.62..........***Paspalidium flavidum***
66. Embryo % in between 52 and 58 .... **67**
67. Dorsal surface have ∩ shaped undulation with thick and elevated walls (under SEM)..........***Panicum antidotale***
67. Dorsal surface slightly undulated with thin and flat walls (under SEM) ....***Paspalidium geminatum***
68. Smooth shiny surface of caryopses.......... **69**
68. Smooth dull or normal surface of caryopses.......... **74**
69. V-shaped scutellum **70**
69. Sickle shaped scutellum.......... **71**

70. Fan shaped hilum .... ***Echinochloa colonum***
70. Oval shaped hilum .... ***Panicum trypheron***
71. Dorsal/lateral striations present................................ ***Cenchrus ciliaris***
71. Dorsal/lateral striations absent.. **72**
72. Circular shaped hilum .... ***Choris Montana***
72. Oval shaped hilum **73**
73. Fusiform shape of caryopses...................................... ***Chloris barbata***
73. Ovate to suborbicular shape of caryopses............. ***Cenchrus biflorus***
74. Dorsal/lateral striations absent.. **75**
74. Dorsal/lateral striations present.. **87**
75. L-embryo type.. **76**
75. N-embryo type **77**
76. Oval shaped hilum .... ***Chionachne koenigii***
76. Fan shaped hilum .... ***Andropogon pumilus***
77. Ventral groove present **78**
77. Ventral groove absent.. **79**
78. Caryopses length is 2.71 mm .... ***Sehima ischaemoides***
78. Caryopses length is 4.22 mm .... ***Sehima sulcatum***
79. Oval shaped hilum **80**
79. Fan or V-shaped hilum **82**
80. Greenish to white color caryopses .... ***Pennisetum pedicellatum***
80. Light brown to brown color caryopses **81**
81. Caryopses breadth is 1.19 mm .... ***Alloteropsis cimicina***
81. Caryopses breadth is 0.57 mm .... ***Tragrus biflorus***
82. Fan shaped hilum **83**
82. V-shaped hilum **85**
83. Caryopses breadth is 0.66 mm ... ***Heteropogon contortus var. contortus sub var. genuinus***
83. Caryopses breadth is 0.39 mm **84**
84. Hilum have ruminate pattern (under SEM) .... ***Heteropogon contortus var. contortus sub var. typicus***
84. Hilum have reticulate pattern (under SEM)....... ***Heteropogon ritchiei***
85. Caryopses shape is ovate.. ***Themeda laxa***
85. Caryopses shape is lanceolate of oblong.. **86**
86. T:B ratio is 75.42.. ***Themeda triandra***
86. T:B ratio is 35.47.. ***Themeda quadrivalvis***

87. Sub basal type of hilum.. **88**
87. Basal type of hilum **90**
88. Linear shaped hilum... ***Eriochloa procera***
88. Oval shaped hilum **89**
89. Caryopses shape is oblong....................................... ***Digitaria ciliaris***
89. Caryopses shape is ellipsoid .... ***Digitaria microbachne***
90. Linear shaped hilum...................................... ***Oplismenus composites***
90. Oval or fan or V-shaped hilum.. **91**
91. V-shaped hilum .... ***Capillipedium hugelii***
91. Oval or fan shaped hilum... **92**
92. Oval shaped hilum **93**
92. Fan shaped hilum **94**
93. Ventral surface have wavy walls (under SEM).....***Cenchrus setigerus***
93. Ventral surface have straight walls (under SEM).....***Cenchrus preurii***
94. Caryopses shape is obovate.............................. ***Bothriochloa pertusa***
94. Caryopses shape is oblong or ovate **95**
95. Caryopses length is 10.20 mm .... ***Heteropogon triticeus***
95. Caryopses length in between 1.8–2.47 mm **96**
96. Hilum have striate pattern (under SEM) ....***Ischaemum pilosum***
96. Hilum have reticulate or ruminate pattern (under SEM) **97**
97. Hilum have ruminate pattern (under SEM)...................................... **98**
97. Hilum have reticulate pattern (under SEM)..................................... **99**
98. Caryopses thickness is 0.41mm .... ***Dicanthium caricosum***
98. Caryopses thickness is 0.81mm .... ***Ischaemum rugosum***
99. Dorsal surface have ∧ shaped undulation (under SEM)... ***Iseilẹma laxum***
99. Dorsal surface not show clearly reticulum but slightly undulating wall seen (under SEM)................................. ***Dicanthium annulatum***

1.6 CLUSTER ANALYSIS

The software used displays a single tree among the possible ones (Figure 48). In the dendrogram based on the morphological characters of the caryopses.

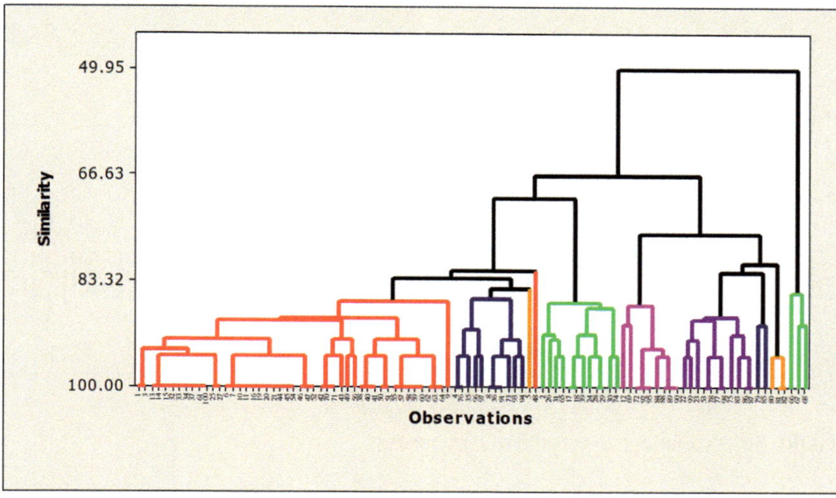

FIGURE 48 A dendrogram showing the clustering of grass species.

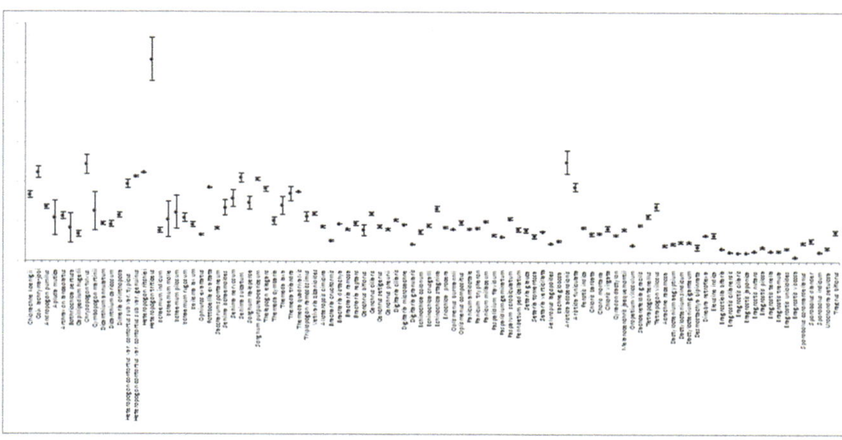

FIGURE 49 Variations in length of caryopses.

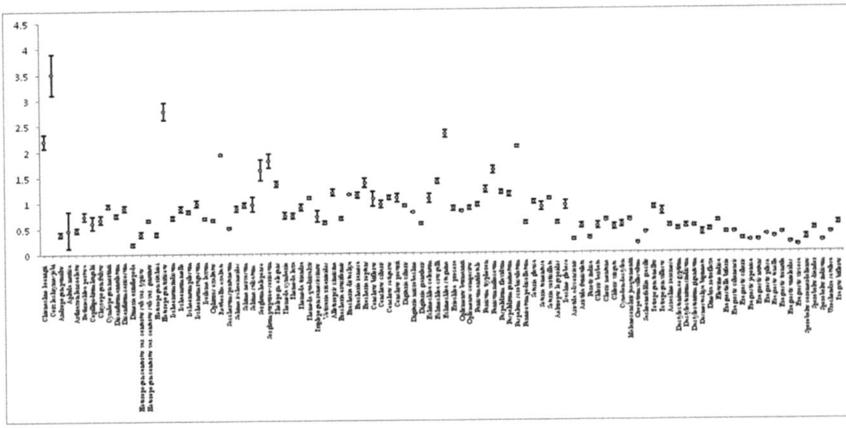

FIGURE 50 Variations in breadth of caryopses.

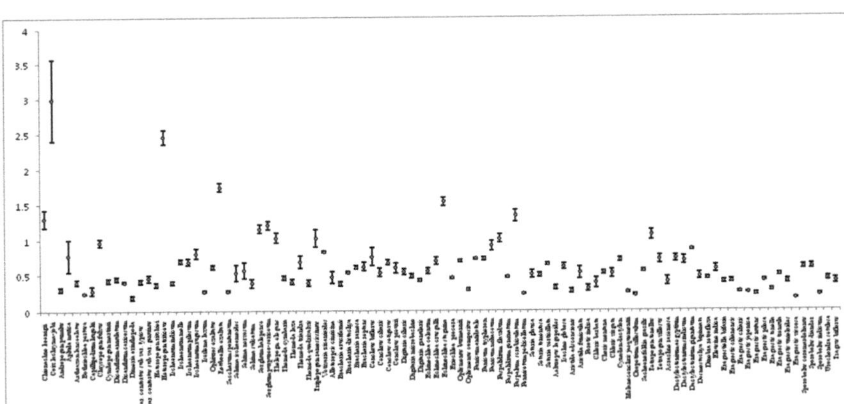

FIGURE 51 Variations in thickness of caryopses.

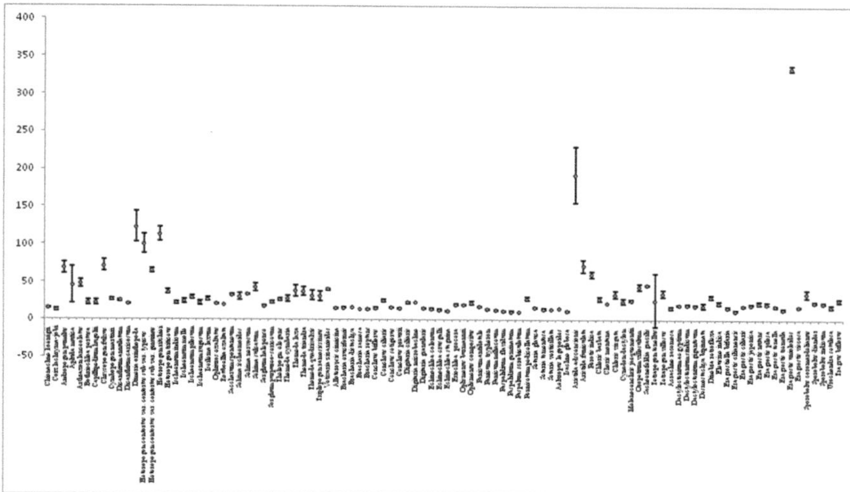

FIGURE 52 Variations in L:B ratio.

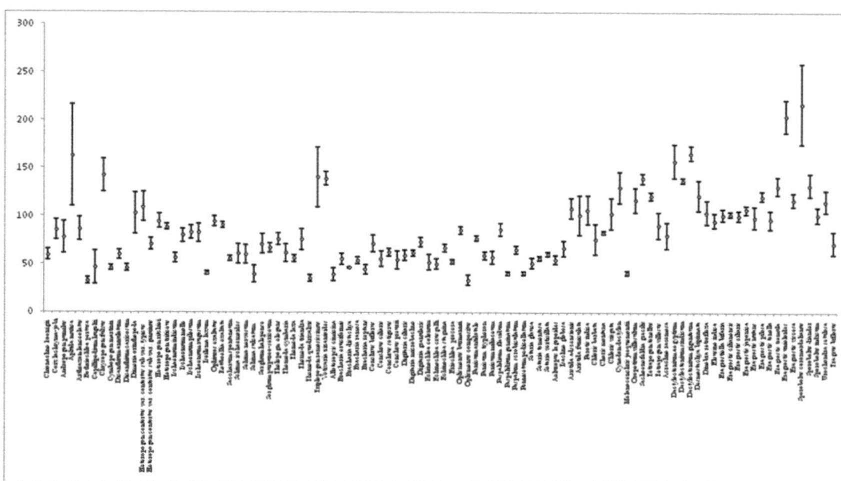

FIGURE 53 Variations in T:B ratio.

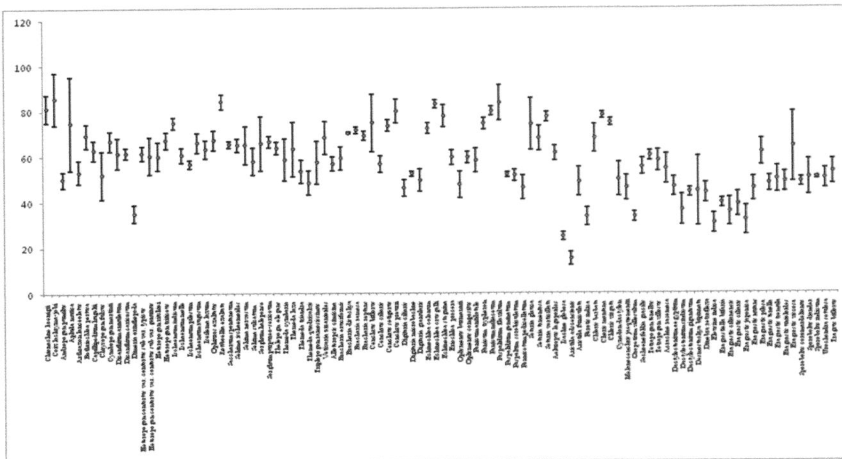

FIGURE 54 Variations in embryo %.

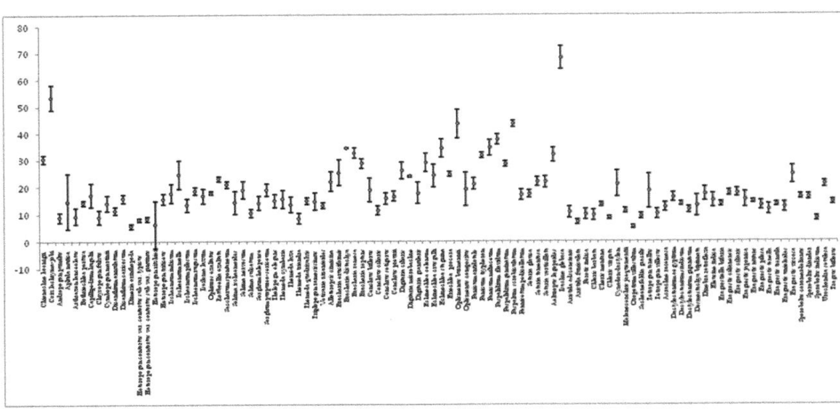

FIGURE 55 Variations in hilum %.

The following features are taken into consideration for the preparation of dendrogram:

I. Texture (4 criteria)
   Smooth shiny (1), Smooth (2), Rough (3), Smooth dull (4)
II. Compression (3 criteria)
   Dorsally compressed (1), Laterally compressed (2), Not compressed
III. Dorsal/Lateral straitions (2 criteria)

Present (1), Absent (2)

IV. Ventral groove (2 criteria)
Present (1), Absent (2)

V. Scutellum shape (2 criteria)
V-shaped (1), Sickle shaped (2)

VI. Embryo type (2 criteria)
N-type (1), L-type (2)

VII. Embryo class (2 criteria)
Large (1), Short (2)

VIII. Hilum type (3 criteria)
Basal (1), Sub basal (2), Linear (3)

IX. Hilum shape (8 criteria)
Circular (1), V-shaped (2), Oval (3), Fan (4), Long (5), Short (6), Linear (7), Semi circular (8)

Majorly two clusters form with the similarity about 50%. Under one major cluster only one sub cluster is present and is cluster 10, which have three species with the similarity about 84%. Other nine clusters are present under the second major group. The observations presented in Table 2.

Clusters 4 and 7 are simplifolius, for example, having only a single species in cluster. *Digitaria granularis* form cluster 4, showed around 82%

TABLE 2 Distribution of Caryopses into Clusters on the Basis of Their Qualitative Features

Cluster number	Number of observation (grass species) per cluster	Value from centroid
1	47	1.35
2	12	1.25
3	11	1.15
4	1	0.0
5	9	1.23
6	11	1.14
7	1	0.0
8	3	1.05
9	2	0.70
10	3	0.67

similarities with the characters while *Arthraxon lanceolatus* form cluster 7, shows 84% similarities of characters with other species.

Cluster 1 showing highest number of species and it shows about 83% of similarity with the other clusters. In this cluster 43 species shows 100% similarity, but they were present in different small clusters. This species further classified on the bases of numerical and scanning electron microscopic characters. Rather than these clusters 2, 5, 6 and 8 also have 4, 6, 6 and 2 species, respectively, which shows 100% similarity with the characters.

According to the dendrogram clusters 6 and 9 are closer and they show around 86% similarity. Cluster 9 had two species, *Acrachne racemosa* and *Eleusine indica* which were belongs to the tribe Eragrosteae and Cluster 6 had 11 species, in which three species from tribe Eragrosteae while others were from tribe Chlorideae, Sporoboleae and Andropogoneae.

Cluster 1 was close to cluster 3 and they had 85% similarity, and had 47 and 11 species respectively. The centroid value of cluster 1 was 1.35 and for cluster 3 was 1.15. Cluster 3 had laterally compressed caryopses with smooth surface and the caryopses sizes varies from 0.2–2.28 mm. Cluster 1 consist of more number of species and all are dorsally compressed caryopses. These species have great variations in their characters. Mainly they were classified on the basis of the morphological character but further numerical data supported.

Cluster 2 show close relationship with clusters 4, 7, 3 and 1. It shows around 84% similarity with other clusters and had 12 species.

KEYWORDS

- **Caryopses**
- **Cluster analysis**
- **Light microscopy**
- **Morphology**
- **Morphometric analysis**
- **Scanning electron microscopy**

REFERENCES

1. Agrawal, D. P. *Metal technology of the Harappans*. pp. 163–168 in *Frontiers of Indus Civilization;* Eds. B. B. Lal, S. P. Gupta; Books & Books: New Delhi, 1984.
2. Banerjee, S. K., Chauhan, K. P. S. Studies on the evolution of seed coat pattern in wheat by scanning electron microscopy identification. *Seed Science and Technology.* 1981, 9(3), 819–822.
3. Barthlott, W. G. Epidermal and seed surface characters of plants: systematic applicability and some evolutionary aspects. *Nordic Journal of Botany* 1981, 1, 345–355.
4. Barthlott, W. G. Microstructural features of seed surface. pp. 95–105 (Heywood, V. H., Moore, D. M., ed.) In: *Current concepts in plant taxonomy.* Academic Press: London, 1984.
5. Bogdan, A. V. Seed morphology of some cultivated African grasses. *Proceedings of the International Seed Testing Association* 1965, 31, 789–799.
6. Chuang, T. I., Heckard, L. R. Seed coat morphology in *Cordylanthus* (Scrophulariaceae) and its taxonomic significance. *American Journal of Botany* 1972, 59, 258–265.
7. Colledge, S. M. Scanning electron microscope studies of the cell patterns of the pericarp layers of some wild wheat's and ryes methods and problems. pp. 225–236 in, S. L. Olsen (ed.) *Microscopy in archaeology, BAR, International series 452* (Oxford). 1988.
8. Costea, M., Tardif, F. J. Taxonomy of the most common weedy European *Echinochloa* species (Poaceae: Panicoidee) with special emphasis on characters of the lemma and caryopsis. *SIDA* 2002, 20, 525–548.
9. Guest, E. *Flora of Iraq*, Vol 1: Introduction. Baghdad: Ministry of Agriculture, Republic of Iraq. 1966.
10. Heywood, V. H. *Scanning electron microscopy: systematic and evolutionary applications.* Academic press, New York, London. 331. 1971.
11. Hillman, F. H. *Distinguishing characters of the seeds of Sudan grass and Johnson grass. USDA, Bulletin 406* (Washington, D. C.). 1916.
12. Hill, R. J. Taxonomic and phylogenetic significance of seed microsculpturing in *Mentzelia* (Loasaceae) in Wyoming and adjacent Western states. *Brittonia* 1976, 28, 86–112.
13. Hilu, K. W., Alice, L. A. *Phylogenetic relationships in subfamily Chloridoideae (Poaceae) based on mark sequences: a preliminary assessment*, pp. 173–179. In Jacobs, S. W. L., Everett, J. [eds.], *Grasses: systematics and evolution.* CSIRO Publishing, Collingwood, Victoria, Australia. 2000.
14. Hilu, K. W., Wright, K. Systematics of Gramineae: A cluster analysis study. *Taxon* 1982, 31, 9–36.
15. Hoagland, R. E., Paul, R. N. A comparative SEM study of red rice and several commercial rice (*Oryza sativa*) varieties. *Weed science* 1978, 26(6), 619–625.
16. Hufford, L. Seed morphology of Hydrangeaceae and its phylogenetic implications. *International Journal of Plant Science* 1995, 156, 555–580.
17. Jensen, L. A. Seed characteristics of certain wild barley, *Hordeum* spp. *Proceedings of the International Seed Testing Association* 1957, 7, 87–91.
18. Joshi, M., Sujatha, K., Harza, S. Effect of TDZ and 2,4-D on peanut somatic embryogenesis and in vitro bud development. *Plant Cell, Tissue and Organ Culture* 2008, 94(1), 85–90.

19. Karcz, J., Weiss, H., Maluszynska, J. Structural and embryological studies of diploid and tetraploid *Arabidopsis thaliana* (L.) Heynh. *Acta Biological Cracoviensia Series Botanica* 2000, 42(2), 113–124.
20. Kislev, M. E., Melamed, Y., Simchoni, O., Marmorstein, M. Computerized key of grass grains of the Mediterranean basin. *Lagascalia* 1997, 19(2), 289–294.
21. Körber-Grohne, U. *Bestimmungsschlüssel für subfossile Juncus-Samen und Gramineen-Früchte.* Probleme der Küsten-forschung im südlichen Nordseegebiet 7. Hildesheim: August Lax. 1964.
22. Koul, K. K., Nagpal, R., Raina, S. N. Seed coat microsculpturing in *Brassica* and Allied genera (Subtribe Brassicinae, Raphanine, Morcandinaceae). *Annals of Botany* 2000, 80, 385–397.
23. Kreitschitz, A., Tadele, Z., Gola, E. M. Slime cells on the surface of *Eragrostis* seeds maintain a level of moisture around the grain to enhance germination. *Seed Science Research* 2009, 19, 27–35.
24. Liu, Q., Zhao, N. X., Hao, G., Hu, X. Y., Liu, Y. X. Caryopsis morphology of the Chloridoideae (Gramineae) and its systematic implications. *Botanical Journal of Linnean Society* 2005b, 148, 57–72.
25. Matthews, J. F., Levins, P. A. The systematic significance of seed morphology in *Portulaca* (Portulacaceae) under scanning electron microscopy. *Systematic Botany* 1986, 11, 302–308.
26. Matsutani, A. Identification of Italian millet from Esashika site by means of scanning electron microscope. *Journal of the Anthropological society of Nippon* 1986, 94 (1), 111–118.
27. Murley, M. R. Seeds of the Cruciferae of North Eastern America. *American Middle Naturalichen* 1951, 46, 1–81.
28. Nesbitt, M. *Identification guide for Near Eastern grass seeds,* Institute of Archaeology, University College London: London. 2006.
29. Osman, A. K., Mohammed, A. Z., Hamed, S. T., Hussein, N. R. Fruit morphology of annual grasses from Egypt. *Asian Journal of Plant Sciences* 2012, 11(6), 268–284.
30. Peterson, P. M., Sánchez Vega, I. *Eragrostis* (Poaceae: Chloridoideae: Eragrostideae: Eragrostidinae) of Peru. *Annals of the Missouri Botanical Gard*ens 2007, 94, 745–790.
31. Qing Liu., Nan-Xian Zhao; Gang Hao; Xiao-Ying Hu; Yun-Xiao Liu. Caryopsis morphology of the Chloridoideae (Gramineae) and its systematic implications. *Botanical Journal of Linnaean Society* 2005, 148, 57–72.
32. Terrell, E. E., Peterson, P. M. Caryopses morphology and classification in the Triticeae (Pooideae: Poaceae). *Smithsonian Contribution to Botany.* 1993, 83, 1–25.
33. Wang, S. J., Guo, P. C., Li, J. H. The major types of caryopses of the Chinese Gramineae in relation to systematics. *Acta Phytotaxon Sin* 1986,24, 327–345.
34. Whiffin, T., Tomb, A. S. The systematic significance of seed morphology in the neotropical capsular-fruited Melastomataceae. *American Journal of Botany* 1972, 59, 411–422.
35. Zhang, Y., Xiaoying, H. U., Yunxiao, L., Qing, L. Caryopsis morphological survey of the genus *Themeda* (Poaceae) and allied spathaceous genera in the Andropogoneae. *Turkish Journal of Botany* 2014, 38, 665–676.

SECTION 2

SEEDLING

CONTENTS

2.1 Introduction 160
2.2 List of Grass Seedling Characterized (Figures 59–107) 162
2.3 Materials and Methods 165
2.4 Illustrated Terminologies 166
2.4.1 Growth Habit 170
2.4.2 Vernation 171
2.4.3 Node 171
2.4.4 Internode 171
2.4.5 Leaf Blade 172
2.4.6 Termination Point 172
2.4.7 Leaf Sheath 173
2.4.8 Ligule 173
2.4.9 Auricle 174
2.4.10 Collar 174
2.5 Illustrated Seedling Description 200
2.6 Identification Key 251
2.7 Cluster Analysis 257
Keywords 259
References 259

2.1 INTRODUCTION

A seedling comes in being when a seed germinates, but the point at which it ceases to be a seedling is much less clear (Fenner, 1987). Seed and seedling traits vary strongly across the tropical forest biome to cope with the variations in the distribution and amount of rainfall, light, temperature and soil nutrient regimes. Seedlings of monocots are much more diverse than those of other angiosperms, often with much derived character. This makes morphological interpretation difficult (Tillich, 2007). Seedlings r and Turner (1933), Hitchcock (1936), Undersnder et al. A seedling is helpful in assessing the natural regeneration of an ecosystem and is of great importance to the forest planners. Recognition of plants by their vegetative characters is essential in the development of a sound pasture-improvement program. In this program, the first step is to recognize the pest. Control measures depend upon accurate identification. Grasses occurring as weeds are difficult to be identified in their vegetative stage. Many grasses do not flower until late in the growing season. By this time they already have had an impact on the associated growing crop yields. Removal of grass seedling at an earlier stage is important because smaller grasses are easy to be removed and they require very small dosage of herbicides. So greater control can be achieved while minimizing control costs.

A seedling is a plant from seed and not by vegetative reproduction. The seedling is in most cases used to refer to very young individuals. The height and size of seedling are often unsuitable for identifying it, but these measures are important for the general impression the plant makes. Seedlings of tree species are very easily identifiable while grasses and herbaceous plants are very difficult to identify at seedling stages, particularly in the one-to-two-leaf seedling growth stage. For the perfect identification of grasses flowering condition is needed (Fishel, 2004).

McAlpine reported identification of grasses by their leaves (1890). The first attempts to study the grasses in their flowerless and anatomical characters were by Jensen (1953) as rightly indicated by Ward (1901). Carrier compiled a key for 48 common grasses in Eastern United States and it proved of more practical value for the fieldwork than other available key (1917). Later on Henning (1930), Whyte (1930), Burr and Turner (1933), Hitchcock (1936), Undersnder et al. (1996), Prosser (2009), Bradley et al. identified grasses at their mature stage before flowering on the basis of the vegetative characters (2010). Nowosad et al. (1942) bulletin on the identification of certain native and naturalized hay and pasture grasses and Phillips, field manual on the identification of grasses describes the identification of few grasses of North Eastern part of Canada by their vegetative characters (1962). The main characteristic features used by them are bud shoot, vernation, collar, ligule and auricle. However in these studies the information has been restricted to only a few numbers of grasses.

Durgan (1999) proposed an identification key for some of the broad leaf and grass weed seedlings. Wintel et al. (2009) studied few seedlings of subtropical perennial grass species of Western Australia, for example, *Panicum maximum, Chloris gayana, Pennisetum dandestinum, Setaria sphacelata, S. splendid, Urochloa brachiaria, Digitaria eriantha, Panicum coloratum*, etc. Jian-Guo et al. (1993) studied seedling characters of 201 species of Gramineae from Australia and China and they grouped these species in 4 clusters.

Our literature reviews and consultation with agrostologist, foresters have not found a compendium of seedlings of native grass species' morphological characteristics. Morphological properties of the seedlings, particularly shape of the first leaf, have been examined systematically by Kuwabara (1960, 1961). Looman (1982) studied 107 grasses of the rangelands of the Prairie Provinces of Canada and proposed key for the identification of these grass species. Harries (1950, 2010) identified and proposed a key to the identification of 50 grass seedlings of Oklahoma grassland. These studies have been supplemented with line diagrams showing the diagnostic features of identification.

From the systematic point of view several workers have analyzed features of grass seedling. To identify the grass seedling, first feature to be checked is to see whether the grass is sod forming (spreading) or bunching (forms clumps).

There are several other features which can be used for the identification of seedlings. First, the life cycle is considered which is very important for collecting seedlings from the field. It will enable an easy identification and collection because some grasses which will not be present at certain time of the year can be easily eliminated from the list during that particular period of collection. Thereafter, there are other features like collar, node, ligule, auricle, culm, leaf surface, leaf tip, stem etc. which can to be considered. The shape of the stem is another feature that is easy to determine. Most grasses have round stem, but some have distinctly flat stems. Most of the identifying characteristic features are confined in collar region of the grasses, which can be seen by carefully pulling the leaf blade back from the stem (Fishel, 2004). Ligule which is present at the base of leaf blade, may or may not be present in all the species. If present they appear as small projection that appears at rim or tuft of hairs or membranous at the base of leaf blade. Besides these features, some grass species have auricles, which are small finger like projection at the base of leaf blade, it clasping the stem at the collar region.

The external surface of seedling organs may contain some hairs, glands, scales, prickles, cystoliths, dots, etc. A spine and a tendril can also be present which may serve as a diagnostic feature of identification.

The present monograph is an effort to evaluate morphological and micromorphology features of 98 common grass species. Caryopses the fruit of grass members have been characterized for their light and scanning electron microscopic features and the seedlings have been characterized solely on their vegetative characteristic features. Photographs depicting the characteristic features observed in the different parts of the studied seedlings are presented along with the description of the grass seedlings. An easy to use field identification key for these seedlings has been prepared.

2.2 LIST OF GRASS SEEDLING CHARACTERIZED (FIGURES 59–107)

1. *Coix lachryma-jobi* L.
2. *Andropogon pumilus* Roxb.
3. *Aplaaauda mutica* L.
4. *Arthraxon lanceolatus* (Roxb.) Hochst.
5. *Bothriochloa pertusa* (L.) A.Camus

6. *Capillipedium hugelii* (Hack.) A.Camus
7. *Chrysopogon fulvus* (Spreng.) Chiov.
8. *Cymbopogon martini* (Roxb.) W. Watson
9. *Dicanthium annulatum* (Forssk.) Stapf
10. *Dicanthium caricosum* (L.) A. Camus
11. *Hackelochloa granularis* (L.) Kuntze
12. *Heteropogon contortus* var. *contortus* sub var. *typicus* Blatt. and McCann
13. *Heteropogon contortus* var. *contortus* sub var. *genuinus* Blatt. and McCann
14. *Heteropogon ritcheii* (Hook.f.) Blatt. and McCann
15. *Heteropogon triticeus* (R.Br.) Stapf ex Craib
16. *Imperata cylindrical* (L.) Raeusch.
17. *Ischaemum indicum* (Houtt.) Merr.
18. *Ischaemum molle* Hook.f.
19. *Ischaemum pilosum* (Willd.) Wight
20. *Ischaemum rugosum* Salisb.
21. *Iseilema laxum* Hack.
22. *Ophiuros exaltatus* (L.) Kuntze
23. *Rottboellia exaltata* Linn. f.
24. *Sehima ischaemoides* Forssk.
25. *Sehima nervosum* (Rottler) Stapf
26. *Sehima sulcatum* (Hack.) A.Camus
27. *Sorghum halepense* (L.) Pers.
28. *Sorghum purpureo-sericeum* (A.Rich.) Schweinf. and Asch.
29. *Thelepogn elegans* Roth
30. *Themeda cymbaria* Hack.
31. *Themeda laxa* A.Camus
32. *Themeda triandra* Forssk.
33. *Triplopogon ramosissimus* (Hack.) Bor
34. *Alloteropsis cimicina* (L.) Stapf
35. *Brachiaria eruciformis* (Sm.) Griseb.
36. *Brachiaria distachya* (L.) Stapf
37. *Brachiaria ramosa* (L.) Stapf
38. *Brachiaria reptans* (L.) C. A. Gardner and C. E. Hubb.
39. *Cenchrus biflorus* Roxb.

40. *Cenchrus ciliaris* L.
41. *Cenchrus setigerus* Vahl
42. *Digitaria ciliaris* (Retz.) Koeler
43. *Digitaria microbachne* (Persl) Hern.
44. *Digitaria granularis* (Trin.) Henrard
45. *Digitaria longiflora* (Retz.) Pers.
46. *Digitaria stircta* Roth
47. *Echinochloa colona* (L.) Link
48. *Echinochloa crusgalli* (L.) P. Beauv.
49. *Echinochloa stagnina* (Retz.) P. Beauv.
50. *Eremopogon foveolatus* (Delile) Stapf
51. *Eriochloa procera* (Retz.) C. E. Hubb.
52. *Oplismenus composites* (L.) P. Beauv.
53. *Panicum antidotale* Retz.
54. *Panicum maximum* Jacq.
55. *Panicum miliaceum* L.
56. *Panicum trypheron* Schult.
57. *Paspalidium flavidum* (Retz.) A. Camus
58. *Paspalidium geminatum* (Forssk.) Stapf
59. *Paspalum scrobiculatum* L.
60. *Paspalum distichum* L.
61. *Paspalum vaginatum* Sw.
62. *Pennisetum setosum* (Sw.) Rich.
63. *Setaria glauca* (L.) P.Beauv.
64. *Setaria tomentosa* (Roxb.) Kunth
65. *Setaria verticillata* (L.) P. Beauv.
66. *Aeluropus lagopoides* (L.) Thwaites
67. *Isachne globosa* (Thunb.) Kuntze
68. *Aristida adscensionis* L.
69. *Aristida funiculata* Trin. and Rupr.
70. *Chloris barbata* Sw.
71. *Chloris virgata* Sw.
72. *Cynadon dactylon* (L.) Pers.
73. *Melanocenchris jaequemontii* Jaub. and Spach
74. *Oropetium villosulum* Stapf ex Bor
75. *Schoenefeldia gracilis* Kunth

76. *Tetrapogon tenellus* (Roxb.) Chiov.
77. *Tetrapogon villosus* Desf.
78. *Dactyloctenium aegyptium* (L.) Willd.
79. *Dactyloctenium sindicum* Boiss.
80. *Desmostachya bipinnata* (L.) Stapf
81. *Dinebra retroflexa* (Vahl) Panz.
82. *Eleusine indica* (L.) Gaertn.
83. *Eleusine verticillata* Roxb.
84. *Eragrostiella bifaria* (Vahl) Bor
85. *Eragrostiella bachyphylla* (Stapf) Bor
86. *Eragrostis cilianensis* (All.) Janch.
87. *Eragrostis ciliaris* (L.) R.Br.
88. *Eragrostis pilosa* (L.) P.Beauv.
89. *Eragrostis tenella* (Linn.) P.Beauv. ex Roem.
90. *Eragrostis tremula* Hochst. ex Steud.
91. *Eragrostis unioloides* (Retz.) Nees ex Steud.
92. *Eragrostis viscosa* (Retz.) Trin.
93. *Sporobolus coromardelianus* (Retz.) Kunth
94. *Sporobolus diander* (Retz.) P. Beauv.
95. *Sporobolus indicus* (L.) R.Br.
96. *Urochondra setulosa* (Trin.) C. E. Hubb.
97. *Tragus biflorus* (Roxb.) Schult. (illegimate name)
98. *Zoysia matrella* (L.) Merr.

2.3 MATERIALS AND METHODS

The seedling study was carried out both in field and in laboratory.

(1) In field: Seedlings of grasses were collected as the new emergents. The collected seedlings and their different parts were photographed by Digital camera in the field itself. Total 10–15 samples for each species were collected from the field and observed for distinctive features.

(2) In laboratory: Total 45–50 seeds were grown in earthen pots and seedlings were raised till the emergence of four to five leaves. Randomly selected seedlings were evaluated for their characteristic features and photographed by digital camera (DSC-T20). All observations were made on seedlings

< 20–25 days old. Characteristic features of seedlings raised in the pots were critically observed and compared with those collected from the field.

Herbarium specimens of the studied species have been deposited in the BARO Herbarium of The M.S. University of Baroda, Vadodara, Gujarat, India.

In grasses, as most of the identifying characteristics are seen on the collar region, this region was examined carefully by pulling the leaf blade back from the stem. Features like the presence of ligule, auricle, characteristics of node, internode, etc. were recorded.

The information included in the morphological data was used to prepare a morphological key. The morphological features were grouped and a dichotomous key was constructed on the basis of the most conspicuous morphological features observed in the field-collected samples. The features were critically observed and compared with the laboratory raised seedling features. In addition, the data was used to perform a preliminary analysis using Minitab software (Version 16) to prepare a dendrogram in order to explore the correspondence between morphological and taxonomic features.

2.4 ILLUSTRATED TERMINOLOGIES

A total of 98 different grass species belonging to group Panicoideae and Pooideae have been morphologically characterized for its diagnostic features of identification. Most of the species belongs to the tribe Andropogoneae and Paniceae. Characters defined by Nowosad et al. (1942) and Phillips (1962) have been used for describing the seedlings. From vernation to ligule the vegetative characteristic feature of mature grass species have been well documented by Nowosad et al. (1942) and Phillips (1962) and so has not been represented in the present study. These features were found to be differing from species to species. The characteristic parts of a grass seedling and the features considered are represented in Figures 56–58.

The following eleven features were taken into the consideration for the study:

X. Growth habit (2 criteria)

FIGURE 56 Different parts of grass seedling.

Annual (1), Perennial (2)

XI. Vernation (2 criteria)

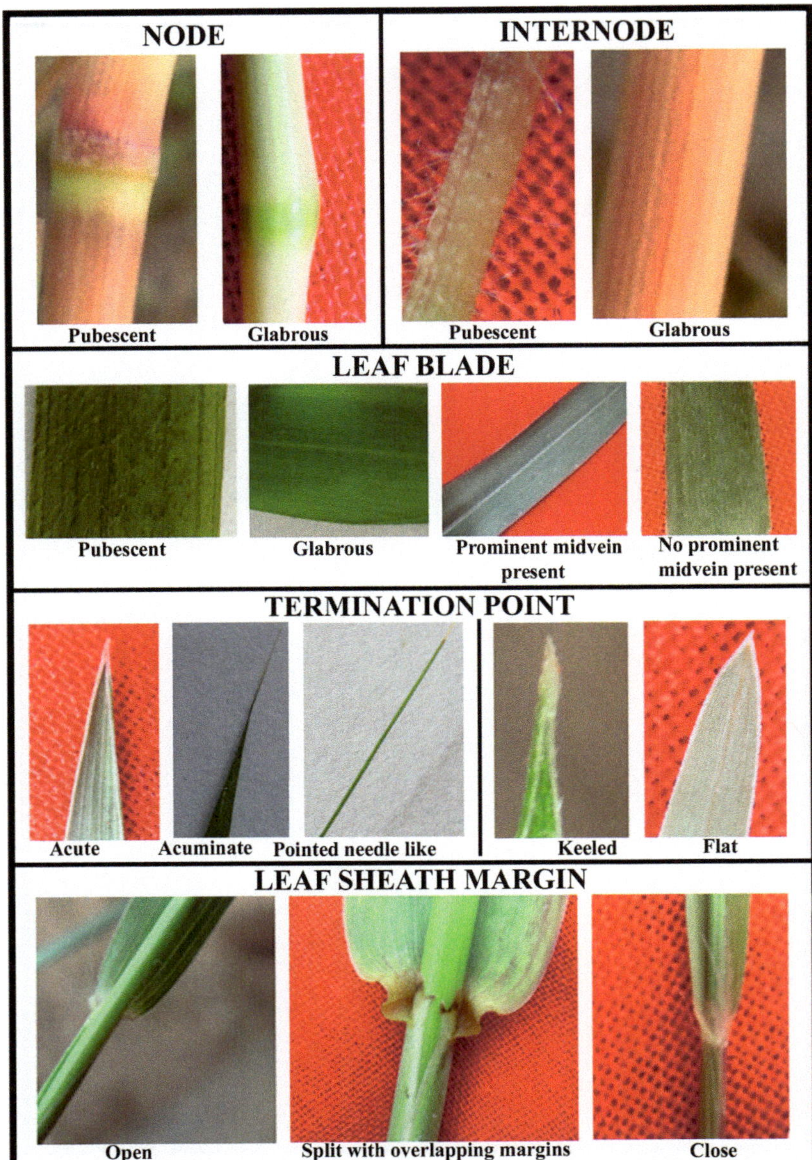

FIGURE 57 Characteristic culm and leaf features of a grass seedling.

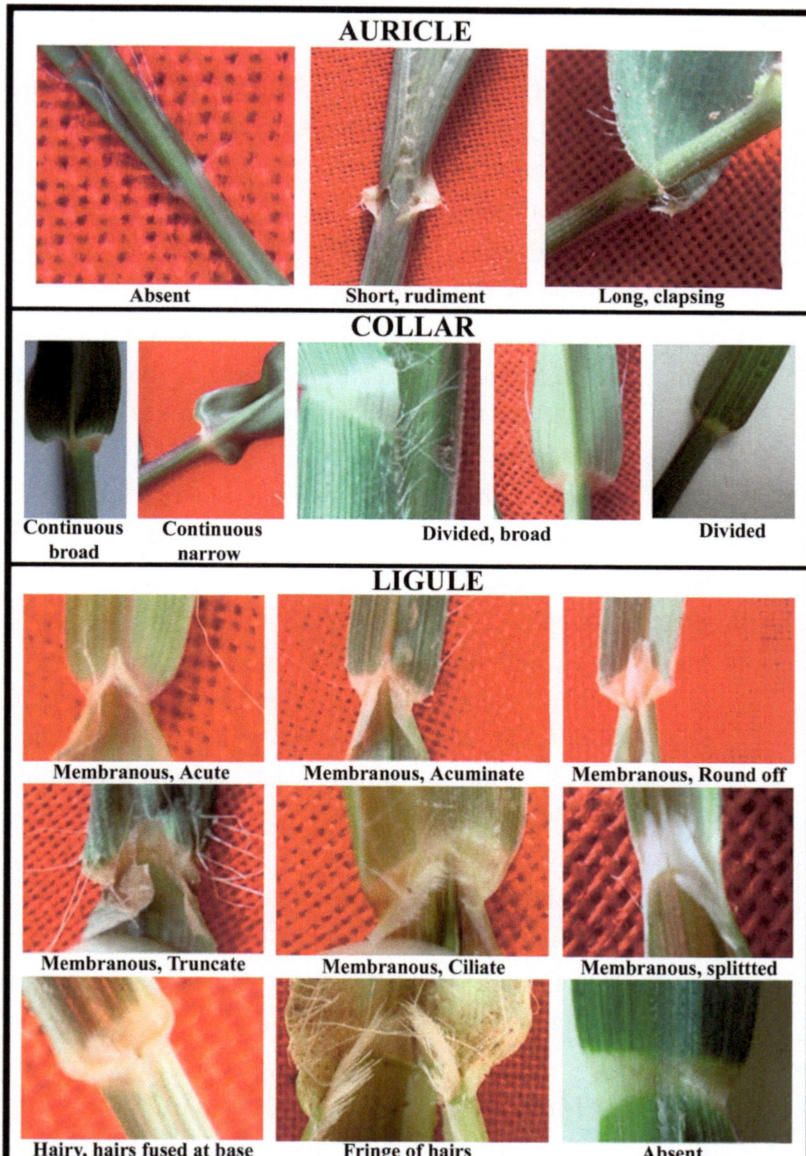

FIGURE 58 Characteristic auricle, ligule and collar features of a grass seedling.

Rolled (1), Folded (2)

XII. Node (2 criteria)

Pubescent (1), Glabrous (2), Glabrous and angled (3), Glabrous and bearded (4), Pubescent and bearded (5)

XIII. Internode (2 criteria)

Pubescent (1), Glabrous (2)

XIV. Leaf Blade (2 criteria)

Pubescent surface (1), Glabrous surface (2)

XV. Termination point (5 criteria)

Acute and flat (1), Acuminate and flat (2), Pointed needle like (3), Acute and keeled (4), Acuminate and keeled (5)

XVI. Leaf sheath margin (3 criteria)

Open (1), Split with open margins (2), Close (3)

XVII. Leaf sheath type (2 criteria)

Round (1), Folded (2)

XVIII. Ligule (10 criteria)

Membranous and truncate (1), Membranous and round off (2), Membranous and ciliate (3), Membranous and acute (4), Fringe of hairs (5), Membranous and obtuse (6), Membranous (7), Absent (8), Fringe of hairs, fused at base (9), Membranous and acuminate (10)

XIX. Auricle (3 criteria)

Absent (1), Short and rudiment (2), Long and clasping (3)

XX. Collar (5 criteria)

Divided (1), Continuous and broad (2), Continuous and narrow (3), Divided and broad (4), Continuous (5)

2.4.1 GROWTH HABIT

The term is used most often to describe the general appearance, growth form, or architecture of a plant. Growth habit may be:

(1) Annual: A plant that completes its life cycle and dies within one year.

(2) Perennial: A plant whose life span extends over several years.

2.4.2 VERNATION

Vernation means the arrangement of the leaves in the bud-shoot. It is used as a first point for the identification of grasses. This arrangement of the leaves in the bud shoot was identified by cutting the shoot across below the ligule of the uppermost leaf and examining the sections. Accordingly, the vernation may be:

(1) Folded (Conduplicate): Which is generally formed in a laterally compressed or flattened shoot. In this the margins meet but do not overlap. It appears elliptical in cross section.
(2) Rolled (Convolute): Which is generally formed in a round or cylindrical shoot. In this the margins meeting overlap. It appears round in cross section and the successive leaves are rolled alternately in clockwise and counter clockwise manner.

2.4.3 NODE

The part of a stem where leaves or branches arise. Node may be:

(1) Glabrous: Without surface ornamentation such as hairs, scales or bristles.
(2) Pubescent: Downy; covered with short, soft, erect hairs.
(3) Bearded: Small appendages that project from the base of the leaf blade and appear to wrap the culm, at least partially.

2.4.4 INTERNODE

The portion of a stem between two nodes.

(1) Glabrous: Without surface ornamentation such as hairs, scales or bristles.
(2) Pubescent: Downy; covered with short, soft, erect hairs.

2.4.5 LEAF BLADE

An outgrowth of a stem, usually flat and green; its main function is food manufacture by photosynthesis. The terms which are use for the explanation of leaf blade are follows:

- Surface:
  (1) Glabrous: Without surface ornamentation such as hairs, scales or bristles.
  (2) Pubescent: Downy; covered with short, soft, erect hairs.
- Midvein: The central, and usually most prominent, vein of a leaf or leaf-like organ;
- Leaf lamina: The blade of a leaf or the expanded upper part of a petal, sepal or bract.
  (1) Dorsal surface: The back surface; in particular, away from the axis in a lateral organ or away from the substratum.
  (2) Ventral surface: The front surface; in particular, towards the axis in a lateral organ or towards substratum.
- Leaf margin: The edge, as in the edge of a leaf blade.
  (1) Serrate margin: Toothed with asymmetrical teeth pointing forward; like the cutting edge of a saw.
  (2) Toothed margin: Margin with a more or less regularly incised margin.
- Tuberculated hairs: Hairs covered in tubercles; warty.
- Amplexicaul: A leaf with its base clasping the stem.
- Scabrous: Rough to touch.

2.4.6 TERMINATION POINT

The termination point is nothing but a leaf blade tip. It is of different types but in grasses generally it is acute or acuminate or pointed needle like.

(1) Acute: Sharply pointed; converging edges making an angle of less than 90°.
(2) Acuminate: Tapering gradually to a point.

Rather than shape of leaf tip other features of leaf tip are also very important. If the leaf tip is not folded it is termed as flat termination point and if it is folded then keeled termination point.

2.4.7 LEAF SHEATH

Leaf sheath is the basal portion of the leaf enveloping the culm or young growing leaves. It is categorized into different types based on the overlapping of margins.

(1) Open: If the margins of the sheath are not overlapping or not joined to each other then it is termed as open sheath margin.
(2) Splitted with overlapping: When the margins are splitted to downward to some extent and then the margins are overlapped on each other that time the sheath is known as split with overlapping margins.
(3) Close: When the margins of the sheaths are joined together then it is known to be closed. Sometimes the midrib of the lamina extends downward into the sheath and the sheath is said to be keeled.

2.4.8 LIGULE

The ligule is a tongue like outgrowth at the junction of the blade and sheath clasping the culm. It may be present or absent. If it is present then it may be membranous or hairy. In membranous, it may be short or long, may be splitted from base or at margin only. Margin of the ligule may be entire or may be toothed. The membranous ligule tip may be acute, acuminate, and truncate, for example, cut off squarely with an abruptly transverse end or round. When hairy then it may be in a fringe of small and short hairs, long hairs or rudimentary.

(1) Acute: Sharply pointed tip; converging edges making an angle of less than 90°.
(2) Acuminate: Tip tapering gradually to a point.
(3) Truncate: Tip cut off squarely; with an abruptly transverse end.
(4) Ciliate: Hairs more or less confined to the margins of the ligule.

(5) Villous: Abounding in or covered with long, soft, straight hairs; shaggy with soft hairs.
(6) Cirrhose: Mucronate like apex ends in a fine thread like structure.

2.4.9 AURICLE

These are appendages projecting from the collar, one on each side. It may be present or absent. When present then it has different types based on the length and shape. It may be it is long and clasping, for example, claw like or it is small and rudimentary.

2.4.10 COLLAR

It is a growth zone marking the division point between the blade and the sheath, distinctly observed from ventral surface of the blade. It may be broad or narrow. It may be continuous from one margin to the other margin of the leaf. It may be divided by a conspicuous midrib.

Important distinguishing characteristics of the field grown specimens did not differ much from the seedlings grown under the laboratory conditions especially for the annual species. But in the perennial species slight variations in the characters were observed from the laboratory grown seedlings. *Apluda mutica, Chrysopogon fulvus, Cymbopogon martini, Heteropogon triticeus, Ophiorus exaltatus, Rottbelia exaltata, Cenchrus biflorus, Urochondra setulosa* and *Themeda triandra* showed differences in their node, internode, leaf and collar hair features. Field collected samples of *Apluda mutica, Chrysopogon fulvus, Cymbopogon martini, Heteropogon triticeus, Rottbelia exaltata* and *Cenchrus biflorus* showed a distinct node and internode, which could not be observed in laboratory raised samples. In *Ophiorus exaltatus,* node, internode and collar regions were well developed in field samples while these characters were not prominent in laboratory-raised seedlings. In *Themeda triandra,* density of hairs at the collar region and on the leaf blade was more in field seedling samples.

Photographs of the characteristics features present in the studied species have been represented in Figures 59–107. A dendrogram prepared with the help of software is presented in Figure 108 and the numerical data is given in Table 3.

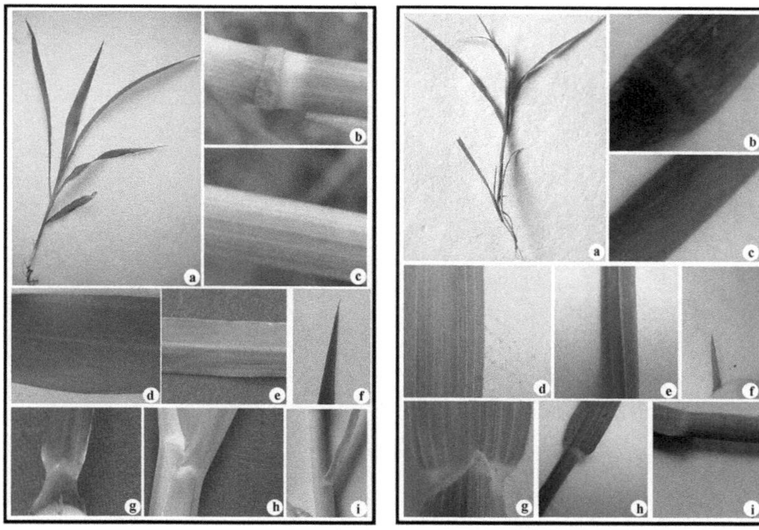

1. *Coix lachrymal-jobi* 2. *Andropogon pumilus*

FIGURE 59 Qualitative features of grass seedlings (A. Habit; B. Node; C. Internode; D. Dorsal surface of lamina; E. Ventral surface of lamina; F. Termination point (Leaf tip); G. Ligule; H. Auricle; I. Collar).

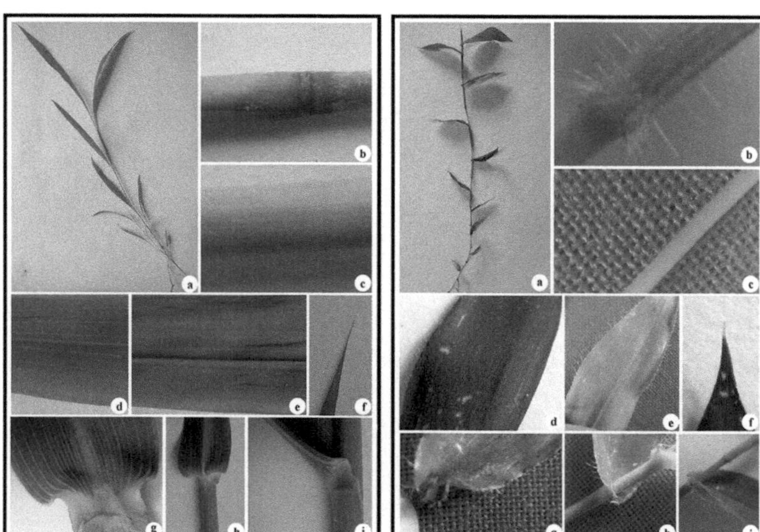

3. *Apluda mutica* 4. *Arthraxon lanceolatus*

FIGURE 60 Qualitative features of grass seedlings (A. Habit; B. Node; C. Internode; D. Dorsal surface of lamina; E. Ventral surface of lamina; F. Termination point (Leaf tip); G. Ligule; H. Auricle; I. Collar).

5. ***Bothriochloa pertusa*** 6. ***Capillipedium huegelii***

FIGURE 61 Qualitative features of grass seedlings (A. Habit; B. Node; C. Internode; D. Dorsal surface of lamina; E. Ventral surface of lamina; F. Termination point (Leaf tip); G. Ligule; H. Auricle; I. Collar).

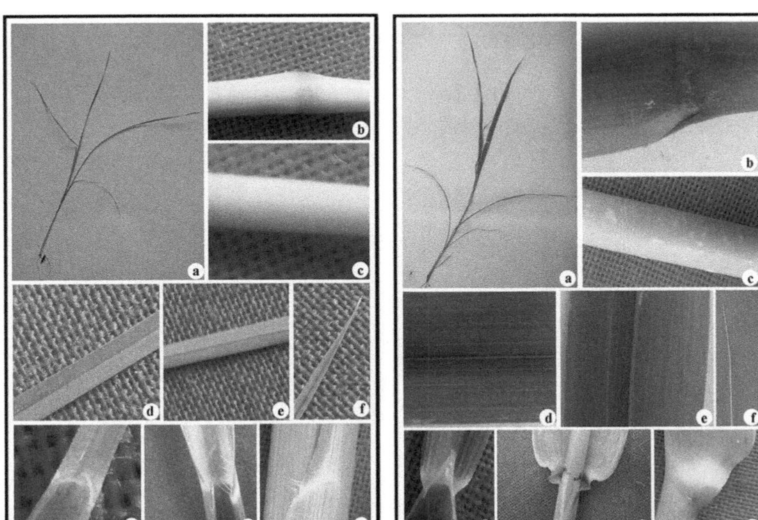

7. ***Chrysopogon fulvus*** 8. ***Cymbopogon martinii***

FIGURE 62 Qualitative features of grass seedlings (A. Habit; B. Node; C. Internode; D. Dorsal surface of lamina; E. Ventral surface of lamina; F. Termination point (Leaf tip); G. Ligule; H. Auricle; I. Collar).

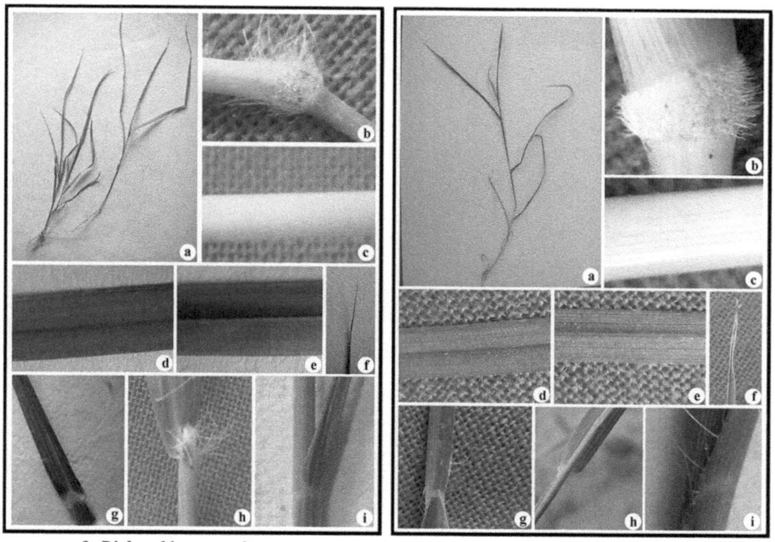

9. ***Dichanthium annulatum*** 10. ***Dichanthium caricosum***

FIGURE 63 Qualitative features of grass seedlings (A. Habit; B. Node; C. Internode; D. Dorsal surface of lamina; E. Ventral surface of lamina; F. Termination point (Leaf tip); G. Ligule; H. Auricle; I. Collar).

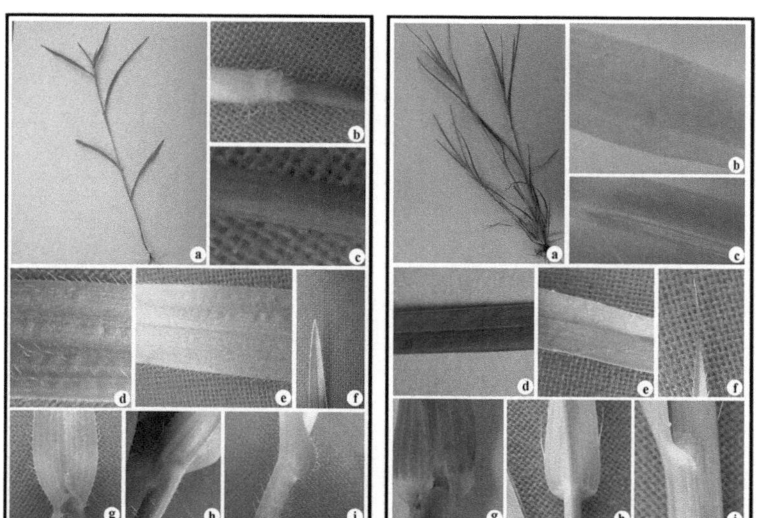

11. ***Hackelochloa granularis*** 12. ***Heteropogon contortus var. contortus sub var. typicus***

FIGURE 64 Qualitative features of grass seedlings (A. Habit; B. Node; C. Internode; D. Dorsal surface of lamina; E. Ventral surface of lamina; F. Termination point (Leaf tip); G. Ligule; H. Auricle; I. Collar).

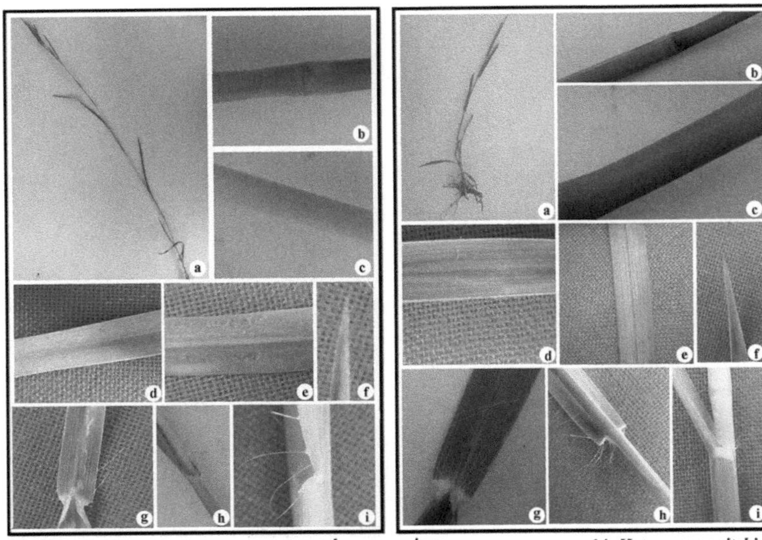

13. ***Heteropogon contortus var. contortus sub var. genuinus*** **14.** ***Heteropogon ritchiei***

FIGURE 65 Qualitative features of grass seedlings (A. Habit; B. Node; C. Internode; D. Dorsal surface of lamina; E. Ventral surface of lamina; F. Termination point (Leaf tip); G. Ligule; H. Auricle; I. Collar).

15. ***Heteropogon triticeus*** **16.** ***Imperata cylindrica***

FIGURE 66 Qualitative features of grass seedlings (A. Habit; B. Node; C. Internode; D. Dorsal surface of lamina; E. Ventral surface of lamina; F. Termination point (Leaf tip); G. Ligule; H. Auricle; I. Collar).

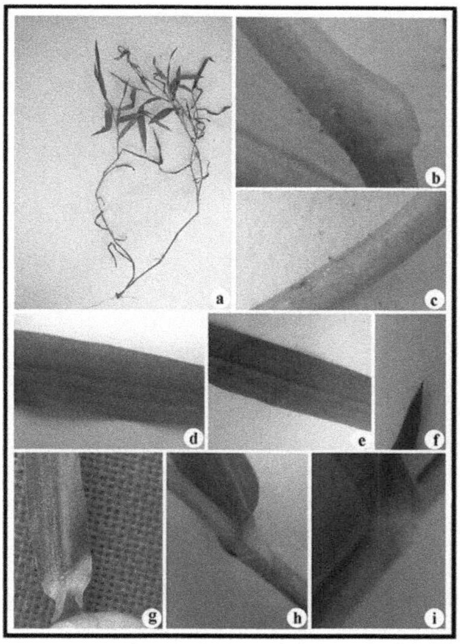

17. *Ischaemum indicum*

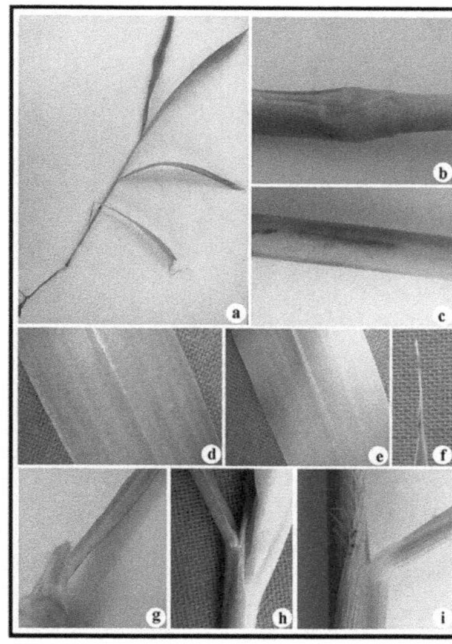

18. *Ischaemum molle*

FIGURE 67 Qualitative features of grass seedlings (A. Habit; B. Node; C. Internode; D. Dorsal surface of lamina; E. Ventral surface of lamina; F. Termination point (Leaf tip); G. Ligule; H. Auricle; I. Collar).

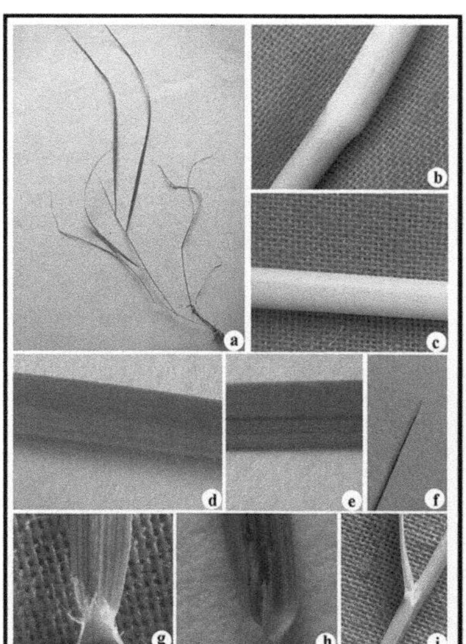

19. *Ischaemum pilosum*

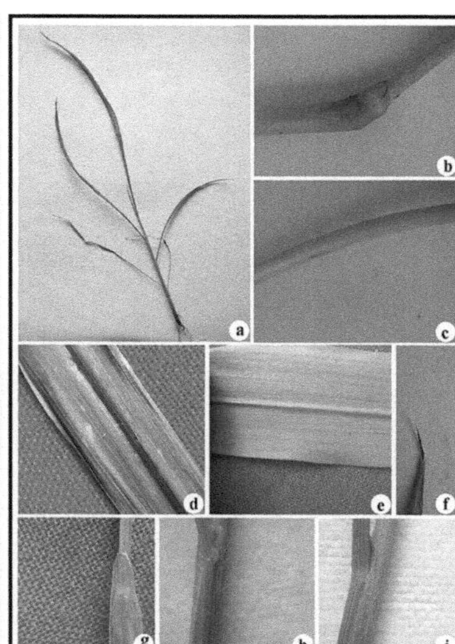

20. *Ischaemum rugosum*

FIGURE 68 Qualitative features of grass seedlings (A. Habit; B. Node; C. Internode; D. Dorsal surface of lamina; E. Ventral surface of lamina; F. Termination point (Leaf tip); G. Ligule; H. Auricle; I. Collar).

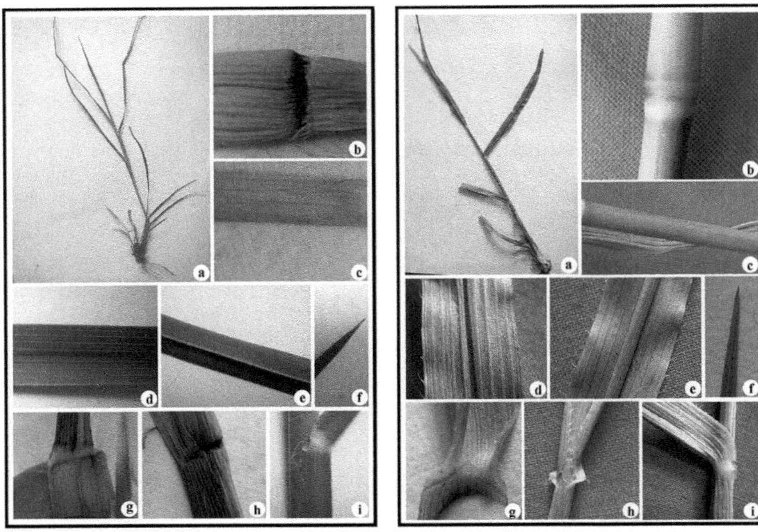

21. ***Iseilema laxum*** **22.** ***Ophiorus exaltatus***

FIGURE 69 Qualitative features of grass seedlings (A. Habit; B. Node; C. Internode; D. Dorsal surface of lamina; E. Ventral surface of lamina; F. Termination point (Leaf tip); G. Ligule; H. Auricle; I. Collar).

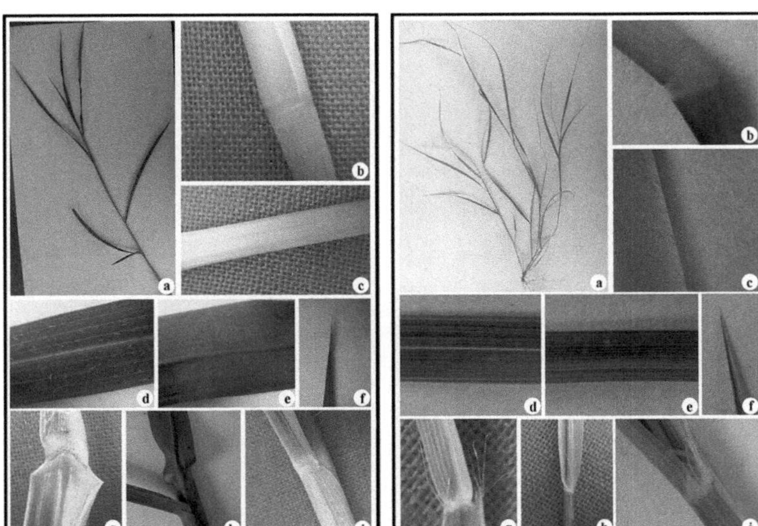

23. ***Rottboellia exaltata*** **24.** ***Sehima ischaemoides***

FIGURE 70 Qualitative features of grass seedlings (A. Habit; B. Node; C. Internode; D. Dorsal surface of lamina; E. Ventral surface of lamina; F. Termination point (Leaf tip); G. Ligule; H. Auricle; I. Collar).

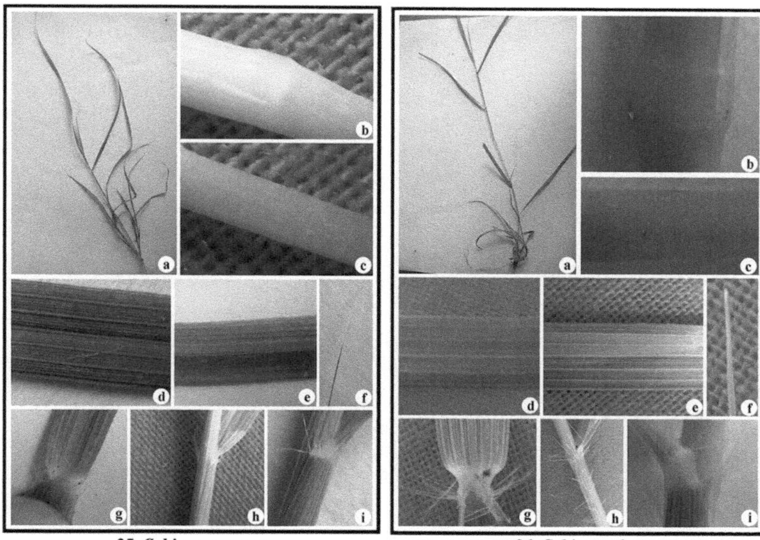

25. *Sehima nervosum* 26. *Sehima sulcatum*

FIGURE 71 Qualitative features of grass seedlings (A. Habit; B. Node; C. Internode; D. Dorsal surface of lamina; E. Ventral surface of lamina; F. Termination point (Leaf tip); G. Ligule; H. Auricle; I. Collar).

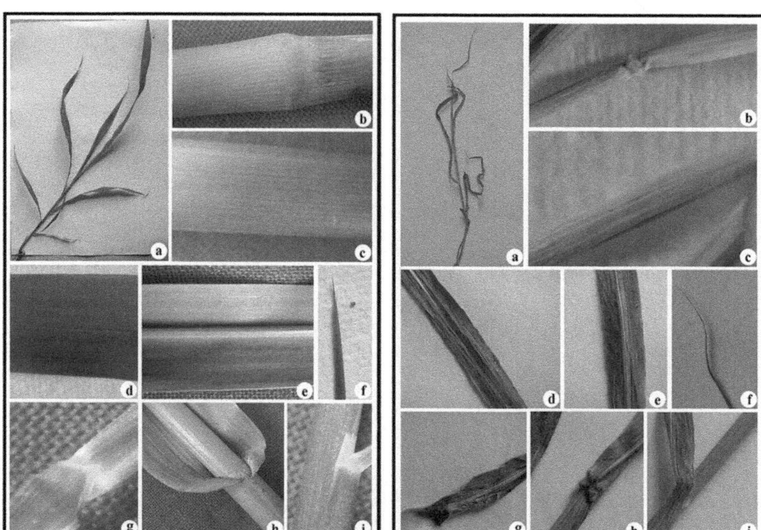

27. *Sorghum halepense* 28. *Sorghum purpureo-sericeum*

FIGURE 72 Qualitative features of grass seedlings (A. Habit; B. Node; C. Internode; D. Dorsal surface of lamina; E. Ventral surface of lamina; F. Termination point (Leaf tip); G. Ligule; H. Auricle; I. Collar).

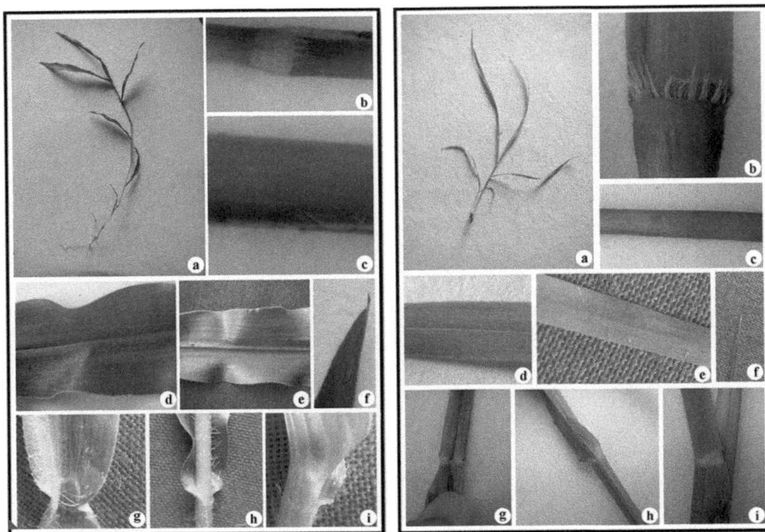

29. *Thelepogon elegans* 30. *Themeda cymbaria*

FIGURE 73 Qualitative features of grass seedlings (A. Habit; B. Node; C. Internode; D. Dorsal surface of lamina; E. Ventral surface of lamina; F. Termination point (Leaf tip); G. Ligule; H. Auricle; I. Collar).

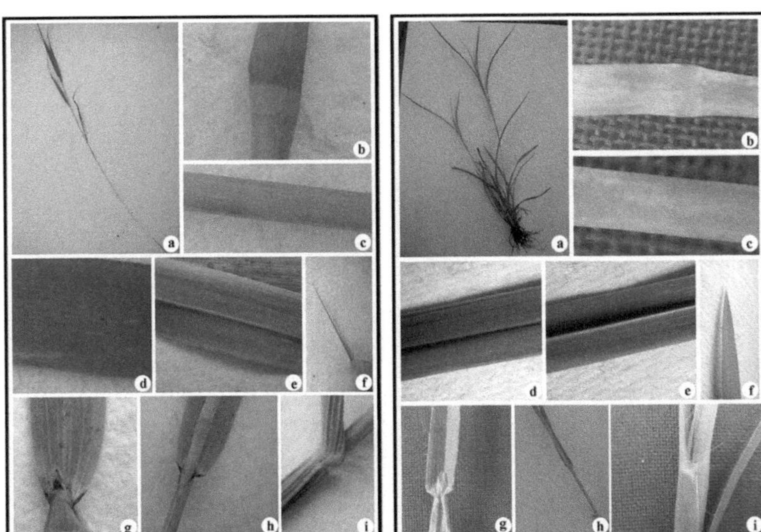

31. *Themeda laxa* 32. *Themeda triandra*

FIGURE 74 Qualitative features of grass seedlings (A. Habit; B. Node; C. Internode; D. Dorsal surface of lamina; E. Ventral surface of lamina; F. Termination point (Leaf tip); G. Ligule; H. Auricle; I. Collar).

33. *Triplopogon ramosissimus*   34. *Alloteropsis cimicina*

FIGURE 75 Qualitative features of grass seedlings (A. Habit; B. Node; C. Internode; D. Dorsal surface of lamina; E. Ventral surface of lamina; F. Termination point (Leaf tip); G. Ligule; H. Auricle; I. Collar).

35. *Brachiaria cruciformis*   36. *Brachiaria distachya*

FIGURE 76 Qualitative features of grass seedlings (A. Habit; B. Node; C. Internode; D. Dorsal surface of lamina; E. Ventral surface of lamina; F. Termination point (Leaf tip); G. Ligule; H. Auricle; I. Collar).

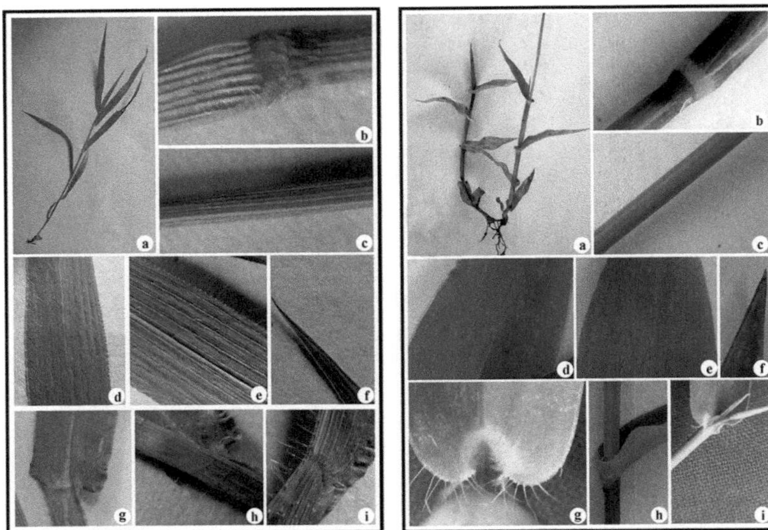

37. *Brachiaria ramosa* 38. *Brachiaria reptans*

FIGURE 77 Qualitative features of grass seedlings (A. Habit; B. Node; C. Internode; D. Dorsal surface of lamina; E. Ventral surface of lamina; F. Termination point (Leaf tip); G. Ligule; H. Auricle; I. Collar).

39. *Cenchrus biflorus* 40. *Cenchrus ciliaris*

FIGURE 78 Qualitative features of grass seedlings (A. Habit; B. Node; C. Internode; D. Dorsal surface of lamina; E. Ventral surface of lamina; F. Termination point (Leaf tip); G. Ligule; H. Auricle; I. Collar).

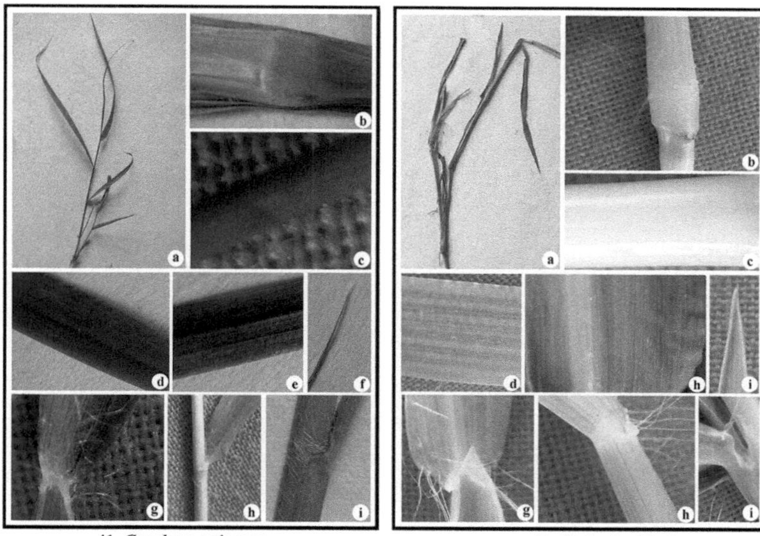

41. Cenchrus setigerus ***42. Digitaria ciliaris***

FIGURE 79 Qualitative features of grass seedlings (A. Habit; B. Node; C. Internode; D. Dorsal surface of lamina; E. Ventral surface of lamina; F. Termination point (Leaf tip); G. Ligule; H. Auricle; I. Collar).

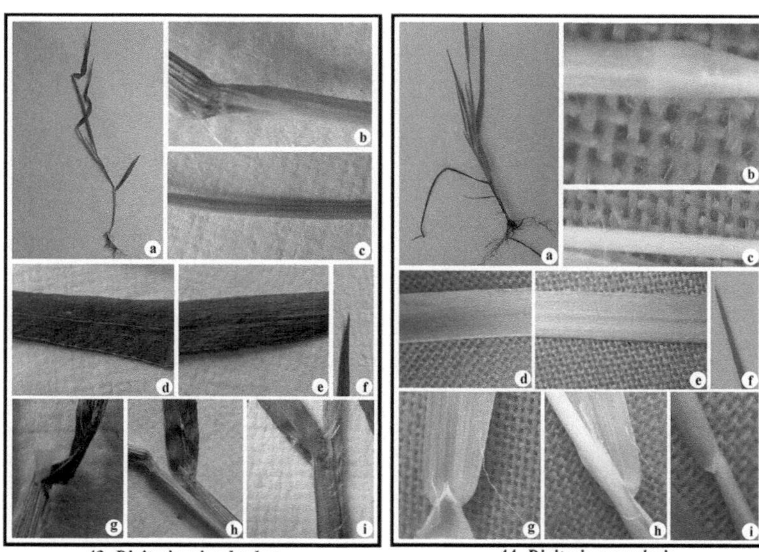

43. Digitaria microbachne ***44. Digitaria granularis***

FIGURE 80 Qualitative features of grass seedlings (A. Habit; B. Node; C. Internode; D. Dorsal surface of lamina; E. Ventral surface of lamina; F. Termination point (Leaf tip); G. Ligule; H. Auricle; I. Collar).

45. ***Digitaria longiflora*** **46.** ***Digitaria stricta***

FIGURE 81 Qualitative features of grass seedlings (A. Habit; B. Node; C. Internode; D. Dorsal surface of lamina; E. Ventral surface of lamina; F. Termination point (Leaf tip); G. Ligule; H. Auricle; I. Collar).

47. ***Echinochloa colonum*** **48.** ***Echinochloa crus-galli***

FIGURE 82 Qualitative features of grass seedlings (A. Habit; B. Node; C. Internode; D. Dorsal surface of lamina; E. Ventral surface of lamina; F. Termination point (Leaf tip); G. Ligule; H. Auricle; I. Collar).

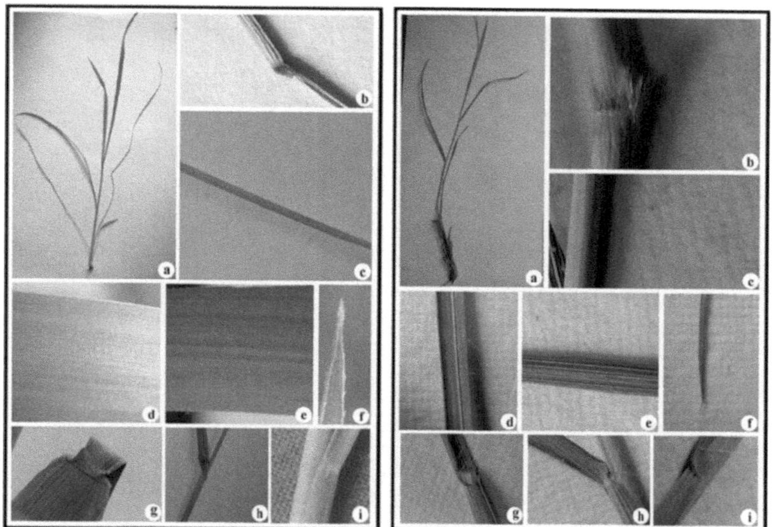

49. *Echinochloa stagnina* 50. *Eremopogon foveolatus*

FIGURE 83 Qualitative features of grass seedlings (A. Habit; B. Node; C. Internode; D. Dorsal surface of lamina; E. Ventral surface of lamina; F. Termination point (Leaf tip); G. Ligule; H. Auricle; I. Collar).

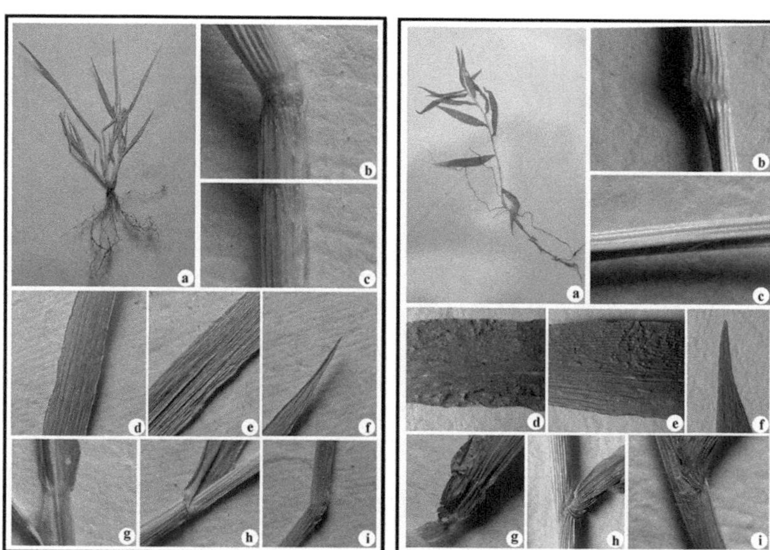

51. *Eriochloa procera* 52. *Oplismenus composites*

FIGURE 84 Qualitative features of grass seedlings (A. Habit; B. Node; C. Internode; D. Dorsal surface of lamina; E. Ventral surface of lamina; F. Termination point (Leaf tip); G. Ligule; H. Auricle; I. Collar).

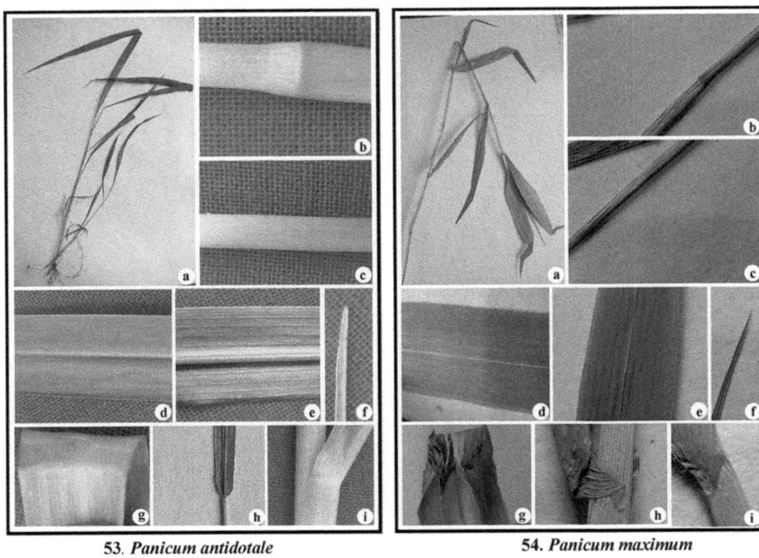

FIGURE 85 Qualitative features of grass seedlings (A. Habit; B. Node; C. Internode; D. Dorsal surface of lamina; E. Ventral surface of lamina; F. Termination point (Leaf tip); G. Ligule; H. Auricle; I. Collar).

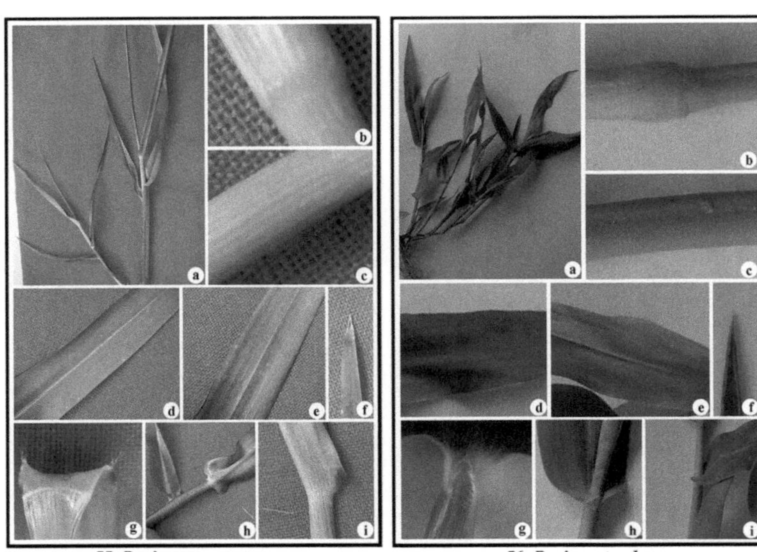

FIGURE 86 Qualitative features of grass seedlings (A. Habit; B. Node; C. Internode; D. Dorsal surface of lamina; E. Ventral surface of lamina; F. Termination point (Leaf tip); G. Ligule; H. Auricle; I. Collar).

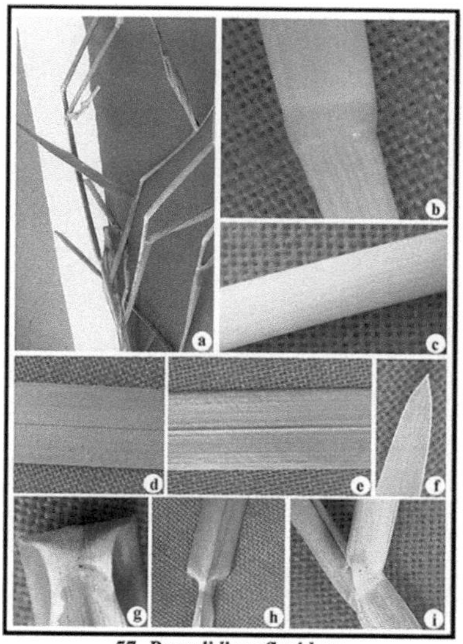

57. *Paspalidium flavidum*

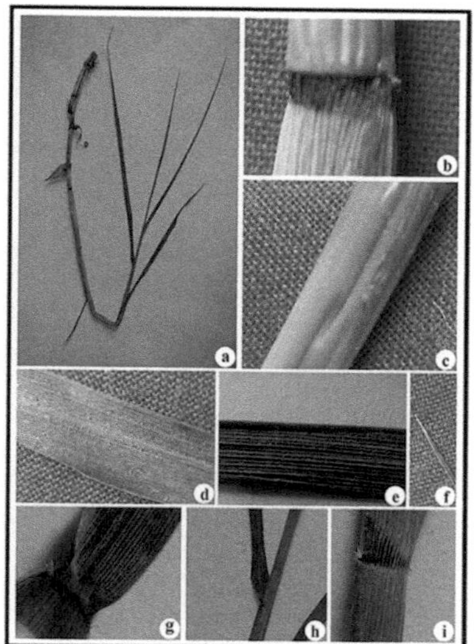

58. *Paspalidium geminatum*

FIGURE 87 Qualitative features of grass seedlings (A. Habit; B. Node; C. Internode; D. Dorsal surface of lamina; E. Ventral surface of lamina; F. Termination point (Leaf tip); G. Ligule; H. Auricle; I. Collar).

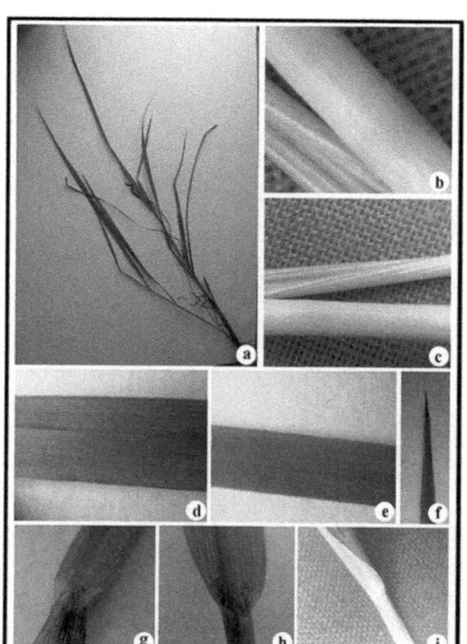

59. *Paspalum scrobiculatum*

60. *Paspalum distichum*

FIGURE 88 Qualitative features of grass seedlings (A. Habit; B. Node; C. Internode; D. Dorsal surface of lamina; E. Ventral surface of lamina; F. Termination point (Leaf tip); G. Ligule; H. Auricle; I. Collar).

61. *Paspalum vaginatum*

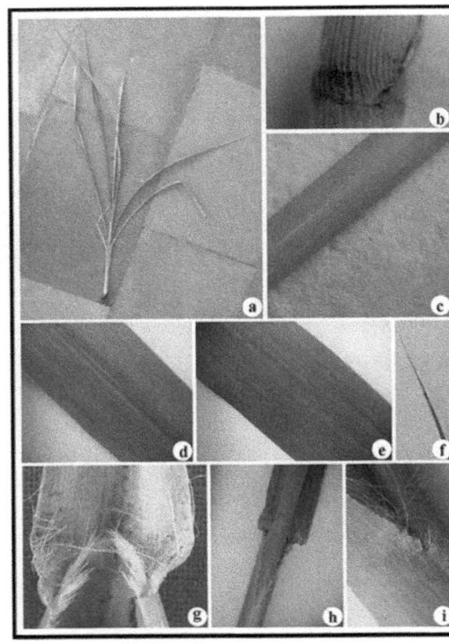

62. *Pennisetum setosum*

FIGURE 89 Qualitative features of grass seedlings (A. Habit; B. Node; C. Internode; D. Dorsal surface of lamina; E. Ventral surface of lamina; F. Termination point (Leaf tip); G. Ligule; H. Auricle; I. Collar).

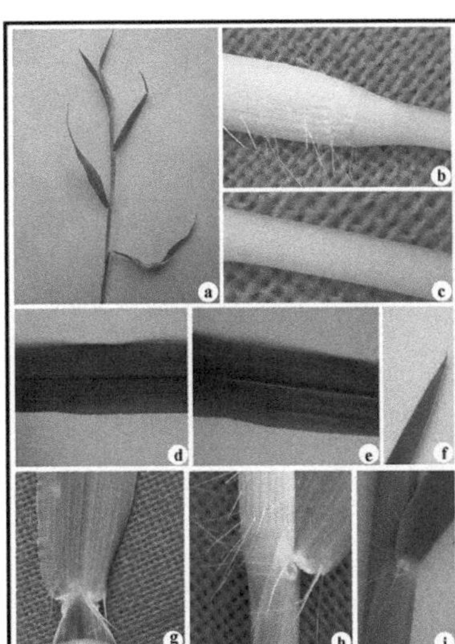

63. *Setaria glauca*

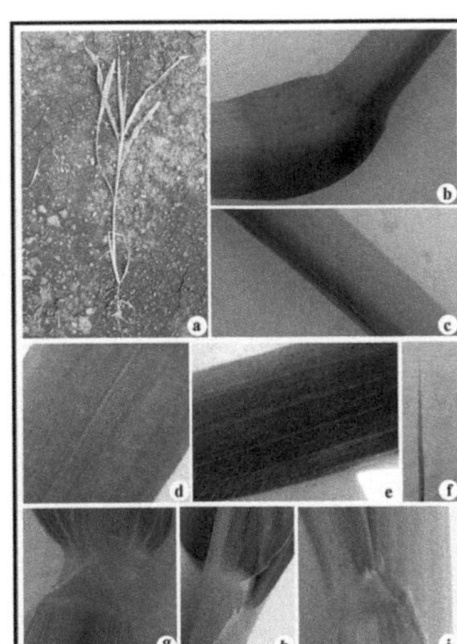

64. *Setaria tomentosa*

FIGURE 90 Qualitative features of grass seedlings (A. Habit; B. Node; C. Internode; D. Dorsal surface of lamina; E. Ventral surface of lamina; F. Termination point (Leaf tip); G. Ligule; H. Auricle; I. Collar).

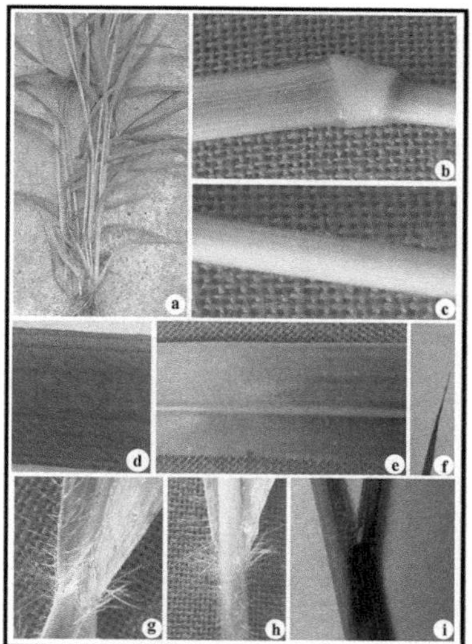

65. Setaria verticillata

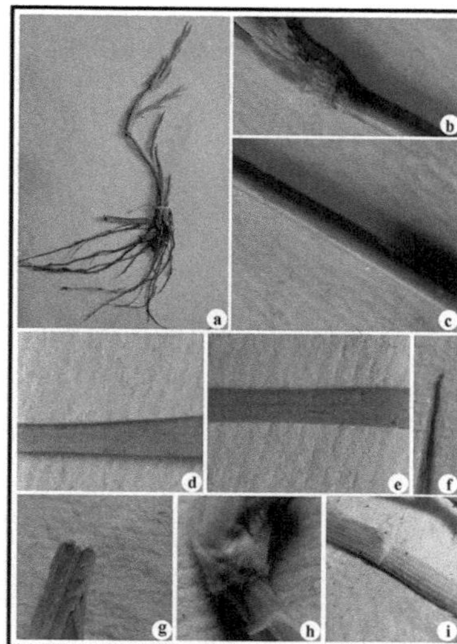

66. Aeluropus lagopoides

FIGURE 91 Qualitative features of grass seedlings (A. Habit; B. Node; C. Internode; D. Dorsal surface of lamina; E. Ventral surface of lamina; F. Termination point (Leaf tip); G. Ligule; H. Auricle; I. Collar).

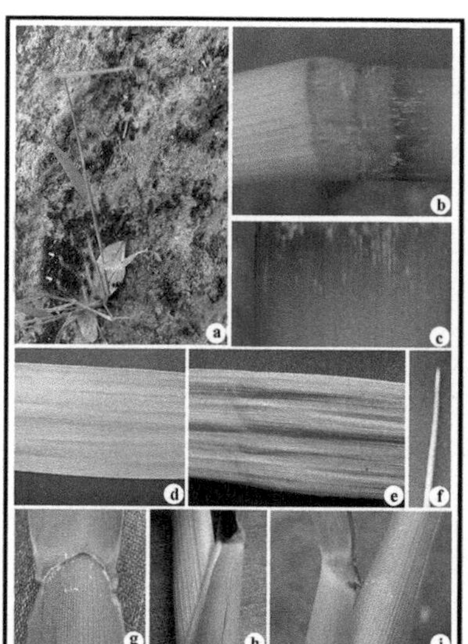

67. Isachne globosa

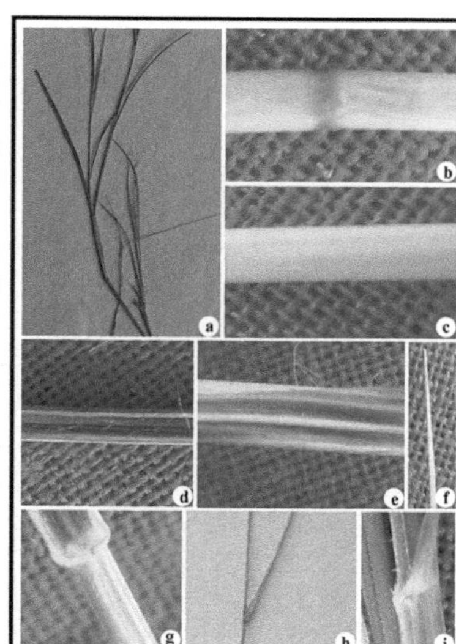

68. Aristida adscensionis

FIGURE 92 Qualitative features of grass seedlings (A. Habit; B. Node; C. Internode; D. Dorsal surface of lamina; E. Ventral surface of lamina; F. Termination point (Leaf tip); G. Ligule; H. Auricle; I. Collar).

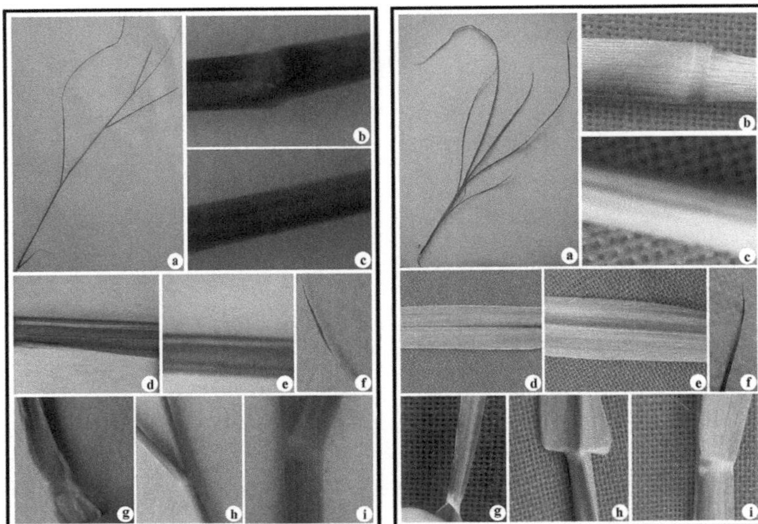

69. *Aristida funiculata* 70. *Chloris barbata*

FIGURE 93 Qualitative features of grass seedlings (A. Habit; B. Node; C. Internode; D. Dorsal surface of lamina; E. Ventral surface of lamina; F. Termination point (Leaf tip); G. Ligule; H. Auricle; I. Collar).

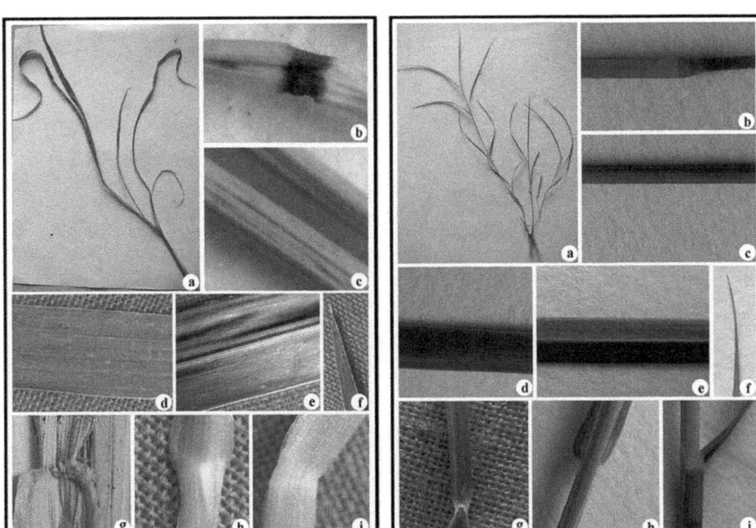

71. *Chloris virgata* 72. *Cynodon dactylon*

FIGURE 94 Qualitative features of grass seedlings (A. Habit; B. Node; C. Internode; D. Dorsal surface of lamina; E. Ventral surface of lamina; F. Termination point (Leaf tip); G. Ligule; H. Auricle; I. Collar).

73. *Melanocenchris jacquemontii*

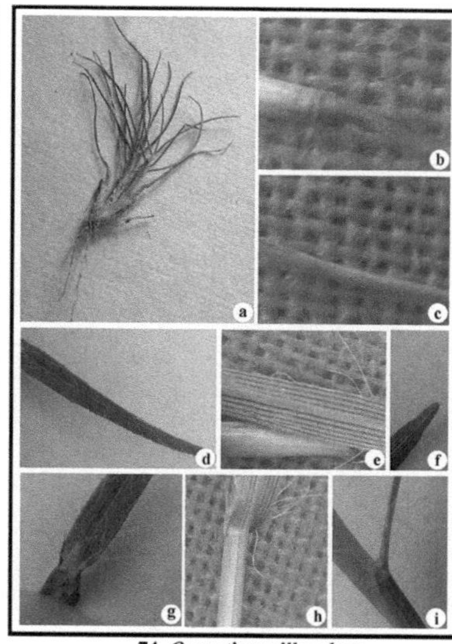

74. *Oropetium villosulum*

FIGURE 95 Qualitative features of grass seedlings (A. Habit; B. Node; C. Internode; D. Dorsal surface of lamina; E. Ventral surface of lamina; F. Termination point (Leaf tip); G. Ligule; H. Auricle; I. Collar).

75. *Schoenefeldia gracilis*

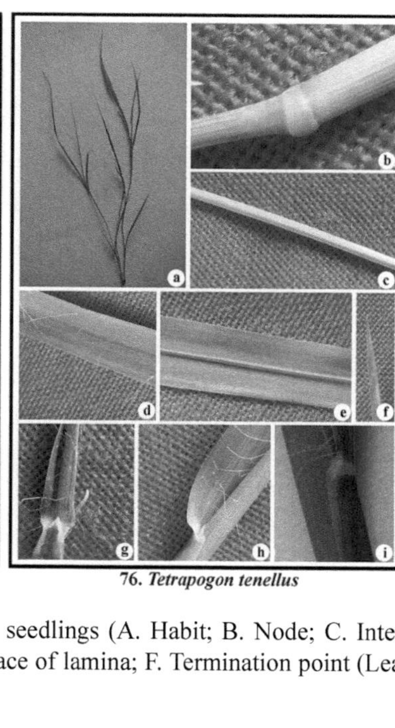

76. *Tetrapogon tenellus*

FIGURE 96 Qualitative features of grass seedlings (A. Habit; B. Node; C. Internode; D. Dorsal surface of lamina; E. Ventral surface of lamina; F. Termination point (Leaf tip); G. Ligule; H. Auricle; I. Collar).

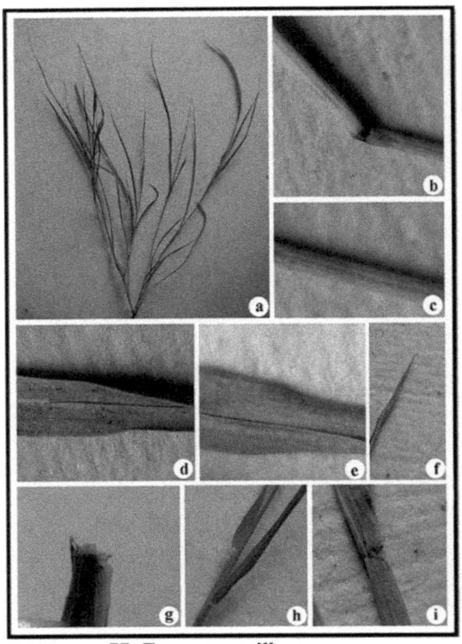

77. Tetrapogon villosus

78. Dactyaloctanium aegyptium

FIGURE 97 Qualitative features of grass seedlings (A. Habit; B. Node; C. Internode; D. Dorsal surface of lamina; E. Ventral surface of lamina; F. Termination point (Leaf tip); G. Ligule; H. Auricle; I. Collar).

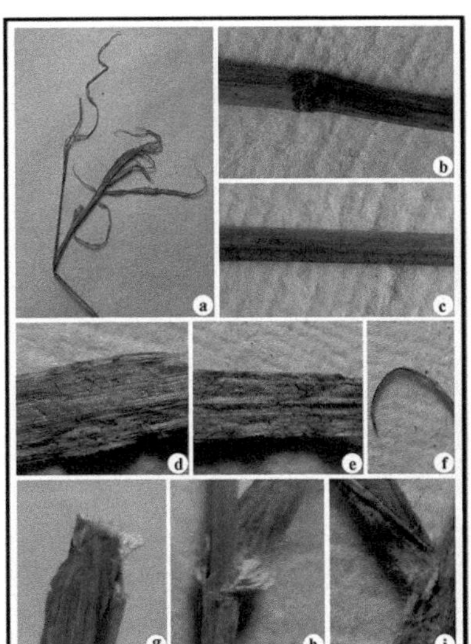

79. Dactyloctenium sindicum

80. Desmostachya bipinnata

FIGURE 98 Qualitative features of grass seedlings (A. Habit; B. Node; C. Internode; D. Dorsal surface of lamina; E. Ventral surface of lamina; F. Termination point (Leaf tip); G. Ligule; H. Auricle; I. Collar).

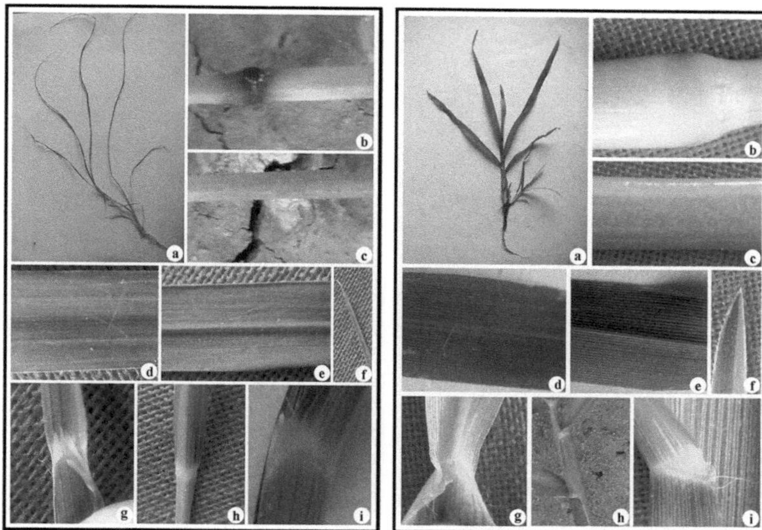

81. ***Dinebra retroflexa*** **82.** ***Eleusine indica***

FIGURE 99 Qualitative features of grass seedlings (A. Habit; B. Node; C. Internode; D. Dorsal surface of lamina; E. Ventral surface of lamina; F. Termination point (Leaf tip); G. Ligule; H. Auricle; I. Collar).

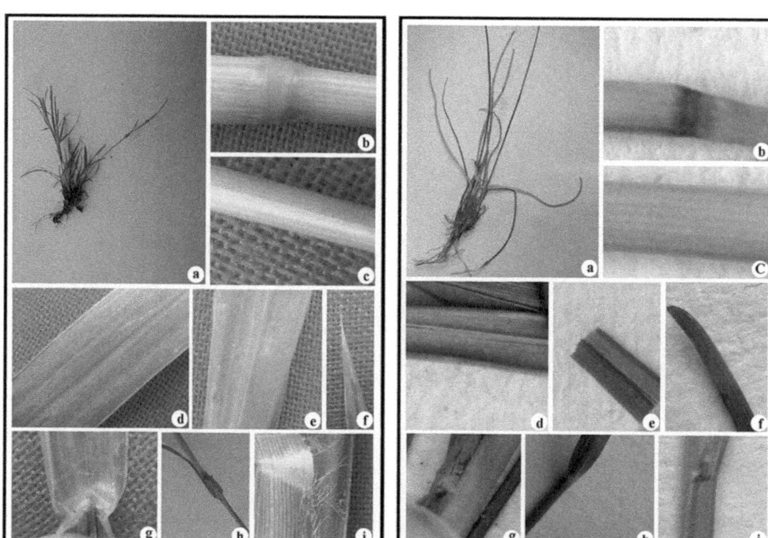

83. ***Acrachne racemosa*** **84.** ***Eragrostiella bifaria***

FIGURE 100 Qualitative features of grass seedlings (A. Habit; B. Node; C. Internode; D. Dorsal surface of lamina; E. Ventral surface of lamina; F. Termination point (Leaf tip); G. Ligule; H. Auricle; I. Collar).

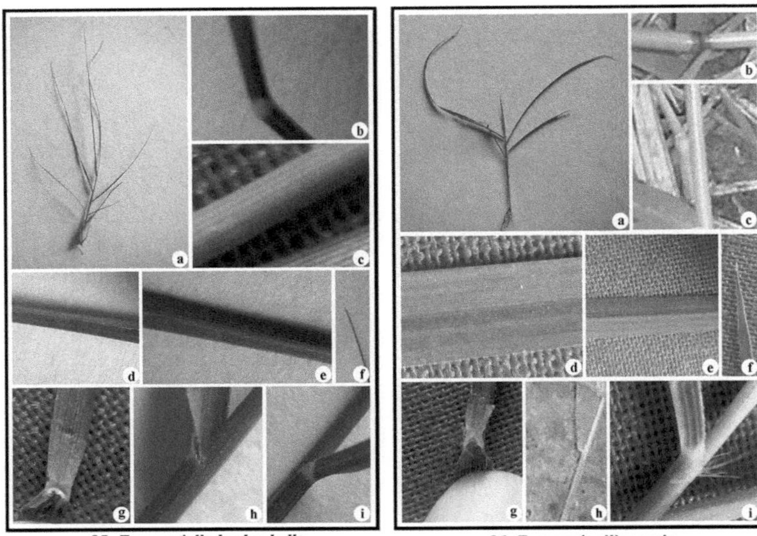

85. *Eragrostiella bachyphylla* 86. *Eragrostis cilianensis*

FIGURE 101 Qualitative features of grass seedlings (A. Habit; B. Node; C. Internode; D. Dorsal surface of lamina; E. Ventral surface of lamina; F. Termination point (Leaf tip); G. Ligule; H. Auricle; I. Collar).

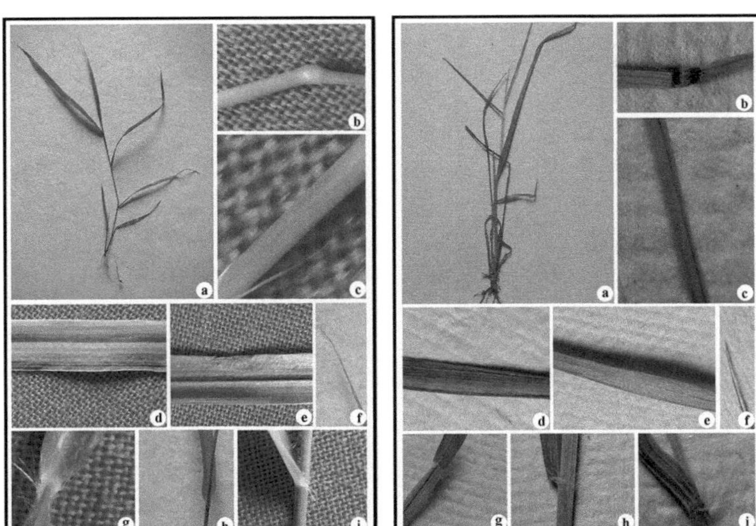

87. *Eragrostis ciliaris* 88. *Eragrostis pilosa*

FIGURE 102 Qualitative features of grass seedlings (A. Habit; B. Node; C. Internode; D. Dorsal surface of lamina; E. Ventral surface of lamina; F. Termination point (Leaf tip); G. Ligule; H. Auricle; I. Collar).

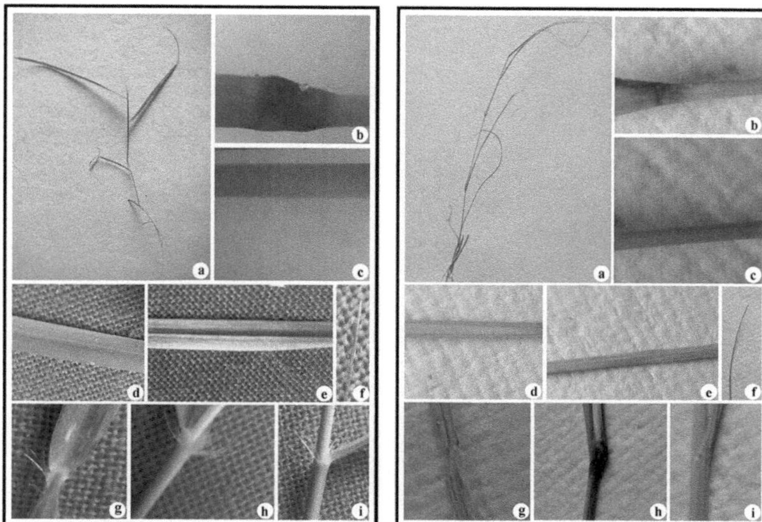

89. *Eragrostis tenella* 90. *Eragrostis tremula*

FIGURE 103 Qualitative features of grass seedlings (A. Habit; B. Node; C. Internode; D. Dorsal surface of lamina; E. Ventral surface of lamina; F. Termination point (Leaf tip); G. Ligule; H. Auricle; I. Collar).

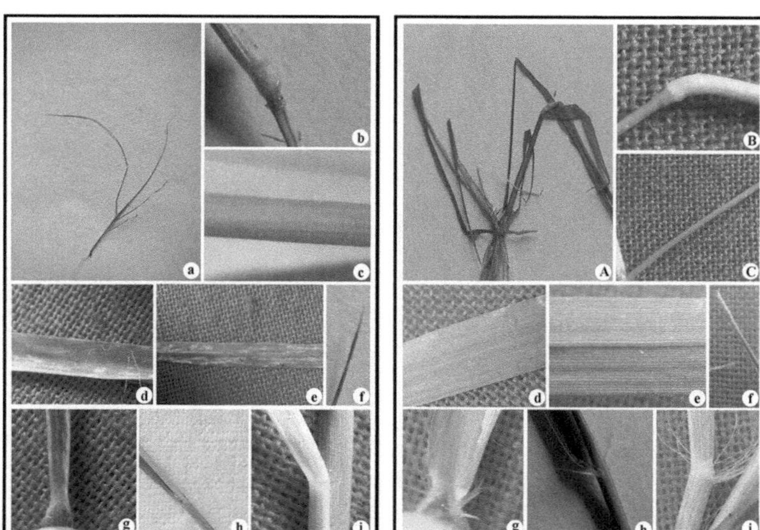

91. *Eragrostis unioloides* 92. *Eragrostis viscosa*

FIGURE 104 Qualitative features of grass seedlings (A. Habit; B. Node; C. Internode; D. Dorsal surface of lamina; E. Ventral surface of lamina; F. Termination point (Leaf tip); G. Ligule; H. Auricle; I. Collar).

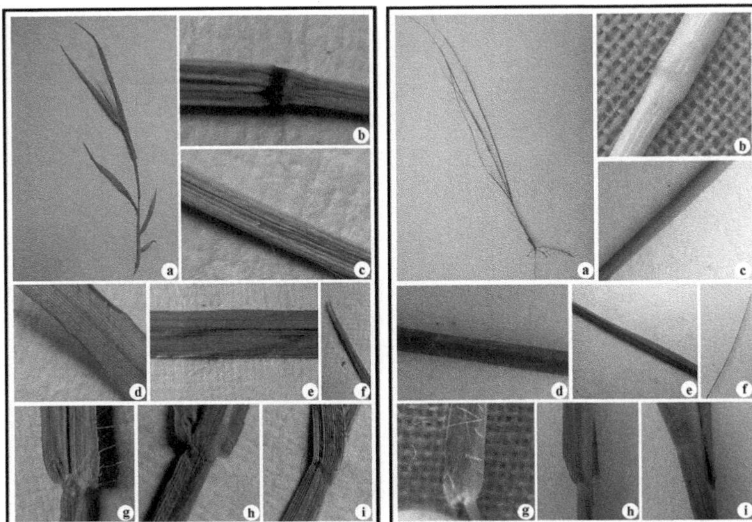

93. *Sporobolus coromardelianus* 94. *Sporobolus diander*

FIGURE 105 Qualitative features of grass seedlings (A. Habit; B. Node; C. Internode; D. Dorsal surface of lamina; E. Ventral surface of lamina; F. Termination point (Leaf tip); G. Ligule; H. Auricle; I. Collar).

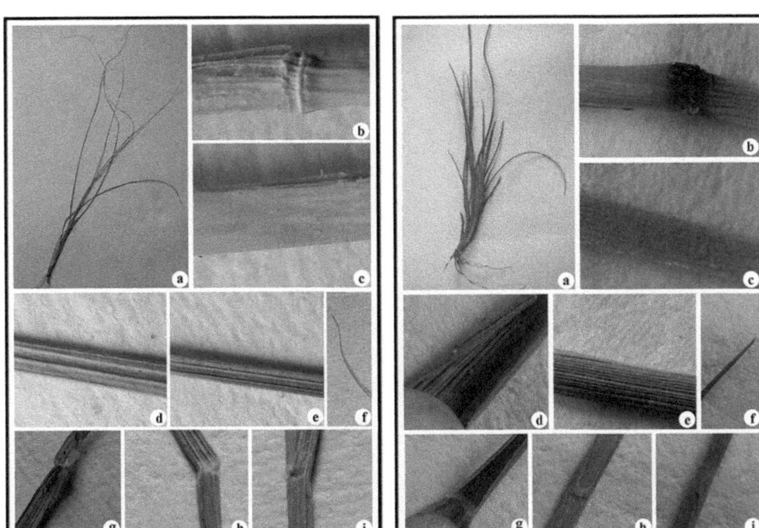

95. *Sporobolus indicum* 96. *Urochondra setulosa*

FIGURE 106 Qualitative features of grass seedlings (A. Habit; B. Node; C. Internode; D. Dorsal surface of lamina; E. Ventral surface of lamina; F. Termination point (Leaf tip); G. Ligule; H. Auricle; I. Collar).

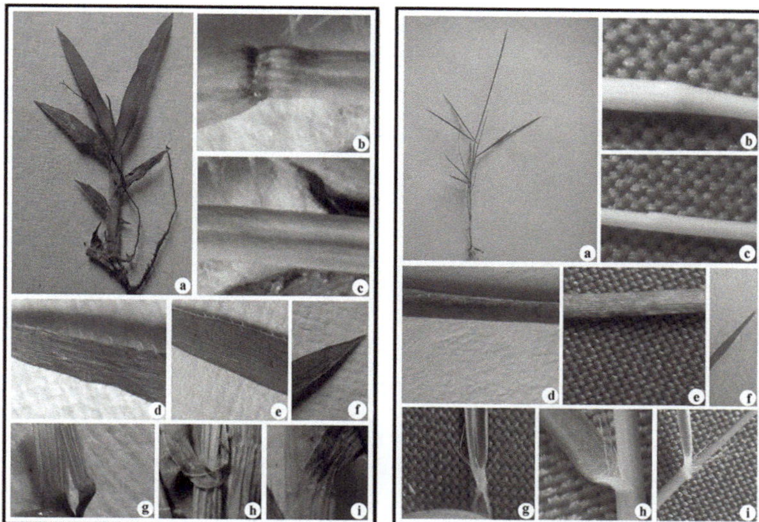

97. *Tragrus biflorus* 98. *Zoysia matrella*

FIGURE 107 Qualitative features of grass seedlings (A. Habit; B. Node; C. Internode; D. Dorsal surface of lamina; E. Ventral surface of lamina; F. Termination point (Leaf tip); G. Ligule; H. Auricle; I. Collar).

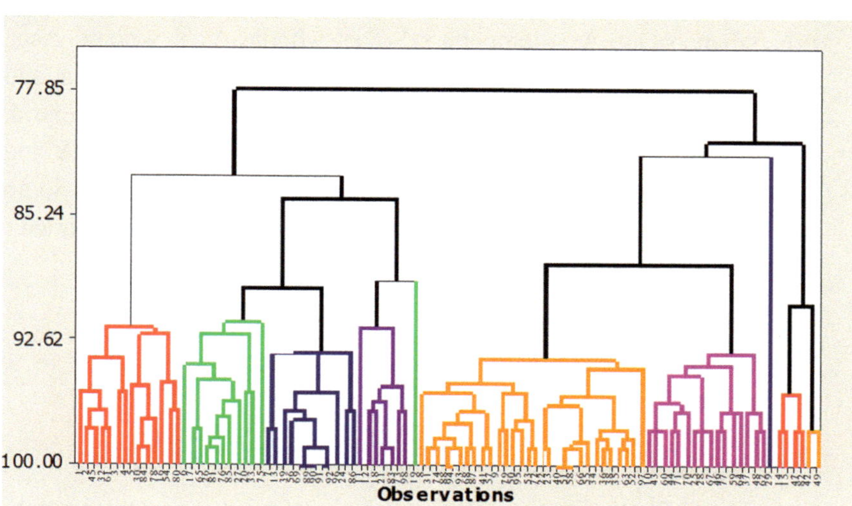

FIGURE 108 A dendrogram showing the clustering of grass species.

TABLE 3 Distribution of Studied Grass Seedlings into Clusters on the Basis of Their Qualitative Features

Cluster number	Number of observation (grass species) per cluster	Value from centroid
1	14	2.124
2	11	2.132
3	12	2.215
4	30	2.235
5	16	2.127
6	7	2.094
7	4	1.322
8	1	0.000
9	1	0.000
10	2	0.707

2.5 ILLUSTRATED SEEDLING DESCRIPTION

1. ***Coix lachryma-jobi* L.**

   Distribution: Europe: southwestern and southeastern. Africa: north, Macaronesia, west tropical, west-central tropical, northeast tropical, east tropical, southern tropical, south, middle Atlantic ocean, and western Indian ocean. Asia-temperate: Soviet Middle Asia, western Asia, China, and eastern Asia. Asia-tropical: India, Indo-China, Malesia, and Papuasia. Australasia: Australia. Pacific: southwestern, south-central, northwestern, and north-central. North America: southeast USA and Mexico. South America: Mesoamericana, Caribbean, northern South America, western South America, Brazil, and southern South America.
   Habitat: Swampy places and near streams, cultivated.

   Seedling description:

   Growth Habit: Annual
   Vernation: Rolled
   Node: Pubescent, purple in tinge
   Internode: Glabrous
   Leaf Blade: Glabrous surface, single midvein is present throughout the lamina

Termination point: Acute, flat
Leaf sheath margin: Open
Leaf sheath type: Flattened **Ligule**: Membranous, truncate
Auricle: Absent **Collar**: Divided

2. ***Andropogon pumilus* Roxb.**

Distribution: Asia-tropical: India and Indo-China.
Habitat: Grasslands, moist plains

Seedling description:

Growth Habit: Annual
Vernation: Rolled
Node: Glabrous, reddish in color
Internode: Smooth
Leaf Blade: Glabrous surface, single midvein with other strongly parallel veins, few hairs present towards the margin
Termination point: Acute, flat
Leaf sheath margin: Split with overlapping margin
Leaf sheath type: Flattened **Ligule**: Membranous, truncate
Auricle: Absent **Collar**: Divided

3. ***Apluda mutica***

Distribution: Africa: northeast tropical. Asia-temperate: Soviet Middle Asia, western Asia, Arabia, China, and eastern Asia. Asia-tropical: India, Indo-China, Malesia, and Papuasia. Australasia: Australia. Pacific: southwestern and northwestern.
Habitat: Very commonly grow in all type of habitat.

Seedling description:

Growth Habit: Annual
Vernation: Rolled
Node: Glabrous, purple color band at node
Internode: Smooth
Leaf Blade: Glabrous surface, many parallel veins are seen
Termination point: Acuminate, tapering towards the apex, flat
Leaf sheath margin: Open
Leaf sheath type: Round **Ligule:** Membranous, round
Auricle: Short, rudiment **Collar:** Continuous, broad

4. *Arthraxon lanceolatus*

Distribution: Africa: south. Asia-tropical: India and Indo-China.
Habitat: Water edges, little moist places, escape from gardens.

Seedling description:

Growth Habit: Annual
Vernation: Folded
Node: Pubescent, long hairs present at and above node
Internode: Smooth
Leaf Blade: Pubescent surface single midvein with other strongly parallel veins, margins is short-serrate, lanceolate leaf
Termination point: Acuminate, flat
Leaf sheath margin: Split with overlapping margin
Leaf sheath type: Round **Ligule:** Membranous, round off
Auricle: Long, clasping **Collar:** Continuous, narrow

5. *Bothriochloa pertusa*

Distribution: Africa: western Indian ocean. Asia-temperate: China. Asia-tropical: India, Indo-China, and Malesia. Australasia: Australia. Pacific: southwestern, south-central, northwestern, and north-central. North America: southeast USA. South America: Mesoamericana, Caribbean, northern South America, and western South America.
Habitat: Grassland on clay soils and open woodland.

Seedling description:

Growth Habit: Annual
Vernation: Rolled
Distance Node: Glabrous
Internode: Purple to greenish color with longitudinal lines
Leaf Blade: Pubescent surface, single prominent midvein is present
Termination point: Mucronate, flat
Leaf sheath margin: Close
Leaf sheath type: Flattened **Ligule:** Membranous, ciliate
Auricle: Absent **Collar:** Divided

6. *Capillipedium huegelii*

Distribution: Asia-tropical: India.

Habitat: Commons in all habitats.

Seedling description:

Growth Habit: Annual
Vernation: Folded
Node: Pubescent, bearded
Internode: Smooth
Leaf Blade: Glabrous surface, single prominent midvein is present, slightly serrate margin
Termination point: Acuminate, flat
Leaf sheath margin: Open
Leaf sheath type: Round
Ligule: Membranous, acute
Auricle: Absent
Collar: Divided, long hairs are present at collar region

7. ***Chrysopogon fulvus***

Distribution: Asia-tropical: India, Indo-China, and Malesia.
Habitat: Dry, gravelly, often rather alkaline soils with grass or semi-desert cover

Seedling description:

Growth Habit: Perennial
Vernation: Rolled
Node: Glabrous, bearded
Internode: Smooth
Leaf Blade: Glabrous surface, single prominent midvein is present
Termination point: Acute, keeled
Leaf sheath margin: Open
Leaf sheath type: Flattened
Ligule: Membranous, acute
Auricle: Absent
Collar: Divided, slightly hairy

8. ***Cymbopogon martinii***

Distribution: Africa: western Indian ocean. Asia-temperate: China. Asia-tropical: India, Indo-China, and Malesia. Australasia:
Habitat: Common in open grasslands, occasional on drier wastelands

Seedling description:

Growth Habit: Perennial

Vernation: Rolled
Node: Glabrous
Internode: Smooth
Leaf Blade: Glabrous surface, single prominent midvein with other parallel vein, subcordate base
Termination point: Pointed needle like, flat
Leaf sheath margin: Split with overlapping margins
Leaf sheath type: Round **Ligule:** Membranous, acute
Auricle: Short, rudiment **Collar:** Divided, broad

9. *Dichanthium annulatum*

Distribution: Africa: north, Macaronesia, west tropical, northeast tropical, east tropical, southern tropical, south, and western Indian ocean. Asia-temperate: western Asia, Arabia, China, and eastern Asia. Asia-tropical: India, Indo-China, Malesia, and Papuasia. Australasia: Australia. Pacific: southwestern, south-central, and north-central. North America: south-central USA and Mexico. South America: Mesoamericana, Caribbean, northern South America.
Habitat: Commonly found everywhere

Seedling description:

Growth Habit: Perennial
Vernation: Rolled
Node: Pubscent, whole node is covered with long hairs
Internode: Smooth
Leaf Blade: Glabrous surface, single prominent midvein with other parallel veins which have dotted structure on dorsal surface
Termination point: Acuminate, flat
Leaf sheath margin: Split with overlapping margins
Leaf sheath type: Round **Ligule:** Membranous, obtuse
Auricle: Short, rudiment **Collar:** Continuous, broad

10. *Dichanthium caricosum*

Distribution: Africa: east tropical and western Indian ocean. Asia-temperate: Arabia and China. Asia-tropical: India, Indo-China, Malesia, and Papuasia. Australasia: Australia. Pacific: southwestern and northwestern. South America: Mesoamericana, Caribbean, and northern South America.

Habitat: Common in all habitats

Seedling description:

Growth Habit: Perennial
Vernation: Rolled
Node: Pubescent, node is covered with hairs
Internode: Smooth
Leaf Blade: Pubescent surface, single prominent midvein with other parallel veins
Termination point: Acuminate, flat
Leaf sheath margin: Open
Leaf sheath type: Round **Ligule:** Membranous, acute
Auricle: Absent **Collar:** Continuous, broad

11. *Hackelochloa granularis*

Distribution: Africa: west tropical, west-central tropical, northeast tropical, east tropical, southern tropical, south, and western Indian ocean. Asia-temperate: Arabia, China, and eastern Asia. Asia-tropical: India, Indo-China, Malesia, and Papuasia. Australasia: Australia. Pacific. South America, Brazil, and southern South America.
Habitat: Occasional in all habitats

Seedling description:

Growth Habit: Annual
Vernation: Folded
Node: Pubescent, presence of long hairs which are somewhat sticky in nature
Internode: Smooth
Leaf Blade: Pubescent surface, single prominent midvein with other parallel veins, in between veins and at margin hairs are present
Termination point: Acute, keeled
Leaf sheath margin: Close
Leaf sheath type: Flattened **Ligule:** Membranous, divided in to segments
Auricle: Short, rudiment **Collar:** Divided

12. *Heteropogon contortus* var. *contortus* subvar. *typicus*

Distribution: Europe: central, southwestern, and southeastern. Africa-southern tropical, south, middle Atlantic ocean, and western Indian ocean.

Asia-temperate: western Asia, Arabia, China, and eastern Asia. Asia-tropical: India, Indo-China, Malesia, and Papuasia. Australasia. Pacific. North America. South America.
Habitat: In open areas, grasslands

Seedling description:

Growth Habit: Perennial
Vernation: Folded
Node: Glabrous
Internode: Smooth
Leaf Blade: Pubescent surface, single prominent midvein with other parallel veins, presence of cilia on the ventral surface, margin toothed
Termination point: Acuminate, keeled
Leaf sheath margin: Split with overlapping margins
Leaf sheath type: Flattened
Ligule: Membranous, splitted margin, round off
Auricle: Absent
Collar: Divided, broad

13. *Heteropogon contortus* var. *contortus* subvar. *genuinus*

Distribution: Europe: central, southwestern, and southeastern. Africa: north, Macaronesia, west tropical, west-central tropical, northeast tropical, east tropical, southern tropical, south, middle Atlantic ocean, and western Indian ocean. Asia-temperate: western Asia, Arabia, China, and eastern Asia. Asia-tropical: India, Indo-China, Malesia, and Papuasia. Australasia: Australia. Pacific: southwestern, south-central, northwestern, and north-central. North America: southwest USA, south-central USA, and Mexico. South America: Mesoamericana, Caribbean, northern South America, western South America, Brazil, and southern South America.
Habitat: In open areas, grasslands

Seedling description:

Growth Habit: Perennial
Vernation: Folded
Node: Glabrous, bearded
Internode: Smooth
Leaf Blade: Pubescent surface, single prominent midvein with other parallel veins, presence of hairs towards base at margin

Termination point: Acute, keeled
Leaf sheath margin: Split with overlapping margins
Leaf sheath type: Flattened **Ligule:** Membranous, acute, entire margin
Auricle: Absent **Collar:** Divided, broad

14. ***Heteropogon ritchiei***

Distribution: Asia-tropical: India.
Habitat: Dry or moist places

Seedling description:

Growth Habit: Perennial
Vernation: Folded
Node: Glabrous
Internode: Smooth
Leaf Blade: Pubescent surface, single prominent midvein with other parallel veins, presence of long cilia on the dorsal surface towards the base only
Termination point: Acute, keeled
Leaf sheath margin: Split with overlapping margins
Leaf sheath type: Round **Ligule:** Membranous, Short
Auricle: Short, rudiment **Collar:** Continuous

15. ***Heteropogon triticeus***

Distribution: Asia-temperate: China. Asia-tropical: India, Indo-China, Malesia, and Papuasia. Australasia: Australia.
Habitat: Moist open localities.

Seedling description:

Growth Habit: Perennial
Vernation: Folded
Node: Glabrous
Internode: Smooth
Leaf Blade: Pubescent surface, narrow, single prominent midvein is present, long cilia like hairs are present on the surface and at margin
Termination point: Acuminate, keeled
Leaf sheath margin: Split with overlapping margins

Leaf sheath type: Flattened **Ligule:** Membranous
Auricle: Short, rudiment **Collar:** Divided, broad

16. *Imperata cylindrica*

Distribution: Europe: southwestern and southeastern. Africa: north, Macaronesia, west tropical, west-central tropical, northeast tropical, east tropical, southern tropical, south, and western Indian ocean. Asia-temperate: Soviet Middle Asia, Caucasus, western Asia, Arabia, China, and eastern Asia. Asia-tropical: India, Indo-China, Malesia, and Papuasia. Australasia: Australia and New Zealand. Pacific: southwestern and northwestern. North America: southeast USA and Mexico.
Habitat: On the banks, in river beds, in damp clayey soils.

Seedling description:

Growth Habit: Perennial
Vernation: Folded
Node: Pubescent, presence of soft hairs
Internode: Smooth
Leaf Blade: Rigid, single prominent midvein with other parallel veins, scabrous
Termination point: Pointed needle like, flat
Leaf sheath margin: Open
Leaf sheath type: Round **Ligule:** Membranous, round off
Auricle: Absent **Collar:** Continuous, broad, presence of long hairs at margin

17. *Ischaemum indicum*

Distribution: Africa: west tropical, west-central tropical, and western Indian ocean. Asia-temperate: China and eastern Asia. Asia-tropical: India, Indo-China, and Malesia. South America: northern South America.
Habitat: Common weed in all habitats

Seedling description:

Growth Habit: Annual
Vernation: Rolled
Node: Glabrous, bearded
Internode: Smooth

Leaf Blade: Glabrous surface, narrow, single prominent midvein is present with other parallel vein
Termination point: Acute, flat
Leaf sheath margin: Split with overlapping margins
Leaf sheath type: Round **Ligule:** Membranous, acute
Auricle: Short, rudiment **Collar:** Continuous, broad

18. *Ischaemum molle*

Distribution: Asia-temperate: Arabia. Asia-tropical: India and Indo-China.
Habitat: Common in all habitats

Seedling description:

Growth Habit: Annual
Vernation: Rolled
Node: Glabrous
Internode: Smooth
Leaf Blade: Glabrous surface, single prominent midvein is present with other parallel veins
Termination point: Acuminate, keeled
Leaf sheath margin: Open
Leaf sheath type: Flattened **Ligule:** Membranous, long, round off
Auricle: Absent **Collar:** Continuous, broad

19. *Ischaemum pilosum*

Distribution: Africa: west tropical, west-central tropical, northeast tropical, east tropical, southern tropical, and south. Asia-tropical: India.
Habitat: Common in wet soils.

Seedling description:

Growth Habit: Annual
Vernation: Rolled
Node: Glabrous, angled
Internode: Smooth
Leaf Blade: Glabrous surface, single prominent midvein is present with other parallel veins
Termination point: Acuminate to pointed, keeled

Leaf sheath margin: Open
Leaf sheath type: Round **Ligule:** Membranous, ciliate
Auricle: Short, rudiment **Collar:** Continuous, broad

20. ***Ischaemum rugosum***

Distribution: west tropical, east tropical, and western Indian ocean. Asia-temperate: China and eastern Asia. Asia-tropical: India, Indo-China, Malesia, and Papuasia. Australasia: Australia. Pacific: southwestern and northwestern. North America: Mexico. South America: Mesoamericana, Caribbean, northern South America, western South America, and Brazil.
Habitat: Common in all habitats

Seedling description:

Growth Habit: Annual
Vernation: Rolled
Node: Glabrous, angled
Internode: Smooth
Leaf Blade: Glabrous surface, single prominent midvein is present with other parallel veins, narrow towards the base
Termination point: Acute to Acuminate, flat
Leaf sheath margin: Open
Leaf sheath type: Flattened **Ligule:** Membranous, long, round off
Auricle: Absent **Collar:** Continuous, narrow

21. ***Iseilema laxum***

Distribution: India, Sri Lanka
Habitat: Common in all habitats

Seedling description:

Growth Habit: Annual
Vernation: Folded
Node: Glabrous
Internode: Smooth
Leaf Blade: Glabrous surface, single prominent midvein is present
Termination point: Acute, keeled
Leaf sheath margin: Open

Leaf sheath type: Flattened **Ligule:** Membranous, short, ciliate
Auricle: **Collar:** Divided, broad

22. ***Ophiuros exaltatus***

Distribution: Asia-temperate: China. Asia-tropical: India, Indo-China, Malesia, and Papuasia. Australasia: Australia.
Habitat: Common in all habitats

Seedling description:

Growth Habit: Perennial
Vernation: Rolled
Node: Glabrous
Internode: Smooth
Leaf Blade: Glabrous surface, single prominent midvein is present with other parallel veins, short margin
Termination point: Acute, flat
Leaf sheath margin: Close
Leaf sheath type: Round **Ligule:** Membranous, short, acute
Auricle: Short, rudiment **Collar:** Continuous, broad

23. ***Rottboellia exaltata***

Distribution: Africa, Northeast Tropical Africa, East Tropical Africa, West-Central Tropical Africa, West Tropical Africa, Asia-temperate: China, Eastern Asia, Asia-tropical: India, Nepal, Sri Lanka, Indo-China, Australasia, Northern America, Southern America.
Habitat: Growing in shade in wooded grassland, along roadsides and an increasingly problematic weed of cultivation.

Seedling description:

Growth Habit: Perennial
Vernation: Rolled
Node: Glabrous
Internode: Smooth
Leaf Blade: Pubescent surface, single prominent midvein is present with other parallel veins, hairs are present on the dorsal surface
Termination point: Acute, flat
Leaf sheath margin: Split with overlapping margins

Leaf sheath type: Round **Ligule:** Membranous, short, acute
Auricle: Short, rudiment **Collar:** Divided, broad

24. ***Sehima ischaemoides***

Distribution: Africa: Macaronesia, west tropical, west-central tropical, northeast tropical, east tropical, southern tropical, and south. Asia-temperate: Arabia. Asia-tropical: India.
Habitat: Dry soil, on slopes of river bands, on hillocks.

Seedling description:

Growth Habit: Annual
Vernation: Rolled
Node: Glabrous, angled
Internode: Smooth
Leaf Blade: Glabrous surface, single prominent midvein is present with other parallel veins
Termination point: Acute, keeled
Leaf sheath margin: Close
Leaf sheath type: Round **Ligule:** Membranous, acute, hairy at margin
Auricle: Absent **Collar:** Continuous, broad

25. ***Sehima nervosum***

Distribution: Africa: northeast tropical, east tropical, and southern tropical. Asia-temperate: Arabia and China. Asia-tropical: India, Indo-China, Malesia, and Papuasia. Australasia: Australia.
Habitat: Dry soil, on slopes of river bands, on hillocks.

Seedling description:

Growth Habit: Perennial
Vernation: Rolled
Node: Glabrous
Internode: Smooth
Leaf Blade: Pubescent surface, single prominent midvein is present with other parallel veins, in between them secondary veins are present, presence of few hairs on dorsal surface
Termination point: Pointed, flat

Leaf sheath margin: Split with overlapping margin
Leaf sheath type: Round **Ligule:** Membranous
Auricle: Absent **Collar:** Continuous, broad

26. ***Sehima sulcatum***

Distribution: Asia-tropical: India and Indo-China.
Habitat: Dry soil, on slopes of river bands, on hillocks.

Seedling description:

Growth Habit: Perennial
Vernation: Rolled
Node: Glabrous, bearded
Internode: Smooth
Leaf Blade: Pubescent surface, single prominent midvein is present with other parallel veins, in between parallel veins secondary veins are present, toothed margin
Termination point: Acuminate, flat
Leaf sheath margin: Split with overlapping margins
Leaf sheath type: Round **Ligule:** Membranous, Short, ciliate
Auricle: Absent **Collar:** Continuous, broad

27. ***Sorghum halepense***

Distribution: Europe: central, southwestern, southeastern, and eastern. Africa: north, Macaronesia, west tropical, west-central tropical, northeast tropical, southern tropical, south, and western Indian ocean. Asia-temperate: Soviet Middle Asia, Caucasus, western Asia, Arabia, China, and eastern Asia. Asia-tropical: India, Indo-China, Malesia, and Papuasia. Australasia: Australia and New Zealand. Pacific: southwestern, south-central, northwestern, and north-central. North America. South America: western South America, Brazil, and southern South America.
Habitat: Common near wet places, in hedges.

Seedling description:

Growth Habit: Perennial
Vernation: Folded
Node: Glabrous, bearded
Internode: Smooth

Leaf Blade: Glabrous surface, single prominent midvein is present with other parallel veins
Termination point: Acuminate, flat
Leaf sheath margin: Split with overlapping margins
Leaf sheath type: Round **Ligule:** Membranous, acute
Auricle: Short, rudiment **Collar:** Continuous

28. *Sorghum purpureosericeum*

Distribution: Africa: west tropical, west-central tropical, northeast tropical, and east tropical. Asia-temperate: Arabia. Asia-tropical: India.
Habitat:

Seedling description:

Growth Habit: Perennial
Vernation: Folded
Node: Pubescent
Internode: Smooth
Leaf Blade: Pubescent surface, single prominent midvein is present with other parallel veins, hairs present on both surfaces
Termination point: Pointed needle like, flat
Leaf sheath margin: Split with overlapping margins
Leaf sheath type: Round **Ligule:** Membranous
Auricle: Absent **Collar:** Divided, broad

29. *Thelepogon elegans*

Distribution: Africa: west tropical, west-central tropical, northeast tropical, east tropical, southern tropical, and south. Asia-tropical: India, Indo-China, and Malesia.
Habitat: Common in open areas.

Seedling description:

Growth Habit: Annual
Vernation: Rolled
Node: Pubescent, bearded
Internode: Smooth
Leaf Blade: Pubescent surface, single prominent midvein is present with other parallel veins; dotted structures are seen on the surface folded, toothed margin

Leaf Blade Acute, flat
Leaf sheath margin: Split with overlapping margins
Leaf sheath type: Round **Ligule:** Membranous, short
Auricle: Absent **Collar:** Continuous, narrow

30. ***Themeda cymbaria***

Distribution: Asia-tropical: India.
Habitat: Common in open area, sometimes also found on hills.

Seedling description:

Growth Habit: Perennial
Vernation: Rolled
Node: Pubescent, long hairs present but not in continuous ring form, it is in segments
Internode: Smooth
Leaf Blade: Glabrous surface, single prominent midvein is present with other parallel veins
Termination point: Acute, flat
Leaf sheath margin: Split with overlapping margins
Leaf sheath type: Flattened **Ligule:** Membranous, short, ciliate
Auricle: Absent **Collar:** Divided

31. ***Themeda laxa***

Distribution: Asia-tropical: India.
Habitat: Drier rocky areas.

Seedling description:

Growth Habit: Annual
Vernation: Rolled
Node: Glabrous
Internode: Smooth
Leaf Blade: Glabrous surface, single prominent midvein is present with other parallel veins, slightly toothed margin
Termination point: Acuminated to pointed, flat
Leaf sheath margin: Split with overlapping margins
Leaf sheath type: Round **Ligule:** Membranous, acute
Auricle: Absent **Collar:** Continuous, narrow

32. ***Themeda triandra***

Distribution: Africa: north, Macaronesia, west tropical, west-central tropical, northeast tropical, east tropical, southern tropical, south, and western Indian ocean. Asia-temperate: western Asia, Arabia, China, and eastern Asia. Asia-tropical: India, Indo-China, Malesia, and Papuasia. Australasia: Australia and New Zealand. Pacific: southwestern.

Habitat: Common in all habitats.

Seedling description:

Growth Habit: Perennial
Vernation: Rolled
Node: Glabrous
Internode: Smooth
Leaf Blade: Glabrous surface, single prominent midvein is present with other parallel veins, leaf base is shorted
Termination point: Acute, flat
Leaf sheath margin: Split with overlapping margins
Leaf sheath type: Round **Ligule:** Membranous, round off
Auricle: Absent **Collar:** Continuous, narrow

33. ***Triplopogon ramosissimus***

Distribution: Asia-tropical: India.

Habitat: Rocky river beds, hill slopes.

Seedling description:

Growth Habit: Perennial
Vernation: Folded
Node: Glabrous, bearded
Internode: Smooth
Leaf Blade: Glabrous surface, single prominent midvein, lanceolate shape
Termination point: Acuminate, flat
Leaf sheath margin: Open
Leaf sheath type: Flattened **Ligule:** Membranous, round off
Auricle: Absent **Collar:** Continuous, narrow

34. ***Alloteropis cimiciana***

Distribution: Africa: west tropical, west-central tropical, northeast tropical, east tropical, southern tropical, south, and western Indian ocean. Asia-temperate: China. Asia-tropical: India, Indo-China, Malesia, and Papuasia. Australasia: Australia. North America: southeast USA.
Habitat: In plains and foot hills

Seedling description:

Growth Habit: Annual
Vernation: Rolled
Node: Pubescent, long hairs are present
Internode: Pubescent
Leaf Blade: Glabrous surface, single prominent midvein is present with other parallel veins, toothed margin
Termination point: Acute, flat
Leaf sheath margin: Split with overlapping margins
Leaf sheath type: Round **Ligule:** Fringe of hairs
Auricle: Absent **Collar:** Divided, narrow

35. ***Brachiaria eruciformis***

Distribution: Europe: southwestern and southeastern. Africa: north, northeast tropical, east tropical, southern tropical, south, and western Indian ocean. Asia-temperate: Soviet Middle Asia, Caucasus, western Asia, Arabia, and China. Asia-tropical: India, Indo-China, Malesia, and Papuasia. Australasia: Australia. Pacific: southwestern and northwestern. South America: Caribbean and southern South America.
Habitat: Common everywhere

Seedling description:

Growth Habit: Annual
Vernation: Rolled
Node: Pubescent, nodal region is totally covered with hairs
Internode: Smooth, longitudinal lines are present
Leaf Blade: Pubescent surface, no single prominent midvein is present, hairs are present on both the surface of blade

Termination point: Acute, flat
Leaf sheath margin: Split with overlapping margins
Leaf sheath type: Round **Ligule:** Fringe of hairs
Auricle: Short, rudiment **Collar:** Continuous, narrow

36. *Brachiaria distachya*

Distribution: Africa: east tropical and western Indian ocean. Asia-tropical: India, Indo-China, Malesia, and Papuasia. Australasia: Australia. Pacific: southwestern, south-central, northwestern, and north-central.
Habitat: Moist shaded places in plains

Seedling description:

Growth Habit: Annual
Vernation: Rolled
Node: Glabrous
Internode: Smooth
Leaf Blade: Glabrous surface, single prominent midvein is present, cilia present on margin, lanceolate
Termination point: Acute, flat
Leaf sheath margin: Split with overlapping margins
Leaf sheath type: Round **Ligule:** Fringe of hairs
Auricle: Absent **Collar:** Continuous, narrow

37. *Brachiaria ramosa*

Distribution: Africa: north, Macaronesia, west tropical, west-central tropical, northeast tropical, east tropical, southern tropical, and western Indian ocean. Asia-temperate: western Asia, Arabia, and China. Asia-tropical: India, Indo-China, and Malesia. Australasia: Australia. North America: southeast USA.
Habitat: Common in all habitats.

Seedling description:

Growth Habit: Annual
Vernation: Rolled
Node: Pubescent, few hairs are present
Internode: Smooth
Leaf Blade: Pubescent surface, single prominent midvein is present with other parallel veins, cilia present on both the surface even at margin

Termination point: Acute, flat
Leaf sheath margin: Split with overlapping margins
Leaf sheath type: Round **Ligule:** Fringe of hairs, fused at base
Auricle: Absent **Collar:** Continuous, narrow

38. ***Brachiaria reptans***

Distribution: Africa: north, west tropical, west-central tropical, northeast tropical, east tropical, southern tropical, and western Indian ocean. Asia-temperate: western Asia, Arabia, China, and eastern Asia. Asia-tropical: India, Indo-China, Malesia, and Papuasia. Australasia: Australia. Pacific. North America. South America, western South America, and Brazil.
Habitat: Common in all habitats

Seedling description:

Growth Habit: Annual
Vernation: Rolled
Node: Glabrous
Internode: Smooth
Leaf Blade: Pubescent surface, single prominent midvein is present with all parallel veins, few hairs present on dorsal surface
Termination point: Acute, flat
Leaf sheath margin: Split with overlapping margins
Leaf sheath type: Round **Ligule:** Fringe of hairs
Auricle: Short, rudiment **Collar:** Continuous, narrow

39. ***Cenchrus biflorus***

Distribution: Africa: north, west tropical, west-central tropical, northeast tropical, east tropical, southern tropical, south, and western Indian ocean. Asia-temperate: western Asia and Arabia. Asia-tropical: India. Australasia: Australia.
Habitat: In sandy soil

Seedling description:

Growth Habit: Annual
Vernation: Rolled
Node: Glabrous, slightly bearded
Internode: Smooth
Leaf Blade: Pubescent surface, single prominent midvein is present with other parallel veins, hairs present at the base of leaf blade

Termination point: Pointed needle like, flat
Leaf sheath margin: Split with overlapping margins
Leaf sheath type: Flattened **Ligule:** Membranous, acute
Auricle: Absent **Collar:** Continuous, narrow

40. ***Cenchrus ciliaris***

Distribution: Europe: southeastern. Africa: north, Macaronesia, west tropical, west-central tropical, northeast tropical, east tropical, southern tropical, south, and western Indian ocean. Asia-temperate: western Asia, Arabia, and eastern Asia. Asia-tropical: India, Indo-China, Malesia, and Papuasia. Australasia: Australia. Pacific: southwestern, south-central, northwestern, and north-central. North America: Mexico. South America: Brazil, and southern South America.
Habitat: Common in dry and sandy habitats, even drier parts of hills

Seedling description:

Growth Habit: Perennial
Vernation: Folded
Node: Glabrous
Internode: Smooth
Leaf Blade: Pubescent surface, single prominent midvein is present with other parallel veins
Termination point: Acute, flat
Leaf sheath margin: Split with overlapping margins
Leaf sheath type: Round **Ligule:** Fringe of hairs
Auricle: Absent **Collar:** Continuous, narrow

41. ***Cenchrus setigerus***

Distribution: Africa: north, west tropical, northeast tropical, and east tropical. Asia-temperate: western Asia, Arabia, and China. Asia-tropical: India and Indo-China. Australasia: Australia. Pacific: north-central. South America: Brazil.
Habitat: In dry shady areas and in fellow fields.

Seedling description:

Growth Habit: Annual
Vernation: Rolled

Node: Glabrous
Internode: Smooth
Leaf Blade: Pubescent surface, single prominent midvein is present with other parallel veins
Termination point: Acuminate, flat
Leaf sheath margin: Split with overlapping margins
Leaf sheath type: Flattened **Ligule:** Fringe of hairs
Auricle: Absent **Collar:** Continuous, narrow, hairy

42. *Digitaria ciliaris*

Distribution: Europe: southeastern. Africa: north, Macaronesia, west tropical, west-central tropical, northeast tropical, east tropical, southern tropical, south, middle Atlantic ocean, and western Indian ocean. Asia-temperate: Caucasus, western Asia, Arabia, China, and eastern Asia. Asia-tropical: India, Indo-China, Malesia, and Papuasia. Australasia: Australia and New Zealand. Pacific: southwestern, south-central, northwestern. North America: Mexico. South America.
Habitat: Common in forest undergrowth, wastelands, road sides, moist places

Seedling description:

Growth Habit: Annual
Vernation: Rolled
Node: Pubescent, numerous long hairs are present
Internode: Smooth
Leaf Blade: Glabrous surface, single prominent midvein is present with other parallel veins, wavy margin
Termination point: Acute, keeled
Leaf sheath margin: Split with overlapping margins
Leaf sheath type: Round **Ligule:** Membranous, acuminate
Auricle: Absent **Collar:** Continuous, broad

43. *Digitaria microbachne*

Distribution: Africa: east tropical and western Indian ocean. Asia-temperate: China and eastern Asia. Asia-tropical: India, Indo-China, Malesia, and Papuasia. Australasia: Australia and New Zealand. Pacific: southwestern, south-central, northwestern, and north-central.

North America: Mexico. South America: Mesoamericana, northern South America, and western South America.

Habitat: Degraded deciduous forests and grasslands, also in wastelands

Seedling description:

Growth Habit: Annual
Vernation: Rolled
Node: Pubescent, few hairs are present
Internode: Smooth
Leaf Blade: Pubescent surface, single prominent midvein is present with other parallel veins
Termination point: Acute, flat
Leaf sheath margin: Open
Leaf sheath type: Round **Ligule:** Membranous, truncate
Auricle: Absent **Collar:** Divided

44. ***Digitaria granularis***

Distribution: Asia-temperate: China. Asia-tropical: India, Indo-China, Malesia, and Papuasia. In India: throughout. In Gujarat: throughout except Kutch.

Habitat: In open fallow fields, in open grasslands

Seedling description:

Growth Habit: Annual
Vernation: Rolled
Node: Glabrous
Internode: Smooth
Leaf Blade: Pubescent surface, single prominent midvein is present with other parallel veins
Termination point: Acuminate, flat
Leaf sheath margin: Open
Leaf sheath type: Round **Ligule:** Membranous, acute
Auricle: Absent **Collar:** Divided

45. ***Digitaria longiflora***

Distribution: Africa: west tropical, west-central tropical, northeast tropical, east tropical, southern tropical, south, and western Indian ocean. Asia-temperate: China and eastern Asia. Asia-tropical: India, Indo-China, Malesia,

and Papuasia. Australasia: Australia. Pacific: southwestern and northwestern. North America: southeast USA. South America: Mesoamericana, Caribbean, northern South America, western South America, and Brazil.
Habitat: In dry areas, in forest undergrowth of rocky habitats

Seedling description:

Growth Habit: Annual
Vernation: Rolled
Node: Glabrous
Internode: Smooth
Leaf Blade: Glabrous surface, single prominent midvein is present with other parallel veins
Termination point: Acute, flat
Leaf sheath margin: Split with overlapping margins
Leaf sheath type: Round **Ligule:** Membranous, truncate
Auricle: Short, rudiment **Collar:** Continuous, broad, hairy

46. *Digitaria stircta*

Distribution: Asia-temperate: Arabia, China, and eastern Asia. Asia-tropical: India and Indo-China.
Habitat: In open wastelands in rocky gravelly habitats.

Seedling description:

Growth Habit: Annual
Vernation: Rolled
Node: Glabrous, angled
Internode: Smooth
Leaf Blade: Pubescent surface, no single prominent midvein is present, long hairs present on the dorsal surface
Termination point: Acute, flat
Leaf sheath margin: Split with overlapping margins
Leaf sheath type: Round **Ligule:** Membranous
Auricle: Absent **Collar:** Continuous, narrow

47. *Echinichloa colonum*

Distribution: Europe: southwestern and southeastern. Africa. Asia-temperate: western Asia, Arabia, China, and eastern Asia. Asia-tropical:

India, Indo-China, Malesia, and Papuasia. Australasia: Australia. Pacific: southwestern, south-central, northwestern, and north-central. North America. South America.
Habitat: Common in all habitats.

Seedling description:

Growth Habit: Annual
Vernation: Rolled
Node: Glabrous
Internode: Smooth
Leaf Blade: Glabrous surface, single prominent midvein is present with other parallel veins
Termination point: Acute, keeled
Leaf sheath margin: Split with overlapping margins
Leaf sheath type: Flattened **Ligule:** Absent
Auricle: Absent **Collar:** Continuous, broad

48. *Echinochloa crusgalli*

Distribution: Europe: northern, central, southwestern, southeastern, and eastern. Africa. Asia-temperate: Siberia, Soviet far east, Soviet Middle Asia, Caucasus, western Asia, Arabia, China, Mongolia, and eastern Asia. Asia-tropical: India, Indo-China, Malesia, and Papuasia. Australasia: Australia and New Zealand. Pacific: southwestern, south-central, northwestern, and north-central. North America. South America.
Habitat: In moist places, water margin, damp places.

Seedling description:

Growth Habit: Annual
Vernation: Rolled
Node: Glabrous
Internode: Smooth
Leaf Blade: Glabrous surface, single prominent midvein with other parallel veins
Termination point: Acuminate, flat
Leaf sheath margin: Split with overlapping margins
Leaf sheath type: Round **Ligule:** Absent
Auricle: Absent **Collar:** Continuous, narrow

49. *Echionochloa stagnina*

Distribution: Africa: north, west tropical, west-central tropical, northeast tropical, east tropical, southern tropical, south, and western Indian ocean. Asia-tropical: India, Indo-China, Malesia, and Papuasia.
Habitat: In moist damp places

Seedling description:

Growth Habit: Annual
Vernation: Rolled
Node: Glabrous
Internode: Smooth
Leaf Blade: Glabrous surface, single prominent midvein is present with other parallel veins, margin toothed
Termination point: Acute, keeled
Leaf sheath margin: Split with overlapping margins
Leaf sheath type: Round
Ligule: Fringe of short hairs, fused at base
Auricle: Absent
Collar: Continuous, narrow

50. *Eremopogon foveolatus*

Distribution: Africa: north, Macaronesia, west tropical, northeast tropical, and east tropical. Asia-temperate: western Asia and Arabia. Asia-tropical: India and Indo-China. In India: throughout. In Gujarat: throughout, common.
Habitat: Grassy hills and open dry habitats

Seedling description:

Growth Habit: Perennial
Vernation: Folded
Node: Pubescent, long hairs present
Internode: Smooth
Leaf Blade: Pubescent surface, single prominent midvein is present with other parallel veins, few long hairs present at margin, margin is toothed
Termination point: Acuminate, flat
Leaf sheath margin: Split with overlapping margins
Leaf sheath type: Round
Ligule: Fringe of long hairs
Auricle: Absent
Collar: Divided

51. *Eriochloa procera*

Distribution: Africa: northeast tropical, east tropical, southern tropical, middle Atlantic ocean, and western Indian ocean. Asia-temperate: China and eastern Asia. Asia-tropical: India, Indo-China, Malesia, and Papuasia. Australasia: Australia. Pacific: southwestern, northwestern, and north-central. South America: northern South America and western South America.

Habitat: Water edges, margin of agricultural fields, waste lands

Seedling description:

Growth Habit: Perennial
Vernation: Folded
Node: Glabrous
Internode: Smooth
Leaf Blade: Glabrous surface, no single prominent midvein is present
Termination point: Acute, flat
Leaf sheath margin: Split with overlapping margins
Leaf sheath type: Round **Ligule:** Fringe of villous hairs
Auricle: Absent **Collar:** Continuous, narrow

52. *Oplismenus composites*

Distribution: Africa: northeast tropical, east tropical, southern tropical, south, and western Indian ocean. Asia-temperate: western Asia, Arabia, China, and eastern Asia. Asia-tropical: India, Indo-China, Malesia, and Papuasia. Australasia: Australia. Pacific. North America. South America.

Habitat: In forest undergrowth

Seedling description:

Growth Habit: Annual
Vernation: Rolled
Node: Pubescent, slightly hairs
Internode: Smooth
Leaf Blade: Pubescent surface, single prominent midvein is present with other parallel veins, dotted structures are present on both surfaces
Termination point: Acute, flat
Leaf sheath margin: Open

Leaf sheath type: Round **Ligule:** Fringe of hairs
Auricle: Short, rudiment **Collar:** Divided

53. *Panicum antidotale*

Distribution: Africa: west tropical, west-central tropical, northeast tropical, east tropical, southern tropical, south, and western Indian ocean. In India: South to Gangetic plains, North West India. In Gujarat: throughout, not common.
Habitat: Common in wet places, grows under bushes and trees and in hedges.

Seedling description:

Growth Habit: Perennial
Vernation: Folded
Node: Glabrous
Internode: Smooth
Leaf Blade: Glabrous surface, single prominent midvein is present with other parallel veins; dotted structures are present on both surfaces
Termination point: Acuminate, flat
Leaf sheath margin: Open
Leaf sheath type: Round **Ligule:** Fringe of minute hairs
Auricle: Absent **Collar:** Continuous, narrow

54. *Panicum maximum*

Distribution: Europe: southeastern. Africa: north, Macaronesia, west tropical, west-central tropical, northeast tropical, east tropical, southern tropical, south, middle Atlantic ocean, and western Indian ocean. Asia-temperate: western Asia, Arabia, China, and eastern Asia. Asia-tropical: India, Indo-China, Malesia, and Papuasia. Australasia: Australia and New Zealand. North America: south-central USA, southeast USA, and Mexico. South America.
Habitat: Generally grows on hilly areas

Seedling description:

Growth Habit: Perennial
Vernation: Folded

Node: Glabrous
Internode: Smooth
Leaf Blade: Glabrous surface, single prominent midvein is present with other parallel veins
Termination point: Acuminate, flat
Leaf sheath margin: Split with overlapping margins
Leaf sheath type: Round **Ligule:** Membranous, Short, ciliate
Auricle: Absent **Collar:** Divided, broad

55. *Panicum miliaceum*

Distribution: Europe: central, southwestern, southeastern, and eastern. Africa: north, Macaronesia, west-central tropical, southern tropical, and western Indian ocean. Asia-temperate: Siberia, Soviet far east, Soviet Middle Asia, Caucasus, western Asia, Arabia, China, Mongolia, and eastern Asia. Asia-tropical: India, Indo-China, and Malesia. Australasia: Australia and New Zealand. North America. South America: Caribbean, northern South America, Brazil, and southern South America.
Habitat: Generally grows on hilly areas

Seedling description:

Growth Habit: Perennial
Vernation: Folded
Node: Glabrous
Internode: Smooth
Leaf Blade: Glabrous surface, single prominent midvein is present with other parallel veins
Termination point: Acute, flat
Leaf sheath margin: Split with overlapping margins
Leaf sheath type: Round **Ligule:** Ciliated rim
Auricle: Short, rudiment **Collar:** Continuous, narrow

56. *Panicum trypheron*

Distribution: Asia-temperate: China and eastern Asia. Asia-tropical: India, Indo-China, Malesia, and Papuasia. In India: Southwards to Gangetic plains. In Gujarat: North to South Gujarat.
Habitat: In wet lands, in sandy waste land, near cultivated fields.

Seedling description:

Growth Habit: Annual
Vernation: Rolled
Node: Glabrous, bearded
Internode: Smooth
Leaf Blade: Glabrous surface, single prominent midvein is present, margin folded
Termination point: Acuminate, keeled
Leaf sheath margin: Close

Leaf sheath type: Round	**Ligule:** Fringe of short hairs, frimbricate
Auricle: Short, rudiment	**Collar:** Continuous, narrow

57. ***Paspalidium flavidum***

Distribution: Africa: western Indian ocean. Asia-temperate: Arabia, China, and eastern Asia. Asia-tropical: India, Indo-China, Malesia, and Papuasia. Australasia: Australia. Pacific: northwestern
Habitat: Common in wet places

Seedling description:

Growth Habit: Annual
Vernation: Folded
Node: Glabrous
Internode: Smooth
Leaf Blade: Glabrous surface, single prominent midvein is present
Termination point: Acuminate, flat
Leaf sheath margin: Split with overlapping margins

Leaf sheath type: Flattened	**Ligule:** Fringe of short hairs
Auricle: Absent	**Collar:** Divided, broad

58. ***Paspalidium geminatum***

Distribution: Africa: north, Macaronesia, west tropical, west-central tropical, northeast tropical, east tropical, southern tropical, south, middle Atlantic ocean, and western Indian ocean. Asia-temperate: western Asia and Arabia. Asia-tropical: India, Indo-China, and Malesia. North America: north-central USA, south-central USA, southeast USA, and Mexico.

South America: Mesoamericana, Caribbean, northern South America, western South America, Brazil, and southern South America.

Habitat: Water edges, river beds, seasonal water bodies

Seedling description:

Growth Habit: Perennial

Vernation: Folded

Node: Glabrous

Internode: Smooth

Leaf Blade: Pubescent surface, single prominent midvein is present with minute parallel veins; Short hairs are present on the dorsal surface

Termination point: Acuminate, flat

Leaf sheath margin: Split with overlapping margins

Leaf sheath type: Round **Ligule:** Fringe of hairs

Auricle: Absent **Collar:** Divided, broad

59. *Paspalum scorbiculatum*

Distribution: Africa: Macaronesia, west tropical, west-central tropical, northeast tropical, east tropical, southern tropical, south, middle Atlantic ocean, and western Indian ocean. Asia-temperate: Arabia, China, and eastern Asia. Asia-tropical: India, Indo-China, Malesia, and Papuasia. Australasia: Australia and New Zealand. Pacific: southwestern, south-central, northwestern, and north-central. North America. South America, Brazil.

Habitat: Water edges

Seedling description:

Growth Habit: Annual

Vernation: Rolled

Node: Glabrous

Internode: Smooth

Leaf Blade: Glabrous surface, single prominent midvein is present

Termination point: Acute, flat

Leaf sheath margin: Split with overlapping margins

Leaf sheath type: Round **Ligule:** Membranous

Auricle: Short, rudiment **Collar:** Continuous, narrow

60. *Paspalum distichum*

Distribution: Europe: southwestern, southeastern, and eastern. Africa: north, Macaronesia, southern tropical, south, middle Atlantic Ocean, and western Indian ocean. Asia-temperate: Soviet Middle Asia, Caucasus, western Asia, Arabia, China, and eastern Asia. Asia-tropical: India, Indo-China, and Malesia. Australasia: Australia and New Zealand. Pacific: southwestern and north-central. North America, Mexico. South America: Brazil, and southern South America.
Habitat: Moist places.

Seedling description:

Growth Habit: Perennial
Vernation: Rolled
Node: Pubescent
Internode: Smooth
Leaf Blade: Glabrous surface, single prominent midvein is present with parallel veins
Termination point: Acute, flat
Leaf sheath margin: Split with overlapping margins
Leaf sheath type: Round **Ligule:** Membranous, splitted
Auricle: Absent **Collar:** Continuous, narrow

61. *Paspalum vaginatum*

Distribution: Europe: southwestern and southeastern. Africa: Macaronesia, west tropical, west-central tropical, northeast tropical, east tropical, southern tropical, south, and western Indian ocean. Asia-temperate: Arabia, China, and eastern Asia. Asia-tropical: India, Indo-China, Malesia, and Papuasia. Australasia: Australia and New Zealand. North America.
Habitat: coastal salt marshes of the tropics and sub-tropics, in brackish marshy areas and mangrove swamps is tolerant of drought, salt and flooding or extended wet periods.

Seedling description:

Growth Habit: Perennial
Vernation: Rolled
Node: Glabrous

Internode: Smooth
Leaf Blade: Pubescent surface, single prominent midvein is present with parallel veins, dotted structures present on the dorsal surface
Termination point: Acute, flat
Leaf sheath margin: Split with overlapping margins
Leaf sheath type: Flattened **Ligule:** Membranous, short, round off
Auricle: Short, rudiment **Collar:** Continuous, narrow

62. *Pennisetum setosum*

Distribution: Asia-temperate: Caucasus, China, and eastern Asia. Asia-tropical: India, Indo-China, and Malesia. Australasia: Australia and New Zealand. In India: throughout. In Gujarat: Kutch, North and South Gujarat.
Habitat: In Open waste lands and fallow fields.
Habit: Annual or perennial

Seedling description:

Growth Habit: Perennial
Vernation: Rolled
Node: Glabrous
Internode: Smooth
Leaf Blade: Glabrous surface, single prominent midvein is present with parallel veins
Termination point: Acuminate to tapering, flat
Leaf sheath margin: Split with overlapping margins
Leaf sheath type: Round **Ligule:** Fringe of long hairs, fused at base
Auricle: Absent **Collar:** Continuous, narrow

63. *Setaria glauca*

Distribution: Europe: eastern. Africa: north, west tropical, west-central tropical, northeast tropical, east tropical, southern tropical, south, and western Indian ocean. Asia-temperate: western Asia, Arabia, and China. Asia-tropical: India and Indo-China. Australasia: Australia. South America: Caribbean. In India: throughout. In Gujarat: throughout.
Habitat: Common in all habitats.

Seedling description:

Growth Habit: Annual
Vernation: Rolled
Node: Pubescent
Internode: Smooth
Leaf Blade: Glabrous surface, single prominent midvein is present with other parallel veins
Termination point: Acute, flat
Leaf sheath margin: Split with overlapping margins
Leaf sheath type: Round **Ligule:** Membranous, acute
Auricle: Short, rudiment **Collar:** Divided

64. *Setaria tomentosa*

Distribution: Africa: northeast tropical, east tropical, and western Indian ocean. Asia-temperate: Caucasus, Arabia, and China. Asia-tropical: India and Indo-China. In India: throughout. In Gujarat: throughout.
Habitat: In hedges, forest undergrowth.

Seedling description:

Growth Habit: Annual
Vernation: Rolled
Node: Glabrous, angled
Internode: Smooth
Leaf Blade: Pubescent surface, single prominent midvein is present with other parallel veins, presence of minute hairs on all over the surface
Termination point: Acuminate, flat
Leaf sheath margin: Split with overlapping margins
Leaf sheath type: Flattened **Ligule:** Membranous, acute
Auricle: Short, rudiment **Collar:** Divided

65. *Setaria verticillata*

Distribution: Europe: central, southwestern, southeastern, and eastern. Africa: north, Macaronesia, west tropical, west-central tropical, northeast tropical, east tropical, southern tropical, south, middle Atlantic ocean, and western Indian ocean. Asia-temperate: Soviet Middle Asia, Caucasus, western Asia, Arabia, China, and eastern Asia. Asia-tropical: India, Indo-China, and Malesia. Australasia: Australia and New Zealand. Pacific: southwestern, south-central, northwestern, and north-central. North America. South America, Brazil, and southern South America.

Habitat: In hedges and shady grounds

Seedling description:

Growth Habit: Annual
Vernation: Rolled
Node: Glabrous
Internode: Smooth
Leaf Blade: Pubescent surface, single prominent midvein is present with other parallel veins, presence of hairs on the surface
Termination point: Acute to Acuminate, flat
Leaf sheath margin: Split with overlapping margins
Leaf sheath type: Flattened **Ligule:** Membranous, ciliolate
Auricle: Short, rudiment **Collar:** Divided

66. *Aleuropus lagopoides*

Distribution:
Habitat:

Seedling description:

Growth Habit: Perennial
Vernation: Folded
Node: Glabrous
Internode: Smooth
Leaf Blade: Pubescent surface, no single prominent midvein is present but many parallel veins are present, rough and tough leaf blade
Termination point: Acute, flat
Leaf sheath margin: Close
Leaf sheath type: Round **Ligule:** Fringe of short hairs
Auricle: Absent **Collar:** Continuous, narrow

67. *Isachne globosa*

Distribution: Asia-temperate: Arabia, China, and eastern Asia. Asia-tropical: India, Indo-China, Malesia, and Papuasia. Australasia: Australia

and New Zealand. Pacific: southwestern, south-central, and northwestern. In India: throughout hotter parts. In Gujarat: throughout.
Habitat: In wet and marshy habitats along the streams, moist grounds.

Seedling description:

Growth Habit: Annual
Vernation: Rolled
Node: Pubescent
Internode: Smooth
Leaf Blade: Glabrous surface, no single prominent midvein is present but many parallel veins are present
Termination point: Pointed needle like, flat
Leaf sheath margin: Split with overlapping margins
Leaf sheath type: Round **Ligule:** Membranous
Auricle: Absent **Collar:** Divided, broad

68. ***Aristida adscensionis***

Distribution: Europe: southwestern and southeastern. Africa: north, Macaronesia, west tropical, west-central tropical, northeast tropical, east tropical, southern tropical, south, middle Atlantic ocean, and western Indian ocean. Asia-temperate: Soviet Middle Asia, Caucasus, western Asia, Arabia, China, and Mongolia. Asia-tropical: India, Indo-China, Malesia, and Papuasia.. North America. Mexico. South America.
Habitat: Common in all habitats

Seedling description:

Growth Habit: Annual
Vernation: Rolled
Node: Glabrous
Internode: Smooth
Leaf Blade: Pubescent surface, single prominent midvein is present with other parallel veins, presence of hairs on dorsal side, margins are curved
Termination point: Pointed needle like, flat
Leaf sheath margin: Close
Leaf sheath type: Round **Ligule:** Fringe of fine, short hairs
Auricle: Absent **Collar:** Continuous, narrow

69. *Aristida funiculata*

Distribution: Africa: north, Macaronesia, west tropical, northeast tropical, and east tropical. Asia-temperate: western Asia and Arabia. Asia-tropical: India and Indo-China. In India: Peninsular and NW India. In Gujarat: throughout.

Habitat: Common on drier and rocky grounds and hills.

Seedling description:

Growth Habit: Annual
Vernation: Rolled
Node: Glabrous, bearded
Internode: Smooth
Leaf Blade: Glabrous surface, no single prominent midvein is present but all parallel veins are present, margins are curved
Termination point: Pointed needle like, flat
Leaf sheath margin: Close
Leaf sheath type: Round **Ligule:** Fringe of fine, short hairs
Auricle: Absent **Collar:** Continuous, narrow

70. *Chloris barbata*

Distribution: Africa: north, Macaronesia, west tropical, west-central tropical, northeast tropical, east tropical, middle Atlantic ocean, and western Indian ocean. Asia-temperate: western Asia, Arabia, and eastern Asia. Asia-tropical: India, Indo-China, Malesia, and Papuasia. Australasia: Australia. North America: Mexico. South America.

Habitat: Common in plains.

Seedling description:

Growth Habit: Perennial
Vernation: Folded
Node: Glabrous
Internode: Smooth
Leaf Blade: Pubescent surface, single prominent midvein is present, few long hairs are seen on ventral surface at margin
Termination point: Acuminate, flat
Leaf sheath margin: Open
Leaf sheath type: Flattened **Ligule:** Membranous, acute, narrow
Auricle: Absent **Collar:** Divided

71. *Chloris virgata*

Distribution: Africa middle Atlantic ocean, and western Indian ocean. Asia-temperate: Soviet far east, Soviet Middle Asia, Caucasus, western Asia, Arabia, China, Mongolia, and eastern Asia. Asia-tropical: India, Indo-China, Malesia, and Papuasia. Australasia: Australia. North America. South America.

Habitat: Common on grassy hills.

Seedling description:

Growth Habit: Annual

Vernation: Folded

Node: Glabrous

Internode: Smooth

Leaf Blade: Glabrous surface, single prominent midvein is present

Termination point: Acute, flat

Leaf sheath margin: Open

Leaf sheath type: Round **Ligule:** Membranous, short

Auricle: Absent **Collar:** Continuous, narrow

72. *Cynadon dactylon*

Distribution: Europe. Africa.: north, Macaronesia, west tropical, west-central tropical, northeast tropical, east tropical, southern tropical, south, middle Atlantic ocean, and western Indian ocean. Asia-temperate: Siberia, Soviet Middle Asia, Caucasus, western Asia, Arabia, China, and eastern Asia. Asia-tropical: India, Indo-China, Malesia, and Papuasia. Australasia: Australia and New Zealand. North America. South America.

Habitat: Common everywhere.

Seedling description:

Growth Habit: Perennial

Vernation: Rolled

Node: Glabrous

Internode: Smooth

Leaf Blade: Glabrous surface, single prominent midvein with other parallel veins

Termination point: Acuminate, flat

Leaf sheath margin: Open

Leaf sheath type: Round **Ligule:** Fringe of fine hairs
Auricle: Absent **Collar:** Continuous, broad

73. *Melanocenchris jaequemontii*

Distribution: Africa: northeast tropical. Asia-temperate: western Asia. Asia-tropical: India. In India: throughout in drier parts, In Gujarat: throughout.
Habitat: Common on rocky soil.

Seedling description:

Growth Habit: Annual
Vernation: Rolled
Node: Pubescent
Internode: Smooth
Leaf Blade: Pubescent surface, no single prominent midvein is present but many parallel veins are present
Termination point: Acute, keeled
Leaf sheath margin: Split with overlapping margins
Leaf sheath type: Round **Ligule:** Membranous, ciliate
Auricle: Absent **Collar:** Continuous, narrow, hairy

74. *Oropetium villosulum*

Distribution: Asia-tropical: India. In India: occasionally seen. In Gujarat: North Gujarat, rare.
Habitat: On rocky soil

Seedling description:

Growth Habit: Annual
Vernation: Rolled
Node: Glabrous
Internode: Smooth
Leaf Blade: Pubescent surface, single prominent midvein is present, long hairs are present on the surface
Termination point: Acuminate, flat
Leaf sheath margin: Split with overlapping margins
Leaf sheath type: Round **Ligule:** Membranous, acute
Auricle: Absent **Collar:** Continuous, narrow

75. *Sachoenefeldia gracilis*

Distribution: Africa: north, west tropical, west-central tropical, northeast tropical, and western Indian ocean. Asia-temperate: Arabia. Asia-tropical: India. In India: throughout. In Gujarat: North to central Gujarat, Saurashtra.
Habitat: In open dry sandy areas.

Seedling description:

Growth Habit: Annual
Vernation: Rolled
Node: Glabrous, bearded
Internode: Smooth
Leaf Blade: Pubescent surface, single prominent midvein is present, long hairs are present on the surface, curved margin
Termination point: Pointed needle like, flat
Leaf sheath margin: Close
Leaf sheath type: Round **Ligule:** Membranous, round off
Auricle: Short, rudiment **Collar:** Continuous, narrow

76. *Tetrapogon tenellus*

Distribution: Africa: northeast tropical, east tropical, southern tropical, and south. Asia-temperate: Arabia. Asia-tropical: India. In India: Peninsular and NW India. In Gujarat: North to South to Central Gujarat.
Habitat: In open dry areas

Seedling description:

Growth Habit: Perennial
Vernation: Rolled
Node: Glabrous, bearded
Internode: Smooth
Leaf Blade: Pubescent surface, single prominent midvein is present with many parallel veins; long tuberculated based hairs are present on the dorsal surface
Termination point: Acuminate, flat
Leaf sheath margin: Split with overlapping margins
Leaf sheath type: Round **Ligule:** Membranous, acute
Auricle: Absent **Collar:** Continuous, broad

77. *Tetrapogon villosus*

Distribution: Africa: north, Macaronesia, west tropical, northeast tropical, and east tropical. Asia-temperate: Soviet Middle Asia, western Asia, and Arabia. Asia-tropical: India. In India: throughout. In Gujarat: throughout except Kutch but not common.
Habitat: In open dry areas.

Seedling description:

Growth Habit: Perennial
Vernation: Rolled
Node: Glabrous, angled
Internode: Smooth
Leaf Blade: Pubescent surface, single prominent midvein is present
Termination point: Acuminate, flat
Leaf sheath margin: Split with overlapping margins
Leaf sheath type: Round **Ligule:** Membranous
Auricle: Absent **Collar:** Continuous

78. *Dactyloctenium aegyptium*

Distribution: Europe: southeastern. Africa: north, Macaronesia, west tropical, west-central tropical, northeast tropical, east tropical, southern tropical, south, middle Atlantic ocean, and western Indian ocean. Asia-temperate: western Asia, Arabia, China, and eastern Asia. Asia-tropical: India, Indo-China, Malesia, and Papuasia. Australasia: Australia. North America: northwest USA, north-central USA, northeast USA, southwest USA, south-central USA, southeast USA, and Mexico. South America.
Habitat: In open plains, water edges, escape from gardens

Seedling description:

Growth Habit: Perennial
Vernation: Rolled
Node: Pubescent
Internode: Smooth
Leaf Blade: Pubescent surface, single prominent midvein is present
Termination point: Acute, flat
Leaf sheath margin: Split with overlapping margins
Leaf sheath type: Round **Ligule:** Membranous, acute
Auricle: Absent **Collar:** Continuous, narrow

79. *Dactyloctenium sindicum*

Distribution: Africa: north, northeast tropical, and east tropical. Asia-temperate: Arabia. Asia-tropical: India. In India: N. W. India. In Gujarat: South to Central Gujarat.
Habitat: Near sea coast, sandy soils.

Seedling description:

Growth Habit: Perennial

Vernation: Rolled

Node: Pubescent

Internode: Smooth

Leaf Blade: Pubescent surface, single prominent midvein is present, hairs present on both the surfaces

Termination point: Pointed needle like, flat

Leaf sheath margin: Split with overlapping margins

Leaf sheath type: Round **Ligule:** Fringe of hairs

Auricle: Short, rudiment **Collar:** Continuous, narrow

80. *Desmostachya bipinnata*

Distribution: Africa: north, west tropical, and northeast tropical. Asia-temperate: western Asia, Arabia, and China. Asia-tropical: India, Indo-China, and Malesia. In India: throughout. In Gujarat: throughout
Habitat: Common in all habitats

Seedling description:

Growth Habit: Perennial

Vernation: Folded

Node: Glabrous

Internode: Smooth

Leaf Blade: Glabrous surface, single prominent midvein is present with other parallel veins are present

Termination point: Pointed needle like, flat

Leaf sheath margin: Close

Leaf sheath type: Round **Ligule:** Membranous, round off

Auricle: Absent **Collar:** Continuous, broad

81. *Dinebra retroflexa*

Distribution: Africa: north, west tropical, west-central tropical, northeast tropical, east tropical, southern tropical, south, middle Atlantic ocean, and western Indian ocean. Asia-temperate: western Asia, Arabia, and China. Asia-tropical: India, Indo-China, and Malesia. Australasia: Australia. In India: throughout. In Gujarat: throughout.
Habitat: Common in all habitats

Seedling description:

Growth Habit: Annual
Vernation: Rolled
Node: Glabrous, bearded
Internode: Pubescent
Leaf Blade: Pubescent surface, single prominent midvein is present; few hairs are present on the surface
Termination point: Acuminate, flat
Leaf sheath margin: Split with overlapping margins
Leaf sheath type: Round **Ligule:** Membranous, ciliate
Auricle: Absent **Collar:** Continuous, broad

82. *Eleusine indica*

Distribution: Europe: northern, southwestern, and southeastern. Africa: north, Macaronesia, west tropical, west-central tropical, northeast tropical, east tropical, southern tropical, south, middle Atlantic ocean, and western Indian ocean. Asia-temperate: Soviet Middle Asia, Caucasus, western Asia, Arabia, China, and eastern Asia. Asia-tropical: India, Indo-China, Malesia, and Papuasia. Australasia: Australia and New Zealand. Pacific. North America and South America.
Habitat: Common in all habitats.

Seedling description:

Growth Habit: Annual
Vernation: Folded
Node: Glabrous
Internode: Smooth
Leaf Blade: Pubescent surface, single prominent midvein is present
Termination point: Acute, slightly keeled

Leaf sheath margin: Split with overlapping margins
Leaf sheath type: Flattened **Ligule:** Membranous
Auricle: Absent **Collar:** Divided

83. *Acrachne racemosa*

Distribution: Africa: north, Macaronesia, west tropical, west-central tropical, northeast tropical, east tropical, southern tropical, and south. Asia-temperate: western Asia, Arabia, and China. Asia-tropical: India and Indo-China. Australasia: Australia. South America: Caribbean. In India: throughout the plains. In Gujarat: throughout.
Habitat: Common in all habitats

Seedling description:

Growth Habit: Annual
Vernation: Folded
Node: Glabrous
Internode: Smooth
Leaf Blade: Glabrous surface, single prominent midvein is present, toothed margin
Termination point: Acuminate, keeled
Leaf sheath margin: Split with overlapping margins
Leaf sheath type: Flattened **Ligule:** Membranous, ciliolate
Auricle: Absent **Collar:** Divided

84. *Eragrostiella bifaria*

Distribution: Africa: northeast tropical and east tropical. Asia-tropical: India and Indo-China. Australasia: Australia. In India: throughout. In Gujarat: North to Central Gujarat.
Habitat: Open rocky areas

Seedling description:

Growth Habit: Perennial
Vernation: Rolled
Node: Glabrous
Internode: Smooth
Leaf Blade: Glabrous surface, single prominent midvein is present, curved margin, rigid

Termination point: Acute, flat
Leaf sheath margin: Split with overlapping margins
Leaf sheath type: Flattened **Ligule:** Membranous, ciliolate
Auricle: Absent **Collar:** Continuous, narrow

85. ***Eragrostiella bachyphylla***

Distribution: Asia-tropical: India, Bangladesh. In India: East India, West India, and South India. In Gujarat: Saurashtra, North to South Gujarat.
Habitat: In Sandy-rocky soils.

Seedling description:

Growth Habit: Perennial
Vernation: Rolled
Node: Glabrous, angled
Internode: Smooth
Leaf Blade: Glabrous surface, single prominent midvein is present
Termination point: Pointed needle like, flat
Leaf sheath margin: Split with overlapping margins
Leaf sheath type: Round **Ligule:** Membranous, ciliolate
Auricle: Absent **Collar:** Continuous, broad

86. ***Eragrostis cilianensis***

Distribution: Europe: central, southwestern, southeastern, and eastern. Africa: north, Macaronesia, west tropical, west-central tropical, northeast tropical, east tropical, southern tropical, south, middle Atlantic ocean, and western Indian ocean. Asia-temperate: Soviet Middle Asia, Caucasus, western Asia, Arabia, China, Mongolia, and eastern Asia. Asia-tropical: India, Indo-China, Malesia, and Papuasia. Australasia: Australia and New Zealand. Pacific: southwestern and north-central. North America. South America.
Habitat: Along margins of water courses.

Seedling description:

Growth Habit: Annual
Vernation: Rolled
Node: Glabrous, angled
Internode: Smooth

Leaf Blade: Glabrous surface, single prominent midvein is present with other parallel veins

Termination point: Acuminate, keeled

Leaf sheath margin: Close

Leaf sheath type: Round **Ligule:** Fringe of short hairs

Auricle: Absent **Collar:** Continuous, narrow

87. ***Eragrostis ciliaris***

Distribution: Africa: north, Macaronesia, west tropical, west-central tropical, northeast tropical, east tropical, southern tropical, south, and western Indian ocean. Asia-temperate: western Asia, Arabia, and eastern Asia. Asia-tropical: India and Indo-China. North America. South America.

Habitat: Common in all habitats.

Seedling description:

Growth Habit: Annual

Vernation: Rolled

Node: Glabrous, angled

Internode: Smooth

Leaf Blade: Glabrous surface, single prominent midvein is present with other parallel veins

Termination point: Pointed needle like, flat

Leaf sheath margin: Close

Leaf sheath type: Round **Ligule:** Fringe of hairs

Auricle: Absent **Collar:** Continuous, narrow, hairy

88. ***Eragrostis pilosa***

Distribution: Europe: central, southwestern, southeastern, and eastern. Africa: north, west tropical, west-central tropical, northeast tropical, east tropical, southern tropical, south, and western Indian ocean. Asia-temperate: Siberia, Soviet far east, Soviet Middle Asia, Caucasus, western Asia, Arabia, China, Mongolia, and eastern Asia. Asia-tropical: India, Indo-China, Malesia, and Papuasia. Australasia: Australia. North America. South America.

Habitat: Common in all habitats

Seedling description:

Growth Habit: Annual

Vernation: Rolled
Node: Glabrous
Internode: Smooth
Leaf Blade: Glabrous surface, no single prominent midvein is present, all are similar parallel veins
Termination point: Pointed needle like, flat
Leaf sheath margin: Split with overlapping margins
Leaf sheath type: Round **Ligule:** Fringe of hairs
Auricle: Absent **Collar:** Continuous, narrow

89. *Eragrostis tenella*

Distribution: Africa: north, west tropical, west-central tropical, northeast tropical, east tropical, southern tropical, south, middle Atlantic ocean, and western Indian ocean. Asia-temperate: western Asia, Arabia, China, and eastern Asia. Asia-tropical: India, Indo-China, Malesia, and Papuasia. Australasia: Australia. Pacific: southwestern, south-central, northwestern, and north-central. North America: south-central USA, southeast USA, and Mexico. South America.
Habitat: Common in all habitats.

Seedling description:

Growth Habit: Annual
Vernation: Rolled
Node: Glabrous, bearded
Internode: Smooth
Leaf Blade: Glabrous surface, no single prominent midvein is present
Termination point: Pointed needle like, flat
Leaf sheath margin: Close
Leaf sheath type: Round **Ligule:** Fringe of hairs
Auricle: Absent **Collar:** Continuous, narrow, long cilia are present at mouth of collar

90. *Eragrostis tremula*

Distribution: Africa: north, Macaronesia, west tropical, west-central tropical, northeast tropical, east tropical, and southern tropical. Asia-temperate: western Asia and Arabia. Asia-tropical: India and Indo-China. In India: throughout. In Gujarat: throughout, common.

Habitat: Common in all habitats.

Seedling description:

Growth Habit: Annual
Vernation: Rolled
Node: Glabrous, bearded
Internode: Smooth
Leaf Blade: Glabrous surface, no single prominent midvein is present
Termination point: Pointed needle like, flat
Leaf sheath margin: Close
Leaf sheath type: Round **Ligule:** Fringe of hairs
Auricle: Absent **Collar:** Continuous, narrow

91. *Eragrostis unioloides*

Distribution: Africa: west tropical, west-central tropical, and western Indian ocean. Asia-temperate: Arabia, China, and eastern Asia. Asia-tropical: India, Indo-China, Malesia, and Papuasia. Australasia: Australia. Pacific: southwestern, northwestern, and north-central. North America: southeast USA. South America.
Habitat: Common in all habitats

Seedling description:

Growth Habit: Annual
Vernation: Rolled
Node: Glabrous, bearded
Internode: Smooth
Leaf Blade: Pubescent surface, single prominent midvein is present, long hairs are present on the dorsal surface
Termination point: Pointed needle like, flat
Leaf sheath margin: Split with overlapping margins
Leaf sheath type: Round **Ligule:** Fringe of short hairs
Auricle: Absent **Collar:** Continuous, narrow

92. *Eragrostis viscosa*

Distribution: Africa: west tropical, west-central tropical, northeast tropical, east tropical, southern tropical, south, and middle Atlantic ocean. Asia-temperate: Arabia. Asia-tropical: India, Indo-China, and Malesia. North America: Mexico. South America: Mesoamericana, northern South America, and western South America.

Habitat: Common in sandy soils

Seedling description:

Growth Habit: Annual
Vernation: Rolled
Node: Glabrous, bearded
Internode: Smooth
Leaf Blade: Glabrous surface, no single prominent midvein is present
Termination point: Pointed needle like, flat
Leaf sheath margin: Split with overlapping margins
Leaf sheath type: Round **Ligule:** Fringe of hairs
Auricle: Absent **Collar:** Continuous, narrow, longcilia are present at mouth of collar

93. *Sporobolus coromardelianus*

Distribution: Africa: northeast tropical, east tropical, southern tropical, south, and western Indian ocean. Asia-temperate: western Asia, Arabia, and China. Asia-tropical: India, Indo-China, Malesia, and Papuasia. Australasia: Australia. In India: throughout in plains and In Gujarat: throughout, common.

Habitat: In shady areas, moist soil, waste lands.

Seedling description:

Growth Habit: Annual
Vernation: Rolled
Node: Glabrous
Internode: Smooth
Leaf Blade: Pubescent surface, single prominent midvein is present, few tuberculated hairs are present on the ventral surface
Termination point: Acuminate, flat
Leaf sheath margin: Split with overlapping margins
Leaf sheath type: Round **Ligule:** Fringe of hairs
Auricle: Absent **Collar:** Continuous

94. *Sporobolus diander*

Distribution: Africa: Macaronesia, northeast tropical, and western Indian ocean. Asia-temperate: Arabia, China, and eastern Asia. Asia-tropical:

India, Indo-China, Malesia, and Papuasia. Pacific: southwestern, south-central, northwestern, and north-central. South America: Mesoamericana. In India: throughout. In Gujarat: throughout.
Habitat: In open grass lands

Seedling description:

Growth Habit: Annual
Vernation: Rolled
Node: Glabrous
Internode: Smooth
Leaf Blade: Pubescent surface, single prominent midvein is present, few hairs are present on the dorsal surface specially towards the base
Termination point: Pointed needle like, flat
Leaf sheath margin: Split with overlapping margins
Leaf sheath type: Round **Ligule:** Fringe of hairs
Auricle: Absent **Collar:** Continuous

95. *Sporobolus indicus*

Distribution: Europe: southwestern and southeastern. Africa: Macaronesia and western Indian ocean. Pacific: southwestern and north-central. North America: southeast USA and Mexico. South America: Mesoamericana, Caribbean, northern South America, western South America, Brazil, and southern South America. Antarctic: Subantarctic islands.
Habitat: along roadsides and in grasslands.

Seedling description:

Growth Habit: Perennial
Vernation: Folded
Node: Glabrous
Internode: Smooth
Leaf Blade: Pubescent surface, single prominent midvein is present
Termination point: Pointed needle like, flat
Leaf sheath margin: Split with overlapping margins
Leaf sheath type: Round **Ligule:** Fringe of hairs
Auricle: Absent **Collar:** Continuous

96. *Urochondra setulosa*

Distribution: Africa: northeast tropical. Asia-temperate: Arabia. Asia-tropical: India.

Habitat: Grow in salty areas.

Seedling description:

Growth Habit: Perennial
Vernation: Folded
Node: Pubescent, bearded, deposition of salt
Internode: Pubescent
Leaf Blade: Pubescent surface, no single prominent midvein is present, all parallel veins are present and in between veins deposition of salt is present
Termination point: Pointed needle like, flat
Leaf sheath margin: Close
Leaf sheath type: Round **Ligule:** Fringe of hairs
Auricle: Absent **Collar:** Continuous, narrow

97. *Tragus biflorus* (According to Kew this name is illegimate name but according to Gujarat flora this name is true and we follow Gujarat flora so we kept this name)

Distribution: Europe: central, southwestern, southeastern, and eastern. Africa: north, Macaronesia, west tropical, west-central tropical, northeast tropical, east tropical, southern tropical, and south. Asia-temperate: Soviet Middle Asia, Caucasus, western Asia, Arabia, and China. North America. South America.

Seedling description:

Growth Habit: Annual
Vernation: Rolled
Node: Pubescent
Internode: Smooth
Leaf Blade: Pubescent surface, no single prominent midvein is present, all veins are parallel, serrate margin
Termination point: Acute, flat
Leaf sheath margin: Close
Leaf sheath type: Round **Ligule:** Fringe of hairs
Auricle: long, clasping **Collar:** Continuous

98. ***Zoysia matrella***

Distribution: Africa: west tropical and western Indian ocean. Asia-temperate: China and eastern Asia. Asia-tropical: India, Indo-China, Malesia, and Papuasia. Australasia: Australia. Pacific: southwestern, south-central, northwestern, and north-central. North America.
Habitat: Lawn grass, escape from garden.

Seedling description:

Growth Habit: Perennial
Vernation: Rolled
Node: Glabrous
Internode: Smooth
Leaf Blade: Pubescent surface, no single prominent midvein is present, all veins are parallel, folded leaf blade
Termination point: Acuminate, keeled
Leaf sheath margin: Split with overlapping margins
Leaf sheath type: Round **Ligule:** Membranous, ciliolate
Auricle: Absent **Collar:** Continuous, broad

2.6 IDENTIFICATION KEY

Based on the characteristic features of identification a diagnostic key for the studied species has been prepared.

Membranous ligule .... Group I
Hairy ligule .... Group II
No ligule.. Group III

Group I

1. Folded vernation **2**
1. Rolled vernation **19**
2. Single prominent midvein absent.................... ***Arthraxon lanceolatus***
2. Single prominent midvein present **3**
3. Flattened leaf sheath type.. **4**
3. Round leaf sheath type.. **12**
4. Pubescent nodal region .... ***Hackelochloa granularis***
4. Glabrous nodal region **5**
5. Short, rudiment auricle.................................. ***Heteropogon triticeous***

5. Auricle absent **6**
6. Glabrous leaf blade surface **7**
6. Pubescent leaf blade surface **9**
7. Narrow collar region .... ***Triplopogon ramosissimus***
7. Divided collar region **8**
8. Termination point is acute .... ***Isilema laxum***
8. Termination point is acuminate .... ***Acrachne racemosa***
9. Leaf sheath margin open .... ***Chloris barbata***
9. Leaf sheath margin is split with overlapping margins **10**
10. Termination point is acuminate .... ***Heteropogon contortus* var. *typicus***
10. Termination point is acute **11**
11. Annual growth habit .... ***Eleusine indica***
11. Perennial growth habit .... ***Heteropogon contortus* var. *geninus***
12. Annual growth habit **13**
12. Perennial growth habit **14**
13. Divided collar region .... ***Capillepedium hugelii***
13. Continuous collar region .... ***Chloris virgata***
14. Keeled termination point .... ***Heteropogon ritcheii***
14. Flat termination point **15**
15. Short, rudiment auricle .... ***Sorghum halepense***
15. Auricle absent **16**
16. Pubescent nodal region **17**
16. Glabrous nodal region **18**
17. Open leaf sheath margin .... ***Imperata cylindrica***
17. Leaf sheath margin is split with overlapping margins .... ***Sorghum purpureoseri-ceum***
18. Pointed needle like termination point .... ***Desmostachya bipinnata***
18. Acuminate termination point .... ***Panicum maximum***
19. Perennial growth habit **20**
19. Annual growth habit **37**
20. Pubescent nodal region **21**
20. Glabrous nodal region **25**
21. Flattened leaf sheath type .... ***Themeda cymbaria***
21. Round leaf sheath type **22**
22. Short, rudiment auricle .... ***Dicanthium annulatum***

22. Auricle absent **23**
23. Acuminate termination point .... ***Dicanthium caricosum***
23. Acute termination point **24**
24. Glabrous leaf blade surface.................................***Paspalum distichum***
24. Pubescent leaf blade surface .... ***Dactyloctenium aegyptium***
25. Flattened leaf sheath type... **26**
25. Round leaf sheath type.. **28**
26. Leaf sheath margin is open .... ***Chrysopogon fulvus***
26. Leaf sheath margin is split with overlapping margins **27**
27. Auricle is absent...***Eragrostiella bifaria***
27. Short, rudiment auricle.................................... ***Paspalum vaginatum***
28. Keeled termination point ....***Zoysia matrella***
28. Flat termination point.. **29**
29. Leaf sheath margin close ....***Ophiuros exaltatus***
29. Leaf sheath margin is split with overlapping margin....................... **30**
30. Glabrous leaf blade surface... **31**
30. Pubescent leaf blade surface **33**
31. Acute termination point .... ***Themeda triandra***
31. Pointed needle like termination point **32**
32. Continuous collar region...........................***Eragrostiella bachyphylla***
32. Divided collar region .... ***Cymbopogon martinii***
33. Short, rudiment auricle..***Rottboelia exaltata***
33. Auricle absent **34**
34. Pointed termination point.......................................***Sehima nervosum***
34. Acute to acuminate termination point... **35**
35. Angled nodal region..***Tetrapogon villosus***
35. Bearded nodal region **36**
36. Ligule shape is acute..***Tetrapogon tenellus***
36. Ligule is ciliate... ***Sehima sulcatum***
37. Pubescent leaf blade surface **38**
37. Glabrous leaf blade surface... **49**
38. Flattened leaf sheath type.. **39**
38. Round leaf sheath type.. **42**
39. Leaf sheath margin open. .... ***Bothriochloa pertusa***
39. Leaf sheath margin is split with overlapping margin....................... **40**
40. Continuous collar region..................................... ***Cenchrus biflorus***

40. Divided collar region **41**
41. Ligule is membranous, acute .... ***Setaria tomentosa***
41. Ligule is membranous, ciliate .... ***Setaria verticillata***
42. Short, rudiment auricle .... ***Sachoenefeldia gracilis***
42. Auricle absent **43**
43. Keeled termination point .... ***Melanocenchris jaequemontii***
43. Flat termination point **44**
44. Pubescent internode .... ***Dinebra retroflexa***
44. Glabrous internode **45**
45. Divide collar region **46**
45. Continuous collar region **47**
46. Pubescent nodal region .... ***Digitaria microbachne***
46. Glabrous nodal region .... ***Digitaria granularis***
47. Acuminate termination point .... ***Oropetium villosum***
47. Acute termination point **48**
48. Pubescent, bearded nodal region .... ***Thelepogon elegans***
48. Glabrous, angled nodal region .... ***Digitaria stricta***
49. Short, rudiment auricle **50**
49. Auricle absent **55**
50. Acuminate to pointed termination point **51**
50. Acute termination point **52**
51. Simple nodal region .... ***Apluda mutica***
51. Angled nodal region .... ***Ischaemum pilosum***
52. Pubescent nodal region .... ***Setaria gluca***
52. Glabrous nodal region **53**
53. Bearded nodal region .... ***Ischaemum indicum***
53. Simple nodal region **54**
54. Pubescent collar region .... ***Digitaria longiflora***
54. Glabrous collar region .... ***Paspalum scorbiculatum***
55. Pubescent nodal region **56**
55. Glabrous nodal region **58**
56. Pointed termination point .... ***Isachane globosa***
56. Acute termination point **57**
57. Leaf sheath margin open .... ***Coix lachryma-jobi***
57. Leaf sheath margin is split with overlapping margin .... ***Digitaria ciliaris***

58. Flattened leaf sheath type **59**
58. Round leaf sheath type **61**
59. Leaf sheath margin is split with overlapping margin .... ***Andropogon pumilus***
59. Leaf sheath margin open **60**
60. Continuous, broad collar region .... ***Ischaemum molle***
60. Continuous narrow collar region .... ***Ischaemum rugosum***
61. Keeled termination point .... ***Sehima ischaemoides***
61. Flat termination point .... ***Themeda laxa***

Group II

1. Pubescent nodal region **2**
1. Glabrous nodal region **9**
2. Glabrous surface of leaf lamina .... ***Alloteropis cimiciana***
2. Pubescent surface of leaf lamina **3**
3. Pubescent internode .... ***Urochondra setulosa***
3. Glabrous internode **4**
4. Pointed termination point .... ***Dactyloctenium sindicum***
4. Acute to acuminate termination point **5**
5. Perennial growth habit and folded vernation .. ***Eremopogn foveolatus***
5. Annual growth habit and rolled vernation **6**
6. No single prominent midvein is present **7**
6. Single prominent midvein is present **8**
7. Leaf sheath margin is split with overlapping margin .... ***Bracharia cruciformis***
7. Leaf sheath margin close .... ***Tragus biflorus***
8. Continuous narrow collar region .... ***Bracharia ramosa***
8. Divided collar region .... ***Oplismenus composites***
9. Short, rudiment auricle **10**
9. Auricle absent **12**
10. Leaf sheath margin close .... ***Panicum trypheron***
10. Leaf sheath margin is split with overlapping margins **11**
11. Glabrous leaf blade surface .... ***Panicum miliaceum***
11. Pubescent leaf blade surface .... ***Bracharia reptans***
12. Pubescent leaf blade surface **13**
12. Glabrous leaf blade surface **21**
13. Leaf sheath margin is close **14**

13. Leaf sheath margin is split with overlapping margins **15**
14. Rolled vernation... ***Aristida adscendionsis***
14. Folded vernation ....***Aleuropus lagopoides***
15. Perennial growth habit **16**
15. Annual growth habit... **18**
16. Pointed needle like termination point ....***Sporobolus indicus***
16. Acute to acuminate termination point.. **17**
17. Presence of hairs on the dorsal surface
    of leaf blade..***Paspalidium geniculatum***
17. Hairs are absent on the dorsal surface
    of leaf blade...***Cenchrus ciliaris***
18. Flattened leaf sheath type...................................***Cenchrus setigerus***
18. Round leaf sheath type **19**
19. Acuminate termination point ....***Sporobolus coromandliens***
19. Pointed termination point.. **20**
20. Bearded nodal region ....***Eragrostis unioloides***
20. Simple nodal region ....***Sporobolus diander***
21. Perennial growth habit **22**
21. Annual growth habit.. **25**
22. No single prominent midvein is present
    on the leaf blade ....***Erichloa procera***
22. Single prominent midvein is present... **23**
23. Folded vernation ....***Panicum antidotalae***
23. Rolled vernation... **24**
24. Leaf sheath margin open .... ***Cynadon dactylon***
24. Leaf sheath margin is split with
    overlapping margins... ***Pennisetum setosum***
25. Folded vernation ....***Paspalidium flavidum***
25. Rolled vernation... **26**
26. Single midvein is present on the leaf blade surface......................... **27**
26. No single midvein is present.. **30**
27. Keeled termination point.................................***Echinochloa stagnina***
27. Flat termination point.. **28**
28. Leaf sheath margin is split with
    overlapping margins..***Bracharia distachya***
28. Leaf sheath margin is close **29**

29. Pointed termination point.. ***Eragrostis ciliaris***
29. Acuminate termination point .... ***Eragrostis cilianensis***
30. Simple nodal region .... ***Eragrostis pilosa***
30. Bearded nodal region **31**
31. Leaf sheath margin is split with overlapping margins... ***Eragrostis viscosa***
31. Leaf sheath margin close.. **32**
32. Curved margin of leaf blade.................................. ***Aristida funiculata***
32. Simple margin of leaf blade **33**
33. Long hairs are present at mouth of collar region .... ***Eragrostis tenella***
33. No hairs are present at mouth of collar region...... ***Eragrostis tremula***

Group III

1. Acute, keeled termination point........................ ***Echinochloa colonum***
1. Pointed, flat termination point .... ***Echinochloa crus-galli***

The present study demonstrated that detailed studies of seedling morphology can provide morphological insights for identification of the grass species.

The proposed key is an easy to use field guide for accurately identifying the commonly growing 98 native grass species at its seedling stage itself. The guide also includes photographs of important characteristics for each species and written descriptions which would help one to compare the samples for positive identification.

2.7 CLUSTER ANALYSIS

Cluster analysis was done to see the relationship between different groups of species. Cluster analysis using Square Euclidean distance showed good resolution of the taxa based on the qualitative characters.

The dendrogram based on analysis of 11 characters were recorded comparatively for 98 grass species. The dendrogram tree was separated into 10 different clusters (Figure 108). The formations of clusters were made on the basis of combination of characters. Clusters 1, 2, 3, 6 and 8 formed the first major group while the other clusters viz. 4, 5, 9, 7 and 10 got clubbed in the second major group.

First group consists 45 species while second group consists 53 species. In first group almost all species (82.22%) have membranous ligule but only 8 species (17.78%) were having hairy ligule. Though these species showed difference in ligule character, the other characters were matching with the rest of the group characters and these species formed cluster 3 in first group (Figure 108). Likewise, second group also showed this type of mixture, for example, second group has total 53 species. Out of that around 36 species, for example, 71% species shows hairy ligule, 15 species, for example, 26% species shows membranous ligule and 2 species, for example, 3% species do not have ligule. The species, which have hairy ligule, were forming cluster 5 of the second group.

Cluster 8 and 9 are simplifolius, for example, having only a single species in cluster. *Ischaemum molle* form cluster 8, showed around 90% similarities with the characters while *Thelepogon elegans* form cluster 9, shows only 80% similarities of characters with other species.

Cluster 1, 2 and 6 constitute of the members with membranous ligule. Cluster 1 has 14 species, which showed 92% similarity level. Among them *Eragrostiella bifaria* and *Themeda cymbaria* showed 99% similarities in characters. Cluster 2 has 11 species, showed 91% similarity level. Among them *Sehima sulcatum* and *Dinebra retroflexa* showed 99% similarities in characters. In this cluster only *Sachoenefeldia gracilis* showed 91% similarity level while others showed more than that. On the basis of dendrogram it can be inferred that, *Panicum trypheron, Aristida funiculata* and *Eragrostis* spp. have hairy ligule and they showed almost 100% similarities in their vegetative characters and formed cluster 3 which is one of the exceptions among the first major group. *Hackelochloa, Iseilema, Melanocenchris* and *Zoysia* showed 91% similarities in their characters and they were forming cluster 6. Among them *Hackelochloa granularis* showed 91% similarity with other species of same cluster which is lowest among the cluster while other species of cluster 6 has 95% similarities.

Cluster 4 has highest number of species (30 species) and it showed around 94% similarities. *Cenchrus ciliaris, Eriochloa procera* and *Paspalidium geminatum* showed 100% similarities while *Oplismenous, Tragus, Eremopogon, Cynadon, Alloteropis, Paspalidium flavidum* and *Sporobolus* spp. in the same cluster showed around 95% similarities in their vegetative characters. In this cluster 7 species have membranous ligule while others have hairy ligule.

Cluster 5 has 16 species and showed 92.7% similarity level. In this cluster 8 species showed membranous ligule while others have hairy ligule and one species, for example, *Echinochloa crus-galli* do not have ligule but in cluster it showed around 98% similarity level. Cluster 7 has 4 species and showed 97.6% similarity level. In this two subclusters were formed and they showed 98.3% similarity level.

Among 10 clusters, clusters 8 and 9 has 0.0 values from the centroid while other clusters have between 1.3 and 2.3 from centroid (Table 3).

KEYWORDS

- **Auricle**
- **Collar**
- **Cluster analysis**
- **Ligule**
- **Morphology**
- **Seedling**
- **Vegetative features**

REFERENCES

1. Blatter, E., McCann, C. *The Bombay Grasses*. Imperial Council of Agriculture Research. 1936.
2. Bradley, K. W., Fishel, F. *Integrated pest management: Identifying grass seedlings*; University of Missouri Extension. 2010.
3. Burr, S. F., Turner, D. M. *British economic grasses, their employment by the leaf anatomy*. Edward Arnold and Co. London. 1933.
4. Carrier, L. The identification of grasses by their vegetative characters. USDA Bulletin. 32, 1917.
5. Durgan, B. *Broadleaf and grass weed seedling identification keys, Communication and educational technology services*, University of Minnesota extension. 1999.
6. Fenner, M. Seedlings. *New phytologist* 1987, 106(1), 35–47.
7. Fishel, F. Identifying Grass seedling, In: *Integrated Pest Management Manual*. University of Missouri Extension Columbia. 5–18, 2004.
8. Harries, W. F. *The identification of some of the more common native Oklahoma grasses by vegetative characters*, Oklahoma native plant record. Vol. 10. 2010.

9. Henning, E. *Table for the identification of grasses and legumes in the nonflowering condition*, Translated from the Swedish by, F. V. Meissner, Springer. Berlin. 1930.
10. Hitchock, C. L. *A key to the grasses of Montana*. Publ. Missoula, Montana. 1936.
11. Jensen, P. Notes on Malaysian grasses. *Reinwardia* 1953, 2, 225–350.
12. Jian-Guo, H., Clifford, H. T., Jia Shen-Xiu, W. P. A cluster analysis of seedling characters of Gramineae. *Acta Pytotax. Sin.*, 1993, 31(6), 517–532.
13. Kuwabara, Y. The first seedling leaf in grass systematics. *J Jap Bot.* 1960, 35, 139–145.
14. Kuwabara, Y. On the shape and the direction of leaves of grass seedlings. *J Jap Bot.* 1961, 36, 368–373.
15. Looman, J. Prairie grasses identified and described by vegetative characters, Agriculture Canada, Pub. 1413. 1982.
16. McAlpine, A. N. *How to know grasses by their leaves*. The Darien press. Edinburgh. 1890.
17. Nowosad, F. S., Newton Swales, D. E., Dore, W. G. The identification of certain native and naturalized hay and pasture grasses by their vegetative characteristics. Macdonald College Technical Bulletin No. 16. Macdonald College, Que. 78 p. 1946.
18. Phillips, C. E. *Some grasses of the Northeast: a key to their identification by their vegetative characters*. Newark, Delaware: University of Delaware, Agriculture Experiment Station. 1962.
19. Prosser, M. Key for the identification of common grass species using vegetative characters. 2009.
20. *The Plant List,* Published on the Internet; http://www.theplantlist.org/ (accessed 1st January). Version 1. 2010.
21. Tillich, H. J. Seedling Diversity and the Homologies of Seedling Organs in the Order Poales (Monocotyledons). *An. Bot.* 2007, 100(7), 1413–1429.
22. Undersander, D., Casler, M., Cosgrove, D. *Identifying pasture grasses*, Cooperative Extension publications, University of Wisconsin-Extension, 1996.
23. Ward, H. M. *Grasses*. Cambridge Univ. Press, London, 1901.
24. Whyte, J. H. The recognition of some agricultural grasses by their vegetative characters. Trans. And Proc. Bot. Soc. Edin. 1930, 30, 206–208.
25. Wintle, B., Moore, G., Nichols, P. *Identifying sub-tropical grass seedlings.* Bulletin No 4775, Department of Agriculture and Food Western Australia, South Perth, Australia, 2009.

APPENDIX

SYNONYMS

A list of the studied grass species along with their known synonyms are given below: (Source: http://www.theplantlist.org/, The Bombay Grasses)

Family: Poaceae

GROUP: PANICOIDEAE

TRIBE: MAYDEAE

1. ***Coix lachryma-jobi* L.**

Synonyms: *Coix ma-yuen* Rom. Caill., *Coix stenocarpa* (Oliv.) Balansa

2. ***Chionachne koenigii*(Spreng.) Thwaites**

Synonyms: *Chionachne gigantea* (J. Koenig) Veldkamp.

TRIBE: ANDROPOGONEAE

3. ***Andropogon pumilus* Roxb.**

Synonyms: *Andropogon demissus* Steud., *Andropogon pachyarthrus* Hack., *Arthrolophis pumilus* (Roxb.) Chiov.,*Sorghum demissum* (Steud.) Kuntze

4. ***Apluda mutica* L.**

Synonyms: *Andropogon aristatus*(L.) Raspail, *Andropogon glaucus* Retz, *Andropogon involucratus* J.Koenig ex Steud., *Apluda aristata* L., *Apluda aristata* var. *ciliata*(Anderson) S.K.Jain, *Apluda aristata* var. *jainii* S.K.Jain, *Apluda blatteri* Sur, *Apluda ciliata* Andersson, *Apluda communis* Nees & Arn., *Apluda communis* Nees, *Apluda cumingii* Buse, *Apluda geniculata* Roxb., *Apluda gigantea* Spreng., *Apluda glauca* Schreb., *Apluda humilis* Kunth, *Apluda inermis* Regel, *Apluda kobila* Buch.-Ham.

Ex Nees, *Apluda microstachya* Nees, *Apluda mucronata* Steud., *Apluda mutica* subsp. *aristata* (L.) Babu, *Apluda mutica* var. *aristata* (L.) Hack. ex K.Bakker, *Apluda mutica* subsp. *aristata* (L.) R.D. Gaur, *Apluda mutica* var. *aristata* (L.) Pilg., *Apluda mutica* var. *major*(Hack.) S.K. Jain, *Apluda pedicellata* Buse, *Apluda rostrata* Arn. & Nees, *Apluda scabra* Andersson, *Apluda varia* Hack., *Apluda varia* subsp. *aristata*(L.) Hack., *Apluda varia* var. *ciliata*(Andersson) Hack., *Apludvaria* var. *humilis*(J. Presl) Hack, *Apluda varia* var. *intermedia* Hack., *Apludvaria* var. *major* Hack., *Apluda villosa* Schreb., *Apluda varia* subsp. *mutica*(L.) Hack., *Calamina gigantea* P.Beauv., *Calamina humilis* J.Presl, *Calamina mutica*(L.) P.Beauv., *Tripsacum giganteum* (P.Beauv.) Raspail, *Xerochloa latifolia* Hassk.

5. *Arthraxon lanceolatus* (Roxb.) Hochst.

Synonyms: *Andropogon appendiculatus* var.*serrulatus* (Link) Nees, *Andropogon lanceolatus* Roxb., *Andropogon serrulatus* Link, *Arthraxon deccanensis* S.K.Jain, *Arthraxonlanceolatus*f.*glaberrimus*Chiov., *Arthraxonlanceolatus*f.*puberulus*Chiov., *Arthraxonlanceolatus*var.*puberulus* (Chiov.) Mattei, *Batratherum lanceolatum* Nees

6. *Bothriochloa pertusa* (L.) A.Camus

Synonyms: *Amphilophis pertusa* (L.) Stapf, *Amphilophis pertusa* var. *barbata* (A. Camus) E.G. Camus & A. Camus, *Andropogon armillaris* Willd. ex Steud., *Andropogon pertusus* (L.) Willd., *Andropogon pertusus* var. *barbatus* A.Camus, *Andropogon pertusus* var. *wightii* Hack., *Andropogon pertusus* var. *maroccanus* Maire, *Bothriochloa nana* W.Z. Fang., *Bothriochloa pertusa* var. *maroccana* (Maire) Maire, *Dichanthium pertusum* (L.) Clayton, *Holcus pertusus* L., *Lepeocercis pertusa* (L.) Hassk., *Sorghum pertusum* (L.) Kuntze

7. *Capillipedium hugelii* (Hack.) A.Camus

Synonyms: *Andropogon huegelii* Hack., *Andropogon huegeli* var. *foetidus* Hack. ex Lisboa, *Andropogon schmidii* Hook.f., *Capillipedium foetidium* (Lisboa) Raiz. & Jain, *Capillipedium foetidum* (Lisboa) Raizada & S.K.Jain, *Capillipedium hugelii* (Hack.) Stapf, *Capillipedium parviflorum* f. *huegelii* (Hack.) Roberty, *Capillipedium schmidii* (Hook.f.) Stapf, *Dichanthium huegelii* (Hack.) S.K.Jain & Deshp., *Sorghum huegelii* (Hack.) Kuntze

8. *Chrysopogon fulvus* (Spreng.) Chiov.

Synonyms: *Andropogon monticola* Schult. & Schult.f., *Andropogon sprengelii* Kunth, *Chrysopogon montanus* Trin., *Chrysopogon monticola* (Schult. & Schult.f.) Haines, *Pollinia fulva* Spreng., *Sorghum monticola* (Schult.) Kuntze

9. *Cymbopogon martini* (Roxb.) W.Watson

Synonyms: *Andropogon martini* Roxb., *Andropogon schoenanthus* var. *martini*(Roxb.) Benth.

10. *Dichanthium annulatum* (Forssk.) Stapf

Synonyms: *Andropogon annulatus* Forssk., *Andropogon annulatus* Forssk. ex f. Schmidt, *Andropogon annulatus* var. *decalvatus* Hack., *Andropogon comosus* Link, *Andropogon garipensis* Steud., *Andropogon ischaemum* Roxb. ex Wight, *Andropogon nodosus* (Willd.) Nash, *Andropogon obtusus* Nees ex Hook. & Arn., *Andropogon scandens* Roxb., *Bothriochloa tuberculata* W.Z.Fang, *Dichanthium annulatum* var. *bullisetosum* B.S.Sun & S.Wang, *Dichanthium annulatum* var *decalvatum*(Hack.) Maire & Weiller, *Dichanthium nodosum* Willemet, *Gymnandropogon annulatum* (Forssk.) Duthie, *Lepeocercis annulata* (Forssk.) Nees, *Sorghum annulatum* (Forssk.) Kuntze

11. *Dichanthium caricosum* (L.) A.Camus

Synonyms: *Andropogon annulatus* var. *subrepens* Hack., *Andropogon binatus* Roxb., *Andropogon caricosus* L., *Andropogon curvatus* Russell ex Wall., *Andropogon depressus* Steud., *Andropogon filiformis* Pers., *Andropogon serratus* Retz., *Andropogon tenellus* Roxb., *Apocopis pallidus* Hook.f., *Dichanthium caricosum* var. *theinlwinii* (Bor) de Wet & Harlan, *Dichanthium pallidum* (Hook.f.) Stapf ex C.E.C.Fisch., *Dichanthium theinlwinii* Bor, *Heteropogon concinnus* Thwaites, *Heteropogon tenellus* (Roxb.) Schult., *Lepeocercis serrata* Trin., *Sorghum caricosum* (L.) Kuntze

12. *Dimeria orinthopoda* Trin.

Synonyms: *Dimeria filiformis* Hochst., *Dimeria stipaeformis* Miq., *Andropogon filiformis* Roxb., *Andropogon roxburghianus* Schult., *Psilostachys filiformis* Dalz. & Gibs.

13. *Hackelochloa granularis* (L.) Kuntze

Synonyms: *Cenchrus granularis* L., *Manisuris granularis* (L.) L.f., *Manisuris granularis* (L.) Sw., *Manisuris granularis* L., *Manisuris polystachya* P.Beauv., *Mnesithea granularis* (L.) de Koning & Sosef, *Rottboellia granularis* (L.) Roberty, *Rytilix glandulosa* Raf., *Rytilix granularis* (L.) Skeels, *Tripsacum granulare* (L.) Raspail

14. *Heteropogon contortus* var. *contortus* sub var. *typicus* Blatt. & McCann

Synonyms: *Andropogon contortus* var. *genuinus* Subvar. *typicus* Hack. Monogr. Androp., *Heteropogon hirtus* Pers.

15. *Heteropogon contortus* var. *contortus* sub var. *genuinus* Blatt. & McCann

Synonyms: *Andropogon contortus* var. *genuinus* Subvar. *hispidissimus* Hack. Monogr. Androp., *Andropogon besukiensis* Steud., *Heteropogon hispidissimus* Hochst.

16. *Heteropogon ritcheii* (Hook.f.) Blatt. & McCann

Synonyms: *Andropogon ritchiei* Hook.f.

17. *Heteropogon triticeus* (R.Br.) Stapf ex Craib

Synonyms: *Andropogon ischyranthus* Steud., *Andropogon liananthus* Steud., *Andropogon lianatherus* Steud., *Andropogon segaenensis* Steud., *Andropogon segaensis* Steud., *Andropogon triticeus* R.Br., *Heteropogon insignis* Thwaites, *Heteropogon ischyranthus* (Steud.) Miq., *Heteropogon lianatherus* (Steud.) Miq., *Sorghum triticeum* (R.Br.) Kuntze

18. *Imperata cylindrica* (L.) Raeusch.

Synonyms: *Calamagrostis lagurus* Koeler, *Imperata allang* Jungh., *Imperata angolensis* Fritsch, *Imperata arundinacea* Cirillo, *Imperata arundinacea* var. *africana* Andersson, *Imperata arundinacea* var. *europaea* Andersson, *Imperata arundinacea* var. *glabrescens* Büse, *Imperata arundinacea* var. *indica* Andersson, *Imperata arundinacea* var. *koenigii* (Retz.) Benth., *Imperata arundinacea* var. *latifolia* Hook.f., *Iperata arundinacea* var. *pedicellata* (Steud.) Debeaux, *Imperata cylindrica* (L.) P. Beauv., *Imperata cylindrica* var. *africana* (Andersson) C.E.Hubb., *Imperata* *cylindrica* var. *europaea* (Andersson) Asch. & Graebn., *Imperata cylindrica* var. *genuina* A. Camus, *Imperata cylindrica* subsp. *koenigii* (Retz.) Tzvelev, *Imperata cylindrica* var. *koenigii* (Retz.) Pilg., *Imperata cylindrica* var. *latifolia* (Hook.f.) C.E.Hubb., *Imperata cylindrica* var. *major* (Nees) C.E.Hubb., *Imperata cylindrica* f. *pallida* Honda, *Imperata cylindrica* var. *parviflora* Batt. & Trab., *Imperata cylindrica* var. *pedicellata* (Steud.) Debeaux, *Imperata cylindrica* var. *thunbergii* (Retz.) T.Durand & Schinz, *Imperata dinteri* Pilg., *Imperata filifolia* Nees ex Steud., *Imperata koenigii* (Retz.) P.Beauv., *Imperata koenigii* var. *major* Nees, *Imperata laguroides* (Pourr.) J.Roux, *Imperata latifolia* (Hook.f.) L.Liou, *Imperata pedicellata* Steud., *Imperata praecoquis* Honda, *Imperata robustior* A.Chev., *Imperata sieberi* Opiz, *Imperata thunbergii* (Retz.) Nees, *Imperata thunbergii* P.Beauv., *Imperata thunbergii* (Retz.) Roem. & Schult., *Lagurus cylindricus* L., *Saccharum cylindricum* (L.) Lam., *Saccharum cylindricum* var. *europaeum* Pers., *Saccharum europaeum* Pers., *Saccharum koenigii* Retz., *Saccharum laguroides* Pourr., *Saccharum sisca* Cav., *Saccharum spicatum* J.Presl, *Saccharum thunbergii* Retz.

19. *Ischaemum indicum* (Houtt.) Merr.

Synonyms: *Andropogon amaurus* Buse, *Andropogon diversiflorus* Steud., *Andropogon firmandus* Steud., *Eulalia amaura* (Buse) Ohwi, *Eulalia nana* Keng & S.L.Chen, *Eulalia praemorsa* (Steud.) Stapf ex Ridl., *Phleum indicum* Houtt., *Pogonatherum amaurum* (Buse) Roberty, *Pollinia diversiflora* (Steud.) Nash, *Pollinia praemorsa* Nees ex Steud., *Pollinia praemorsa* Nees, *Polytrias amaura* (Buse) Kuntze, *Polytrias amaura* var. *nana* (Keng & S.L.Chen) S.L.Chen, *Polytrias diversiflora* (Steud.) Nash, *Polytrias indica* (Houtt.) Veldkamp, *Polytrias praemorsa* (Steud.) Hack., *Polytrias racemosa* (Nees) Hack.

20. *Ischaemum molle* Hook.f.

No synonyms recorded

21. *Ischaemum pilosum* (Willd.) Wight

Synonyms: *Andropogon afer* J.F.Gmel., *Andropogon brachyatherus* Hochst., *Andropogon fazoglensis* (Chiov.) Chiov., *Andropogon intumescens* Pilg., *Andropogon matteodanus* Chiov., *Andropogon pilosus* Klein ex Willd., *Arthrolophis fazoglensis* Chiov., *Ischaemum afrum* (J.F.Gmel.)

Dandy, *Ischaemum brachyatherum* (Hochst.) Fenzl ex Hack., *Ischaemum glaucostachyum* Stapf, *Spodiopogon pilosus* (Willd.) Nees ex Steud.

22. *Ischaemum rugosum* Salisb.

Synonyms: *Andropogon arnottianus* Steud., *Andropogon griffithsiae* Steud., *Andropogon monostachyus* Steud., *Andropogon rugosus* Steud., *Andropogon segetus* (Trin.) Steud., *Andropogon tong-dong* Steud., *Apluda rugosa* Russell ex Wall., *Colladoa distachya* Cav., *Ischaemum colladoa* Spreng., *Ischaemum royleanum* Miq., *Ischaemum rugosum* Gaertn., *Ischaemum rugosum* var. *distachyum* (Cav.) Merr., *Ischaemum rugosum* var. *nanum* A.Camus, *Ischaemum rugosum* var. *segetum* (Trin.) Hack., *Ischaemum segetum* Trin., *Ischaemum tashiroi* Honda, *Meoschium arnottianum* Nees, *Meoschium griffithii* Nees & Arn., *Meoschium griffithsiae* Nees ex Steud., *Meoschium rugosum* (Salisb.) Nees, *Meoschium rugosum* Wall., *Tripsacum distachyum* (Cav.) Poir., *Tripsacum distichum* Raspail

23. *Iseilema laxum* Hack.

Synonyms: *Iseilema jainiana* P.Umam. & P.Daniel

24. *Ophiuros exaltatus* (L.) Kuntze

Synonyms: *Aegilops exaltata* L., *Mnesithea exaltata* (L.) Skeels, *Ophiuros corymbosus* (L.f.) C.F.Gaertn., *Ophiuros tongcalingii* (Elmer) Henrard, *Rottboellia corymbosa* L.f., *Rottboellia punctata* Retz., *Rottboellia tongcalingii* Elmer

25. *Rottboellia exaltata* Linn. f.

Synonyms: *Aegilops exaltata* L., *Rottboellia exaltata* var. *genuine* Schweinf., *Rottboellia exaltata f. arundinacea* Hack., *Rottboellia arundinacea* Hochst ex A.Rich., *Stegosia cochinchinensis* Lour. Fl., *Stegosia exaltata* Nash, *Manisuris exaltata* (L.) Kuntze, *Rottboellia cochinchinensis(Lour.) W.D. Clayton*

26. *Saccharum spontanum* L.

Saccharum spontanum var. *aegyptiacum* Hack., *Saccharum semidecumbens* Roxb., *Saccharum canaliculatum* Roxb., *Saccharum chinense* Nees Hook, *Saccharum spontanum* Beauv., *Saccharum biflorum* Forsk., *Imperata spontanea* Beauv.

27. *Sehima ischaemoides* Forssk.

Synonyms: *Andropogon inscalptus* (Hochst.) Schweinf., *Andropogon lineatus* Steud., *Andropogon rhynchophorus* Stapf, *Andropogon sehima* (Spreng.) Steud., *Calamina themeda* P.Beauv., *Ischaemum inscalptum* (Hochst.) A.Rich., *Ischaemum laxum* var. *inscalptum* (Hochst.) Hack., *Ischaemum sehima* Spreng, *Sehima inscalptum* Hochst., *Sehima kotschyi* Hochst., *Sehima kotschyi* var. *schangulicum* Hochst.

28. *Sehima nervosum* (Rottler) Stapf

Synonyms: *Andropogon brownii* Kunth, *Andropogon macrostachyus* (A.Rich.) Schweinf., *Andropogon nervosus* Rottler, *Andropogon nervosus* Rottler ex Roem. & Schult., *Andropogon philippinensis* Merr., *Andropogon robertianus* Steud., *Andropogon sorghum* var. *nervosus* (Rottler ex Roem. & Schult.) Stapf, *Andropogon striatus* Klein ex Willd., *Andropogon tacazensis* Steud, *Hologamium nervosum* (Rottler) Nees, *Ischaemum laxum* R.Br., *Ischaemum macrostachyum* Hochst. ex A.Rich., *Ischaemum nervosum* (Rottler) Thwaites, *Ischaemum striatum* (Willd.) Domin, *Ischaemum striatum* var. *stenophyllum* Domin, *Pollinia striata* (Willd.) Spreng., *Schizachyrium striatum* (Willd.) Nees, *Sehima macrostachyum* Hochst. ex Hack.

29. *Sehima sulcatum* (Hack.) A.Camus

Synonyms: *Ischaemum sulcatum* Hack.

30. *Sorghum halepense* (L.) Pers.

Synonyms: *Andropogon arundinaceus* Scop., *Andropogon arundinaceus* Scop. ex Schrad., *Andropogon avenaceus* Kunth, *Andropogon crupina* Kunth, *Andropogon decolorans* Kunth, *Andropogon dubitatus* Steud., *Andropogon dubius* K.Koch ex B.D.Jacks., *Andropogon halepensis* (L.) Brot., *Andropogon halepensis* subsp. *anatherus* Piper, *Andropogon halepensis* var. *anatherus* Piper, *Andropogon halepensis* var. *effusus* Stapf, *Andropogon halepensis* var. *genuinus* Stapf, *Andropogon halepensis* var. *muticus* (Hack.) Asch. & Graebn., *Andropogon halepensis* subsp. *muticus* (Hack.) Piper, *Andropogon laxum* Roxb., *Andropogon sorghum* var. *exiguum* (Forssk.) Piper, *Andropogon sorghum* subsp. *exiguus* (Forssk.) Piper, *Andropogon sorghum* var. *glaberrimus* Hack., *Andropogon sorghum* subsp. *halepensis* (L.) Hack., *Andropogon sorghum* var. *halepensis* (L.)

Hack., *Andropogon sorghum* var. *perennis* Bertoni, *Blumenbachia halepensis* (L.) Koeler, *Holcus decolorans* Willd., *Holcus exiguus* Forssk., *Holcus halepensis* L., *Holcus sorghum* subsp. *exiguus* (Forssk.) Hitchc., *Milium halepense* (L.) Cav., *Rhaphis halepensis* (L.) Roberty, *Sorghum bicolor* subsp. *halepense* Barkworth *et al.*, *Sorghum bicolor* subsp. *halepense* (L.) de Wet & Huckabay, *Sorghum crupina* Link, *Sorghum decolorans* (Willd.) Roem. & Schult., *Sorghum dubium* K.Koch, *Sorghum halepense* var. *anatherum* Barkworth *et al.*, *Sorghum halepense* var. *latifolium* Willk. & Lange, *Sorghum halepense* f. *muticum* (Hack.) C.E.Hubb., *Sorghum halepense* var. *muticum* (Hack.) Grossh., *Sorghum halepense* var. *muticum* (Hack.) Hayek, *Sorghum halepense* var. *muticum* (Hack.) Parodi, *Sorghum saccharatum* var. *halepense* (L.) Kuntze, *Sorghum schreberi* Ten., *Trachypog avenaceus* Nees

31. *Sorghum purpureo-sericeum* (A.Rich.) Schweinf. & Asch.

Synonyms: *Andropogon pappii* Gand., *Andropogon purpureosericeus* Hochst. ex A.Rich., *Andropogon purpureosericeus* var. *calomelas* Hack., *Andropogon purpureosericeus* var. *pallidior* Hack., *Sarga purpureosericea* (Hochst. ex A.Rich.) Spangler, *Sorghum deccanense* Stapf ex Raizada, *Sorghum dimidiatum* Stapf, *Sorghum purpureosericeum* subsp. *deccanense* Garber, *Sorghum purpureosericeum* subsp. *dimidiatum* (Stapf) Garber

32. *Thelepogn elegans* Roth

Synonyms: *Andropogon princeps* A.Rich., *Jardinea abyssinica* Steud., *Meoschium elegans* (Roem. & Schult.) Arn. & Nees, *Rhiniachne princeps* (A.Rich.) Hochst. ex Steud., *Rhytachn princeps* (A.Rich.) T.Durand & Schinz, *Sehima elegans* (Roem. & Schult.) Roberty

33. *Themeda cymbaria* Hack.

Synonyms: *Themeda serratifolia* Roberty

34. *Themeda laxa* A.Camus

Synonyms: *Anthistiria laxa* Andersson, *Themeda forskalii* var. *laxa* (Andersson) Hack., *Themeda triandra* var. *laxa* (Andersson) Noltie

35. *Themeda triandra* Forssk.

Synonyms: *Andropogon tenuipedicellatus* Steud., *Anthistiria argentea* Nees, *Anthistiria arguens* var. *japonica* (Willd.) Andersson, *Anthistiria*

australis R.Br., *Anthistiria brachyantha* Boiss., *Anthistiria caespitosa* Andersson, *Anthistiria ciliata* var. *brachyantha* (Boiss.) Boiss., *Anthistiria ciliata* var. *burchellii* (Hack.) Hack., *Anthistiria ciliata* var. *mollicoma* Nees, *Anthistiria ciliata* var. *syriaca* (Boiss.) Boiss., *Anthistiria cuspidata* Andersson, *Anthistiria depauperata* Andersson, *Anthistiria desfontainii* Kunth, *Anthistiria forskalii* Kunth, *Anthistiria glauca* Desf., *Anthistiria imberbis* Retz., *Anthistiria imberbis* var. *argentea* (Nees) Stapf, *Anthistiria imberbis* var. *burchellii* (Hack.) Stapf, *Anthistiria imberbis* var. *mollicoma* (Nees) Stapf, *Anthistiria imberbis* var. *roylei* Hook.f., *Anthistiria japonica* Willd., *Anthistiria paleacea* (Poir.) Ball, *Anthistiria polystachya* Roxb., *Anthistiria puberula* Andersson, *Anthistiria punctata* Hochst. ex A.Rich., *Anthistiria subglabra* Buse, *Anthistiria syriaca* Boiss., *Anthistiria vulgaris* Hack., *Calamina imberbis* (Retz.) P.Beauv., *Calamina imberbis* (Retz.) Roem. & Schult., *Stipa arguens* var. *japonica* Houtt., *Stipa paleacea* Poir., *Themeda australis* (R.Br.) Stapf, *Themeda barbinodis* B.S.Sun & S.Wang, *Themeda brachyantha* (Boiss.) Trab., *Themeda forskalii* Hack., *Themeda forskalii* (Kunth) Hack. ex Duthie, *Themeda forskalii* var. *argentea* (Nees) Hack., *Themeda forskalii* var. *brachyantha* (Boiss.) Hack, *Themeda forskalii* var. *burchellii* Hack., *Themeda forskalii* var. *glauca* Hack., *Themeda forskalii* var. *imberbis* (Retz.) Hack., *Themeda forskalii* var. *major* Hack., *Themeda forskalii* var. *mollissima* Hack., *Themeda forskalii* var. *paleacea* (Poir.) T.Durand & Schinz, *Themeda forskalii* var. *punctata* (Hochst. ex A. Rich.) Hack., *Themeda forskalii* var. *syriaca* (Boiss.) Hack., *Themeda forskalii* var. *vulgaris* (Hack.) Hack., *Themeda glauca* (Desf.) Batt. & Trab., *Themeda imberbis* (Retz.) T.Cooke, *Themeda japonica* (Houtt.) Tanaka, *Themeda japonica* var. *viridiflora* Honda, *Themeda polygama* J.F.Gmel., *Themeda triandra* var. *brachyantha* (Boiss.) Hack., *Themeda triandra* var. *bracteosa* Peter, *Themeda triandra* var. *burchellii* (Hack.) Domin, *Themeda triandra* var. *glauca* (Hack.) Thell., *Themeda triandra* var. *hispida* Stapf, *Themeda triandra* var. *imberbis* (Retz.) Hack., *Themeda triandra* subsp. *japonica* (Houtt.) T.Koyama, *Themeda triandra* var. *japonica* (Houtt.) Makino, *Themeda triandra* var. *punctata* (A.Rich.) Stapf, *Themeda triandra* var. *roylei* (Hook. f.) Domin, *Themeda triandra* var. *sublaevigata* Chiov., *Themeda triandra* var. *syriaca* (Boiss.) Hack., *Themeda triandra* var. *trachyspathea* Gooss., *Themeda triandra* var. *vulgaris* (Hack.) Domin, *Themeda triandra* var. *vulgaris* auctt. non Hackel

36. *Themeda quadrivalvis* (L.) Kuntze

Synonyms: *Themeda quadrivalvis* var. helferi (Hack.) Bor

37. *Vetiveria zizanoides* (L.) Nash

Synonyms: *Vetiveria odorata* Virey, *Phalaris zinanioides* L., *Andropogon muricatus* Retz., *Andropogon festucoides* Presl., *Andropogon squarrosus* Hack., *Andropogn squarrosus* Cook, *Agrostis verticillata* Lam., *Anatherum muricatum* Beauv.

38. *Triplopogon ramosissimus* (Hack.) Bor

Synonyms: *Ischaemum ramosissimum* Hack., *Ischaemum spathiflorum* Hook.f., *Sehima ramosissimum* (Hack.) Roberty, *Sehima spathiflorum* (Hook.f.) Blatt. & McCann, *Triplopogon spathiflorua* (Hook.f.) Bor, *Triplopogon spathiflorus* (Hook. f.) Bor

TRIBE: PANICEAE

39. *Alloteropsis cimicina* (L.) Stapf

Synonyms: *Agrostis cimicina* (L.) Poir., *Agrostis digitata* Lam., *Alloteropsis latifolia* (Peter) Pilg., *Alloteropsis quintasii* (Mez) Pilg., *Axonopus cimicinus* (L.) P.Beauv., *Axonopus latifolius* Peter, *Coridochloa cimicina* (L.) Nees, *Coridochloa cimicina* (L.) Nees ex B.D. Jacks., *Coridochloa cimicina* (L.) Nees ex Chase, *Eriachne melicaceae* var. *fragrans* F.M.Bailey, *Melica cimicina* (L.) Salisb., *Milium cimicinum* L., *Oplismenus fasciculatus* Roem. & Schult., *Panicum cimicinum* (L.) Retz., *Panicum fasciculatum* Lam., *Panicum fasciculiflorum* Steud., *Urochloa cimicina* (L.) Kunth, *Urochloa fasciculata* (Roem. & Schult.) Kunth, *Urochloa quintasii* Mez

40. *Brachiaria eruciformis* (Sm.) Griseb.

Synonyms: *Brachiaria cruciformis* Griseb., *Brachiaria eruciformis* var. *divaricata* Basappa & Muniy., *Brachiaria isachne* (Roth ex Roem. & Schult.) Stapf, *Echinochloa eruciformis* (Sm.) Rchb., *Echinochloa eruciformis* (Sm.) Koch, *Milium alternans* Bubani, *Moorochloa eruciformis* (Sm.) Veldkamp, *Panicum caucasicum* Trin., *Panicum cruciforme* Steud., *Panicum cruciforme* Sibth. ex Roem., *Panicum eruciforme* Sm., *Panicum isachne* Roth, *Panicum isachne* var. *mexicana* Beal, *Panicum isachne* var. *mexicanum* Vasey ex Beal, *Panicum pubinode* Hochst. ex A.Rich.,

Panicum wightii Nees, *Urochloa eruciformis* (Sm.) C.Nelson, Sutherl & Fern.Casas

41. *Brachiaria distachya* (L.) Stapf

Synonyms: *Brachiaria subquadrifida* Stehlé, *Digitaria distachya* (L.) Pers., *Panicum distachyon* L., *Panicum radicans* Llanos, *Urochloa distachya* (L.) T.Q.Nguyen

42. *Brachiaria ramosa* (L.) Stapf

Synonyms: *Brachiaria chennaveeraiana* Basappa & Muniy., *Brachiaria multispiculata* H.Scholz, *Brachiaria ramosa* var. *pubescens* Basappa & Muniy., *Brachiaria regularis* var. *nidulans* (Mez) Täckh., *Echinochloa ramosa* (L.) Roberty, *Panicum arvense* Kunth, *Panicum brachylachnum* Steud., *Panicum breviradiatum* Hochst., *Panicum canescens* Roth, *Panicum cognatissimum* Steud., *Panicum grossarium* J.Koenig, *Panicum nidulans* Mez, *Panicum ozogonum* Steud., *Panicum pallidum* Peter, *Panicum patens* Bojer, *Panicum petiveri* var. *puberulum* Chiov., *Panicum ramosum* L., *Panicum sorghum* Steud., *Panicum supervacuum* C.B. Clarke, *Setaria canescens* (Roem. & Schult.) Kunth, *Urochloa ramosa* (L.) T.Q.Nguyen, *Urochloa ramosa* (L.) R.D. Webster, *Urochloa supervacua* (Clarke) Noltie

43. *Brachiaria reptans* (L.) C.A.Gardner & C.E.Hubb.

Synonyms: *Brachiaria balansae* Henrard, *Brachiaria prostrata* (Lam.) Griseb., *Brachiaria reptans* var. *hispida* Basappa & Muniy., *Digitaria umbrosa* (Retz.) Pers., *Echinochloa reptans* (L.) Roberty, *Echinochloa subcordata* Roem. & Schult., *Panicum aurelianum* Hale ex Alph. Wood, *Panicum barbipedum* Hayata, *Panicum brachythyrsum* Peter, *Panicum caespitosum* Sw., *Panicum calaccanzense* Steud., *Panicum extensum* Desv., *Panicum grossarium* L., *Panicum grossarium* Roxb., *Panicum insularum* Steud., *Panicum luxurians* Willd. ex Nees, *Panicum nilagiricum* Steud., *Panicum parvum* Buse, *Panicum patulum* Mez, *Panicum procumbens* Nees, *Panicum procumbens* var. *umbrosum* (Retz.) Nees, *Panicum prostratum* Lam., *Panicum prostratum* var. *marquisense* F.Br., *Panicum prostratum* var. *pilosum* Eggers, *Panicum reptans* L., *Panicum sieberi* Link, *Panicum subcordatum* (Roem. & Schult.) Roth,

Panicum taitense Steud., *Panicum umbrosum* Retz., *Panicum virescens* Poir., *Setaria subcordata* (Roem. & Schult.) Kunth, *Setaria umbrosa* (Retz.) P.Beauv., *Urochloa reptans* (L.) Stapf, *Urochloa reptans* var. *glabra* S.L.Chen & Y.X.Jin

44. *Cenchrus biflorus* Roxb.

Synonyms: *Cenchrus annularis* Andersson, *Cenchrus barbatus* Schumach., *Cenchrus catharticus* Delile, *Cenchrus catharticus* Schltdl., *Cenchrus leptacanthus* A.Camus, *Cenchrus niloticus* Fig. & De Not., *Cenchrus perinvolucratus* Stapf & C.E.Hubb., *Cenchrus rajasthanensis* Kanodia & P.C.Nanda, *Cenchrus triflorus* Aitch., *Elymus caput-medusae* Forssk.

45. *Cenchrus ciliaris* L.

Synonyms: *Cenchrus bulbosus* Fresen. ex Steud., *Cenchrus bulbosus* Fresen., *Cenchrus ciliaris* Fig. & De Not., *Cenchrus ciliaris* var. *anachoreticus* (Chiov.) Pirotta, *Cenchrus ciliaris* var. *genuina* Chiov., *Cenchrus ciliaris* var. *genuinus* (Leeke) Maire & Weiler, *Cenchrus ciliaris* subsp. *ibrahimii* Chrtek & Osb.-Kos., *Cenchrus ciliaris* var. *leptostachys* (Leeke) Maire & Weiller, *Cenchrus ciliaris* var. *nubicus* Fig. & De Not, *Cenchrus ciliaris* var. *nubicus* T. Durand & Schinz, *Cenchrus ciliaris* var. *pallens* (Leeke) Maire & Weiller, *Cenchrus ciliaris* var. *robustior* Penz., *Cenchrus ciliaris* var. *villiferus* Fig. & De Not, *Cenchrus ciliaris* var. *villiferus* T. Durand & Schinz, *Cenchrus glaucus* C.R.Mudaliar & Sundararaj, *Cenchrus longifolius* Hochst. ex Steud., *Cenchrus melanostachyus* A.Camus, *Cenchrus pennisetiformis* Hochst. & Steud., *Cenchrus pennisetiformis* var. *rigidifolia* (Fig. & De Not.) Chiov., *Cenchrus pennisetiformis* var. *typica* Chiov., *Cenchrus pubescens* L. ex B.D. Jacks., *Cenchrus rufescens* Desf., *Pennisetum cenchroides* Rich., *Pennisetum ciliare* (L.) Link, *Pennisetum ciliare* var. *anachoreticum* Chiov., *Pennisetum ciliare* f. *brachystachys* PETER, *Pennisetum ciliare* var. *ciliare*, *Pennisetum ciliare* var. *genuina* Leeke, *Pennisetum ciliare* var. *leptostachys* Leeke, *Pennisetum ciliare* f. *longifolium* PETER, *Pennisetum ciliare* var. *pallens* Leeke, *Pennisetum ciliare* var. *robustior* Penz., *Pennisetum distylum* Guss., *Pennisetum incomptum* Nees ex Steud., *Pennisetum oxyphyllum* Peter, *Pennisetum pachycladum* Stapf, *Pennisetum panormitanum* Lojac., *Pennisetum petraeum* Steud., *Pennisetum polycladum* Chiov., *Pennisetum prieurii* A. Chev., *Pennisetum* *rangei* Mez, *Pennisetum rufescens* (Desf.) Spreng., *Pennisetum rufescens* Hochst. ex Steud., *Pennisetum teneriffae* Steud.

46. *Cenchrus preurii* (Kunth) Maire

Synonyms:*Pennisetum prieurii* Kunth, *Pennisetum breviflorum* Steud., *Cenchrus macrostachyus* Hochst. ex Steud.

47. *Cenchrus setigerus* Vahl

Synonyms: *Cenchrus ciliaris* var. *setigerus* (Vahl) Maire & Weiller, *Cenchrus setigerus* Steud., *Pennisetum ciliare* var. *setigerum* (Vahl) Leeke, *Pennisetum setigerum* (Vahl) Wipff, *Pennisetum vahlii* Kunth

48. *Digitaria ciliaris* (Retz.) Koeler

Synonyms: *Asperella digitaria* Lamk., *Panicum adscendens* H.B.K., *Digitaria chinensis* Horn., *Panicum ornithopus* Trin., *Digitaria marginata* Link., *Digitaria marginata* var. *fimbriata* (Link) Stapf, *Paspalum inaequalis* Link., *Digitaria commutate* Schult., *Digitaria inaequalis* (Link) Sperng., *Digitaria australis* Willd., *Digitaria brevifolia* Link., *Digitaria fimbriata* Link., *Digitaria adscendens* (H.B.K.) Henr., *Panicum fimbriatum* (Link) Kunth, *Panicum glaucescens* Nees, *Panicum neesii* Kunth, *Panicum brachyphyllum* Steud., *Paniucum sanguinale* Linn., *Syntherisma fimbriata* (Link) Nash, *Digitaria henryi* Rendle, *Syntherisma marginatum* (Link) Nash, *Digitaria sanguinalis* var. *marginata* (Link) Fern., *Systherisma henryi* (Rendle) Newbold

49. *Digitaria microbachne* (Persl) Hern.

Synonyms: *Panicum microbachne* J.S. Presl ex C.B. Presl, *Digitaria setigera* Roth ex Roem. & Schult., *Digitaria pruriens* (Fisch. ex Trin.) Büse var. *microbachne* (J. Presl) Fosberg, *Digitaria pruriens* (Fisch. ex Trin.) Büse

50. *Digitaria granularis* (Trin.) Henrard

Synonyms: *Digitaria abludens* (Roem. & Schult.) Veldkamp, *Digitaria pedicellaris* (Hook.f.) Prain, *Helopus sanguinalis* (Roxb.) Nees, *Milium sanguinale* Roxb., *Panicum abludens* Roem. & Schult., *Panicum pedicellare* (Hook.f.) Hack., *Paspalum granulare* Trin., *Paspalum pedicellare* Trin. ex Hook.f., *Paspalum pedicellare* Trin. ex Steud., *Paspalum pedicellatum* Nees ex Duthie

51. *Digitaria longiflora* (Retz.) Pers.

Synonyms: *Paspalum longiflorum* Retz., *Panicum longiflorum* Gmel., *Paspalum brevifolius* Flugge, *Panicum propinquum* R.Br, *Digitaria propinqus* (R.Br.) P.Beauv., *Digitaria roxburghii* Sperng., *Panicum parvulum* Trin., *Panicum pseudo-durva* Nees, *Paspalum filiculme* Nees, *Syntherisma longiflora* Skeels, *Digitaria caespitosa* Ridl., *Digitaria filiculmis* (Nees) Ohwi

52. *Digitaria stircta* Roth

Synonyms: *Agrostis pilosa* Retz., *Digitaria denudata* Link, *Digitaria glabrescens* (Bor) L.Liou, *Digitaria puberula* Link, *Digitaria royleana* (Nees ex Hook.f.) Prain, *Digitaria stricta* var. *denudata* (Link) Henrard, *Digitaria stricta* var. *denudata* (Link) Bor, *Digitaria stricta* var. *glabrescens* Bor, *Panicum concinnum* Schrad., *Panicum denudatum* (Link) Kunth, *Panicum kunthii* Steud., *Panicum orthum* Voigt, *Panicum pseudosetaria* Steud., *Panicum puberulum* (Link) Kunth, *Panicum violascens* var. *denuta* (Link) Döll, *Paspalum concinnum* Steud., *Paspalum royleanum* Nees ex Hook.f., *Paspalum royleanum* Nees ex Thwaites, *Reimaria puberula* (Link) Link, *Setaria stricta* (Roth ex Roem. & Schult.) Kunth, *Syntherisma puberulum* (Link) Newbold, *Syntherisma royleana* (Nees) Newbould

53. *Echinochloa colona* (L.) Link

Synonyms: *Brachiaria longifolia* Gilli, *Digitaria cuspidata* (Roxb.) Schult., *Echinochloa colona* var. *equitans* (Hochst. ex A.Rich.) Cufod., *Echinochloa colona* var. *glauca* (Sickenb.) Simpson, *Echinochloa colona* var. *glaucum* (Sickenb.) Simps., *Echinochloa colona* var. *leiantha* Boiss., *Echinochloa colona* var. *repens* (Sickenb.) Simpson, *Echinochloa colona* f. *vivipara* Beetle, *Echinochloa colona* f. *viviparum* Beetle, *Echinochloa colona* f. *zonalis* (Guss.) Wiegand, *Echinochloa colona* var. *zonalis* (Guss.) Wooton & Stand., *Echinochloa colonum* var. *zonalis* (Guss.) Wooton & Standl., *Echinochloa crus-galli* subsp. *colona* (L.) Honda, *Echinochloa crus-galli* var. *longiseta* (Trin.) Hara, *Echinochloa divaricata* Andersson, *Echinochloa equitans* (Hochst. ex A.Rich.) C.E.Hubb., *Echinochloa subverticillata* Pilg., *Echinochloa zonalis* (Guss.) Parl., *Milium colonum* (L.) Moench, *Milium colonum* (L.) Kunth, *Oplismenus colonus*

(L.) Kunth, *Oplismenus colonus* var. *zonalis* (Guss.) Schrad., *Oplismenus crus-galli* var. *colonus* (L.) Coss. & Durieu, *Oplismenus cuspidatus* (Roxb.) Kunth, *Oplismenus daltonii* (Parl. ex Webb.) J.A.Schmidt, *Oplismenus margaritaceus* (Link) Kunth, *Oplismenus muticus* Phil., *Oplismenus pseudocolonus* (Roem. & Schult.) Kunth, *Oplismenus pseudocolonus* (Roem. & Schult.) Kunth, *Orthopogon dichotomus* Llanos, *Orthopogon subverticillatus* Llanos, *Panicum brachiariaeforme* Steud., *Panicum brizoides* L., *Panicum caesium* Hook. & Arn., *Panicum colonum* L., *Panicum colonum* var. *angustatum* Peter, *Panicum colonum* var. *atroviolaceum* Hack., *Panicum colonum* var. *equitans* (A.Rich.) T.Durand & Schinz, *Panicum colonum* var. *glaucum* Sickenb., *Panicum colonum* var. *haematodes* (C. Presl) Richt., *Panicum colonum* var. *humile* Nees, *Panicum colonum* f. *maculatum* Arechav., *Panicum colonum* var. *pseudocolonum* (Roth) Nees, *Panicum colonum* var. *repens* Sickenb., *Panicum colonum* var. *zonale* (Guss.) Dewey, *Panicum crus-galli* subsp. *colonum* (L.) Makino & Nemoto, *Panicum crus-galli* subsp. *colonum* (L.) K.Richt., *Panicum crus-galli* var. *colonum* Cosson ex Richter, *Panicum crus-galli* var. *colonum* (L.) Fiori, *Panicum crus-galli* var. *minor* Thwaites, *Panicum cumingianum* Steud., *Panicum cuspidatum* Roxb., *Panicum daltonii* Parl. ex Webb, *Panicum echinochloa* T.Durand & Schinz, *Panicum equitans* Hochst. ex A.Rich., *Panicum equitans* f. *aquaticum* Chiov., *Panicum equitans* f. *terrestris* Chiov., *Panicum haematodes* C.Presl, *Panicum hookeri* Parl., *Panicum margaritaceum* Link, *Panicum musei* Steud., *Panicum numidianum* C.Presl, *Panicum petiveri* Kotschy ex Griseb., *Panicum prorepens* Steud., *Panicum pseudocolonum* Roth, *Panicum tetrastichum* Forssk., *Panicum zonale* Guss., *Setaria brachiariaeformis* (Steud.) T.Durand & Schinz

54. *Echinochloa crusgalli* (L.) P.Beauv.

Synonyms: *Digitaria hispidula* (Retz.) Willd., *Echinochloa caudate Roshev*, *Echinochloa commutata* Schult., *Echinochloa crus-corvi* (L.) P.Beauv., *Echinochloa crus-galli* f. *aristata* (Honda) M.Hiroe, *Echinochloa crus-galli* f. *aristata* (Vasinger) Morariu, *Echinochloa crus-galli* var. *aristata* Gray, *Echinochloa crus-galli* f. *atra* (Kuntze) Soó, *Echinochloa crus-galli* var. *austrojaponensis* Ohwi, *Echinochloa crus-galli* f. *breviseta* (Döll) Morariu, *Echinochloa crus-galli* var. *breviseta* (Döll) Neilr.,

Echinochloa crus-galli f. *breviseta* (Döll) Pinto de Silva, *Echinochloa crus-galli* var. *caudata* (Roshev.) Kitag., *Echinochloa crus-galli* f. *echinata* (Willd.) Morariu, *Echinochloa crus-galli* var. *echinata* (Willd.) Honda, *Echinochloa crus-galli* f. *exigua* (Holub) Soó, *Echinochloa crus-galli* var. *formosensis* Ohwi, *Echinochloa crus-galli* var. *kasaharae* Ohwi, *Echinochloa crus-galli* f. *longiseta* (Trin.) Farw., *Echinochloa crus-galli* var. *longiseta* (Döll) Podp., *Echinochloa crus-galli* f. *mitis* (Pursh) Farw., *Echinochloa crus-galli* var. *mitis* (Pursh) Peterm., *Echinochloa crus-galli* f. *mixta* (A.F.Schwarz) Soó, *Echinochloa crus-galli* f. *mutica* (Vasinger) Morariu, *Echinochloa crus-galli* var. *mutica* (Elliott) Rydb., *Echinochloa crus-galli* var. *mutica* Wooton & Standl., *Echinochloa crus-galli* var. *praticola* Ohwi, *Echinochloa crus-galli* f. *purpurea* (Pursh) Farw., *Echinochloa crus-galli* f. *rohlenae* (Domin) Soó, *Echinochloa crus-galli* f. *submutica* (Neilr.) Morariu, *Echinochloa crus-galli* var. *submutica* Neilr., *Echinochloa crus-galli* f. *vittata* F.T.Hubb., *Echinochloa crus-galli* f. *zelayensis* (Kunth) Farw., *Echinochloa crus-galli* subsp. *zelayensis* (Kunth) Shinners, *Echinochloa crus-galli* var. *zelayensis* (Humb., Bonpl. & Kunth) Hitchc., *Echinochloa crus-pavonis* var. *austrojaponensis* (Ohwi) S.L.Dai, *Echinochloa crus-pavonis* var. *breviseta* (Döll) S.L.Dai, *Echinochloa crus-pavonis* var. *praticola* (Ohwi) S.L.Dai, *Echinochloa crusgalli* subsp. *submutica* (Meyer) Honda, *Echinochloa disticha* St-Lag., *Echinochloa dubia* Roem. & Schult., *Echinochloa echinata* (Willd.) Nakai, *Echinochloa formosensis* (Ohwi) S.L.Dai, *Echinochloa glabrescens* Kossenko, *Echinochloa glabrescens* var. *glabra* Kossenko, *Echinochloa glabrescens* var. *pilosa* Kossenko, *Echinochloa hispida* (E.Forst.) Schult., *Echinochloa macrocarpa* var. *aristata* Vasinger, *Echinochloa macrocarpa* var. *mutica* Vasinger, *Echinochloa macrocorvi* Nakai, *Echinochloa madagascariensis* Mez, *Echinochloa micans* Kossenko, *Echinochloa muricata* var. *occidentalis* Wiegand, *Echinochloa occidentalis* (Wiegand) Rydb., *Echinochloa paracorvi* Nakai, *Echinochloa persistentia* Z.S.Diao, *Echinochloa pungens* var. *occidentalis* (Wiegand) Fernald & Griscom, *Echinochloa spiralis* Vasinger, *Echinochloa zelayensis* (Kunth) Schult., *Milium crus-galli* (L.) Moench, *Oplismenus crus-galli* (L.) Dumort., *Oplismenus crus-galli* var. *muticus* (Elliott) Alph.Wood, *Oplismenus dubius* (Roem. & Schult.) Kunth, *Oplismenus echinatus* (Willd.) Kunth, *Oplismenus limosus* J.Presl, *Oplismenus zelayensis* Kunth, *Orthopogon crus-galli* (L.) Spreng.,

Orthopogon echinatus (Willd.) Spreng., *Panicum alectorocnemum* St.-Lag., *Panicum alectromerum* Dulac, *Panicum corvi* Thunb., *Panicum corvipes* Stokes, *Panicum cristagalli* Gromov ex Trautv., *Panicum crus-galli* L., *Panicum crus-galli* var. *angustifolium* Döll, *Panicum crus-galli* var. *aristatum* Pursh, *Panicum crus-galli* f. *atra* Kuntze, *Panicum crus-galli* var. *brevisetum* Döll, *Panicum crus-galli* var. *brevisetum* Gaudich. ex Ducommun, *Panicum crus-galli* var. *echinatum* (Willd.) Döll, *Panicum crus-galli* f. *exiguum* Holuby, *Panicum crus-galli* f. *hispidum* Kuntze, *Panicum crus-galli* var. *longisetum* Trin., *Panicum crus-galli* var. *longisetum* Döll, *Panicum crus-galli* var. *mite* Pursh, *Panicum crus-galli* f. *mixtum* A.F.Schwarz, *Panicum crus-galli* var. *muticum* Elliott, *Panicum crus-galli* var. *muticum* Hack. ex Nakai, *Panicum crus-galli* var. *prostratum* Kuntze, *Panicum crus-galli* var. *pumilum* Goiran, *Panicum crus-galli* var. *purpureum* Pursh, *Panicum crus-galli* var. *pygmaeum* Kuntze, *Panicum crus-galli* var. *rohlenae* Domin, *Panicum crus-galli* f. *rohlenae* Domin, *Panicum crus-galli* var. *submuticum* (Neilr.) Hayek, *Panicum crus-galli* var. *zelayense* (Kunth) Stratman, *Panicum cruscorvi* L., *Panicum crus-galli* var. *submuticum* Meyer, *Panicum echinatum* Willd., *Panicum goiranii* Rouy, *Panicum grossum* Salisb., *Panicum hispidum* G.Forst., *Panicum limosum* J.Presl ex Nees, *Panicum oryzetorum* Sickenb., *Panicum scindens* Nees ex Steud., *Panicum zelayense* (Kunth) Steud., *Pennisetum crus-galli* (L.) Baumg.

55. *Echinochloa stagnina* (Retz.) P.Beauv.

Synonyms: *Echinochloa barbata* Vanderyst, *Echinochloa crus-galli* var. *stoloniferum* (Schweinf. & Muschl.) Chevalier, *Echinochloa hostii* Steven ex Link, *Echinochloa lelievrei* (A.Chev.) Berhaut, *Echinochloa oryzetorum* (A.Chev.) A.Chev., *Echinochloa scabra* (Lam.) Roem. & Schult., *Oplismenus scaber* (Lam.) Kunth, *Oplismenus stagninus* (Retz.) Kunth, *Orthopogon stagninus* (Retz.) Spreng., *Panicum burgu* A.Chev., *Panicum crus-galli* var. *leiostachyum* Franch., *Panicum crus-galli* var. *maximum* Franch., *Panicum crus-galli* var. *sieberiana* Asch. & Schweinf., *Panicum crus-galli* var. *sieberianum* Asch. & Schweinf., *Panicum crus-galli* var. *stagninum* (Retz.) Trimen, *Panicum crus-galli* var. *stagninum* (Retz.) T. Durand & Schinz, *Panicum crus-galli* var. *stagninum* (Retz.) Ridl., *Panicum crus-galli* var. *stoloniferum* Schweinf. & Muschl., *Panicum* *crus-galli* var. *submuticum* Franch., *Panicum galli*Thunb., *Panicum lelievrei* A.Chev., *Panicum oryzetorum* (A.Chev.) A.Chev., *Panicum scabrum* Lam., *Panicum scabrum* subsp. *burgu* (A.Chev.) A.Chev., *Panicum scabrum* var. *leiostachyum* (Franch.) A.Chev., *Panicum scabrum* subsp. *lelievrei* A Chev., *Panicum scabrum* subsp. *oryzetorum* A.Chev., *Panicum scabrum* subsp. *stagninum* (Retz.) A.Chev., *Panicum scabrum* var. *submuticum* (Franch.) A.Chev., *Panicum sieberianum* (Asch. & Schweinf.) Sickenb., *Panicum stagninum* Retz., *Panicum stagninum* var. *burgu* (A.Chev.) Chev., *Panicum subaristatum* Peter

56. *Eremopogon foveolatus* (Delile) Stapf

Synonyms: *Andropogon foveolatus* Delile, *Andropogon foveolatus* var. *plumosus* N.Terracc., *Andropogon foveolatus* var. *strictus* (Roxb.) Hack., *Andropogon monostachyus* Spreng., *Andropogon orthos* Schult., *Andropogon strictus* Roxb., *Cymbopogon strictus* (Roxb.) Bojer, *Dichanthium foveolatum* (Delile) Roberty, *Eremopogon strictus* A.Camus, *Hypogynium foveolatum* (Delile) Haines, *Sorghum foveolatum* (Delile) Kuntze

57. *Eriochloa procera* (Retz.) C.E.Hubb.

Synonyms: *Agrostis procera* Retz., *Agrostis ramosa* (Retz.) Poir., *Eriochloa annulata* (Flüggé) Kunth, *Eriochloa hackelii* Honda, *Eriochloa polystachya* Hook. f., *Eriochloa polystachya* var. *annulata* (Flüggé) Maiden & Betche, *Eriochloa procera* var. *involucrata* (Hack.) Jansen, *Eriochloa procera* var. *procera*, *Eriochloa ramosa* (Retz.) Kuntze, *Eriochloa ramosa* var. *involucrata* Hack., *Eriochloa sundaica* Miq., *Helopus annulatus* (Flüggé) Nees, *Helopus javanicus* Steud., *Helopus pilosus* (Retz.) Trin., *Milium ramosum* Retz., *Milium zonatum* Llanos, *Panicum annulatum* (Flüggé) A.Rich., *Paspalum annulatum* Flüggé, *Piptatherum annulatum* (Nees) J.Presl, *Thysanolaena procera* (Retz.) Mez, *Vilfa procera* (Retz.) P.Beauv.

58. *Oplismenus burmanni* (Retz.) P.Beauv

Synonyms: *Oplismenus burmannii* f. cristata (J. Presl) Hier. ex Peter.

59. *Oplismenus composites* (L.) P.Beauv.

Synonyms: *Andropogon undatus* Jacq., *Digitaria composita* Willd., *Digitaria elatior* Willd., *Echinochloa aristata* Raspail, *Echinochloa*

lanceolata (Retz.) P.Beauv., *Echinochloa lanceolata* (Retz.) Roem. & Schult., *Hippagrostis composita* (L.) Kuntze, *Hippagrostis setaria* (Lam.) Kuntze, *Oplismenus africanus* J.M. Wood, *Oplismenus burmannii* var. *intermedius* Honda, *Oplismenus compositus* var. *angustifolius* L.C.Chia, *Oplismenus compositus* var. *compositus*, *Oplismenus compositus* var. *formosanus* (Honda) S.L.Chen & Y.X.Jin, *Oplismenus compositus* f. *glabratus* F.Br., *Oplismenus compositus* var. *intermedius* (Honda) Ohwi, *Oplismenus compositus* var. *lasiorhachis* Hack., *Oplismenus compositus* var. *patens* (Honda) Ohwi, *Oplismenus compositus* f. *pubescens* F.Br., *Oplismenus compositus* var. *rariflorus* (K.B.Presl) U.Scholz, *Oplismenus compositus* var. *setarius* (Lam.) F.M.Bailey, *Oplismenus compositus* var. *submuticus* S.L.Chen & Y.X.Jin, *Oplismenus compositus* var. *sylvaticus* Hildebr., *Oplismenus compositus* f. *vittatus* (Bailey) Beetle, *Oplismenus compositus* var. *vittatus* L.H.Bailey, *Oplismenus decompositus* Nees, *Oplismenus elatior* (L.f.) P.Beauv., *Oplismenus elatius* (L. f.) P.Beauv., *Oplismenus formosanus* Honda, *Oplismenus hirtellus* subsp. *setarius* (Lam.) Mez ex Ekman, *Oplismenus hirtiflorus* C. Presl, *Oplismenus indicus* Roem. & Schult., *Oplismenus jacquinii* Kunth, *Oplismenus junghuhnii* Boerl., *Oplismenus lanceolatus* (Retz.) Kunth, *Oplismenus liebmannii* E.Fourn., *Oplismenus owatarii* Honda, *Oplismenus patens* Honda, *Oplismenus patens* var. *angustifolius* (L.C.Chia) S.L.Chen & Y.X.Jin, *Oplismenus patens* var. *yunnanensis* S.L.Chen & Y.X.Jin, *Oplismenus polliniaefolius* Honda, *Oplismenus pratensis* (Spreng.) Schult. & Schult. f., *Oplismenus rariflorus* J.Presl, *Oplismenus setarius* (Lam.) Roem. & Schult., *Oplismenus setarius* f. *sterilis* F.Br., *Oplismenus sylvaticus* (Lam.) Roem. & Schult., *Oplismenus thiebautii* E.Fourn., *Orthopogon compositus* (L.) R.Br., *Orthopogon compositus* var. *glabrescens* Büse, *Orthopogon gonyrrhizus* (Steud.) Miq., *Orthopogon hirtellus* Eaton & J. Wright, *Orthopogon junghuhnii* Nees, *Orthopogon longeracemosum* (Steud.) Miq., *Orthopogon longiracemosus* (Steud.) Miq., *Orthopogon pratensis* Spreng., *Orthopogon remotus* Trin., *Orthopogon setarius* (Lam.) Spreng., *Orthopogon sylvaticus* (Lam.) Miq., *Panicum aristatum* Retz., *Panicum bidentatum* Steud, *Panicum bidentulum* Steud., *Panicum certificandum* Steud., *Panicum compositoproximum* Rottler ex Willd., *Panicum compositum* L., *Panicum compositum* var. *fimbriusculum* Döll, *Panicum elatius* L.f., *Panicum gonyrrhizum* Steud., *Panicum lanceolatum*

Retz., *Panicum longiracemosum* Steud., *Panicum parciflorum* Steud., *Panicum peninsulanum* Steud., *Panicum peninsularum* Steud., *Panicum pratense* (Spreng.) Steud., *Panicum setarium* Lam., *Panicum sylvaticum* Lam., *Panicum undatum* (Jacq.) Steud., *Pollinia undata* (Jacq.) Spreng., *Urochloa aristata* (Retz.) B.D.Jacks.

60. *Panicum antidotale* Retz.

Synonyms: *Chasea prolifera* (Lam.) Nieuwl., *Milium effusum* Lour., *Panicum miliare* Lam., *Panicum proliferum* Lam., *Panicum sparsum* Rottler ex Steud., *Panicum subalbidum* Hochst. ex T. Durand & Schinz, *Paspalum miliare* (Lam.) K.Schum. & Hollrung

61. *Panicum maximum* Jacq.

Synonyms: *Megathyrsus maximus* (Jacq.) B.K.Simon & Jacobs, *Megathyrsus maximus* var. *coloratus* (C.T.White) B.K.Simon & Jacobs, *Megathyrsus maximus* var. *pubiglumis* (K.Schum.) B.K.Simon & Jacobs, *Panicum bivonianum* Brullo, Miniss., Scelsi & Spamp., *Panicum compressum* Biv., *Panicum eburneum* Trin., *Panicum giganteum* Mez, *Panicum heynii* Roth, *Panicum hirsutissimum* Steud., *Panicum jumentorum* Pers., *Panicum laeve* Lam., *Panicum maximum* var. *altissimum* Kuntze, *Panicum maximum* var. *coloratum* C.T.White, *Panicum maximum* subsp. *commune* (Nees) Peter, *Panicum maximum* var. *commune* Nees, *Panicum maximum* var. *confine* Chiov., *Panicum maximum* var. *congoensis* Vanderyst, *Panicum maximum* var. *glaucum* Nees, *Panicum maximum* var. *heterotrichum* Peter, *Panicum maximum* var. *hirsutissimum* (Steud.) Oliv., *Panicum maximum* var. *hirsutum* Peter, *Panicum maximum* var. *laeve* Nees, *Panicum maximum* var. *laevis* Nees, *Panicum maximum* var. *maximum*, *Panicum maximum* subsp. *pubescens* M.Sharma, *Panicum maximum* var. *pubiglume* K.Schum., *Panicum maximum* f. *pubiglume* (K. Schum.) K. Schum. ex Peter, *Panicum maximum* var. *trichoglume* Robyns, *Panicum maximum* var. *trichoglume* Eyles, *Panicum pamplemoussense* Steud., *Panicum poiforme* Willd. ex Spreng., *Panicum polygamum* Sw., *Panicum praelongum* Steud., *Panicum scaberrimum* Lag., *Panicum sparsum* Schumach., *Panicum teff* Desv., *Panicum tephrosanthum* Hack., *Panicum trichocondylum* Steud., *Urochloa maxima* (Jacq.) R.D.Webster, *Urochloa maxima* var. *trichoglume* (Robyns) R.D.Webster

62. *Panicum miliaceum* L.

Synonyms: *Leptoloma miliacea* (L.) Smyth, *Milium esculentum* Moench, *Milium panicum* Mill., *Panicum asperrimum* Fisch., *Panicum asperrimum* Fischer ex Jacq., *Panicum densepilosum* Steud., *Panicum miliaceum* var. *aerugineum* Krassavin & Uljanova, *Panicum miliaceum* subsp. *agricola* H.Scholz & Mikoláš, *Panicum miliaceum* var. *anthracinum* Krassavin & Uljanova, *Panicum miliaceum* var. *aquilum* Krassavin & Uljanova, *Panicum miliaceum* var. *atrobrunneum* Krassavin & Uljanova, *Panicum miliaceum* var. *coffeatum* Krassavin & Uljanova, *Panicum miliaceum* var. *corsinum* Krassavin & Uljanova, *Panicum miliaceum* var. *densobrunneum* Agaf., *Panicum miliaceum* var. *fuscum* Krassavin & Uljanova, *Panicum miliaceum* var. *glaucum* Krassavin & Uljanova, *Panicum miliaceum* var. *miliaceum*, *Panicum miliaceum* var. *nicotianum* Krassavin & Uljanova, *Panicum miliaceum* subsp. *ruderale* (Kitag.) Tzvelev, *Panicum miliaceum* var. *ruderale* Kitag., *Panicum miliaceum* var. *spontaneum* Barkworth *et al.*, *Panicum miliaceum* var. *subvitellinotephrum* Agaf., *Panicum miliaceum* var. *virescens* Krassavin & Uljanova, *Panicum milium* Pers., *Panicum ruderale* (Kitag.) D.M.Chang

63. *Panicum trypheron* Schult.

Synonyms: *Panicum curvatum* var. *tenellum* T.Durand & Schinz, *Panicum curviflorum* var. *suishaense* (Hayata) Veldkamp, *Panicum curviflorum* Hornem., *Panicum deccanense* Naik & Patunkar, *Panicum neesianum* Wight & Arn. ex Steud., *Panicum papuanum* Mez, *Panicum roxburghii* Spreng., *Panicum suishaense* Hayata, *Panicum tenellum* Roxb., *Panicum trypheron* var. *suishaense* (Hayata) P.S.Hsu

64. *Paspalidium flavidum* (Retz.) A.Camus

Synonyms: *Panicum flavidum* Retz., *Panicum flavidum* var. *orarium* Domin, *Panicum floridum* Royle, *Panicum granulare* Lam., *Paspalum villosum* Náves ex Fern.-Vill., *Setaria flavida* (Retz.) Veldkamp

65. *Paspalidium geminatum* (Forssk.) Stapf

Synonyms: *Digitaria adpressa* Bojer, *Digitaria affinis* Roem. & Schult., *Digitaria affinis* Opiz ex Bercht., *Digitaria appressa* (Lam.) Pers., *Echinochloa geminata* (Forssk.) Roberty, *Panicum affine* (Roem. & Schult.)

Nees, *Panicum appressum* Forssk., *Panicum appressum* (Lam.) Döll, *Panicum beckmanniiforme* J.C.Mikan ex Trin., *Panicum briziforme* J.Presl, *Panicum brizoides* Lam., *Panicum emergens* Hochst., *Panicum fluitans* Retz., *Panicum geminatum* Forssk., *Panicum geminatum* Hochst., *Panicum glomeratum* Buckley, *Panicum paspaloides* Pers., *Panicum paspaloides* Raddi, *Panicum truncatum* Trin., *Paspalidium geminatum* var. *geminatum*, *Paspalidium paludivagum* (Hitchc. & Chase) Henrard, *Paspalidium paludivagum* (Hitchc. & Chase) Pilg., *Paspalidium paludivagum* (Hitchc. & Chase) Herter, *Paspalum adpressum* Pers. ex B.D.Jacks., *Paspalum appressum* Lam., *Setaria geminata* (Forssk.) Veldkamp

66. *Paspalum scrobiculatum* L.

Synonyms: *Paspalum adelogaeum* Steud., *Paspalum akoense* Hayata, *Paspalum amazonicum* Trin., *Paspalum auriculatum* J.Presl, *Paspalum barbatum* Schumach., *Paspalum borbonicum* Steud., *Paspalum boscianum* Flüggé, *Paspalum brunneum* Bosc ex Flüggé, *Paspalum cartilagineum* J.Presl, *Paspalum cartilagineum* var. *biglumaceum* Fosberg & Sachet, *Paspalum commersonii* Lam., *Paspalum commersonii* var. *hirsutum* Jansen, *Paspalum commersonii* var. *polystachyum* (R.Br.) Stapf, *Paspalum commersonii* var. *turgidum* (Buse) Jansen, *Paspalum commutatum* Nees, *Paspalum confertum* J.Le Conte, *Paspalum coromandelinum* Lam., *Paspalum deightonii* (C.E.Hubb.) Clayton, *Paspalum dissectum* Nees, *Paspalum dissectum* var. *grande* Nees, *Paspalum firmum* Trin., *Paspalum horneri* Henrard, *Paspalum jardinii* Steud., *Paspalum kora* Willd., *Paspalum ledermannii* Mez, *Paspalum longifolium* var. *pseudo-orbiculare* Jansen, *Paspalum mauritanicum* Nees ex Steud., *Paspalum metabolon* Steud., *Paspalum metzii* Steud., *Paspalum moratii* Toutain, *Paspalum orbiculare* G.Forst., *Paspalum orbiculare* var. *cartilagineum* (J.Presl) Summerh. & C.E.Hubb., *Paspalum polo* F.M.Bailey, *Paspalum polystachyum* R.Br., *Paspalum puberulum* Roem. & Schult., *Paspalum pubescens* R.Br., *Paspalum purpurascens* Elliott, *Paspalum scrobiculatum* var. *auriculatum* (J.Presl & C.Presl) Merr., *Paspalum scrobiculatum* var. *bispiculatum* Hack., *Paspalum scrobiculatum* var. *commersoni* (Lam.) Stapf, *Paspalum scrobiculatum* var. *deightonii* C.E.Hubb., *Paspalum scrobiculatum* var. *gracillimum* Domin, *Paspalum scrobiculatum* var. *horneri* (Henrard) de Koning & Sosef, *Paspalum* *scrobiculatum* var. *jardinii* (Steud.) Franch., *Paspalum scrobiculatum* var. *orbiculare* (E.Forst.) Hack., *Paspalum scrobiculatum* var. *polystachyum* (R.Br.) A.Chev., *Paspalum scrobiculatum* var. *polystachyum* (R. Br.) Stapf, *Paspalum scrobiculatum* var. *turgidum* Buse, *Paspalum scrobiculatum* var. *velutinum* Hack., *Paspalum serpens* J.Presl ex Trin., *Paspalum thunbergii* var. *minus* Makino, *Paspalum virgatum* Walter, *Paspalum virgatum* var. *latifolium* Alph.Wood, *Paspalum virgatum* var. *purpurascens* (Elliott) Alph.Wood, *Paspalum zollingeri* Steud.

67. *Paspalum distichum* L.

Synonyms: *Anastrophus paspalodes* (Michx.) Nash, *Digitaria disticha* (L.) Fiori & Paol., *Digitaria paspalodes* Michx., *Dimorphostachys oaxacensis* (Steud.) E.Fourn. ex Hemsl., *Milium distichum* (L.) Muhl., *Milium paspalodes* (Michx.) Elliott, *Panicum digitaria* (Poir.) Latirr., *Panicum fernandezianum* Colla, *Panicum paspaliforme* J.Presl, *Panicum polyrrhizum* J.Presl, *Paspalum berterianum* Balb. ex Colla, *Paspalum chepica* Steud., *Paspalum digitaria* Poir., *Paspalum distichum* var. *digitaria* (Poir.) Hack., *Paspalum distichum* var. *indutum* Shinners, *Paspalum distichum* var. *longirepens* Domin, *Paspalum distichum* var. *microstachyum* Domin, *Paspalum distichum* subsp. *paspalodes* (Michx.) Thell., *Paspalum distichum* var. *paspalodes* (Michx.) Thell., *Paspalum elliottii* S.Watson, *Paspalum fernandezianum* Colla, *Paspalum michauxianum* Kunth, *Paspalum oajacense* Hemsl., *Paspalum oaxacense* Steud., *Paspalum paspaliforme* J.Presl, *Paspalum paspalodes* (Michx.) Scribn., *Paspalum paspalodes* var. *paspalodes*, *Paspalum paucispicatum* Vasey, *Paspalum polyrrhizum* J.Presl, *Paspalum schaffneri* Griseb. ex E Fourn., *Paspalum schaffneri* Griseb. ex Hemsl., *Paspalum vaginatum* Döll, *Paspalum vaginatum* var. *pubescens* Döll

68. *Paspalum vaginatum* Sw.

Synonyms: *Digitaria foliosa* Lag., *Digitaria paspalodes* var. *longipes* (Lange) Willk. & Lange, *Digitaria tristachya* (Lecomte) Schult., *Digitaria vaginata* (Sw.) Philippe, *Digitaria vaginata* (Sw.) Magnier, *Panicum littorale* (R.Br.) Kuntze, *Panicum vaginatum* (Sw.) Godr., *Paspalum boryanum* J.Presl, *Paspalum distichum* var. *anpinense* Hayata, *Paspalum distichum* var. *littorale* (R.Br.) F.M.Bailey, *Paspalum distichum* var. *nanum*

(Döll) Stapf, *Paspalum distichum* var. *tristachyum* (Schult.) Alph.Wood, *Paspalum distichum* subsp. *vaginatum* (Sw.) Maire, *Paspalum distichum* var. *vaginatum* (Sw.) Griseb., *Paspalum foliosum* (Lag.) Kunth, *Paspalum furcatum* var. *fissum* Döll, *Paspalum gayanum* É.Desv., *Paspalum inflatum* A.Rich., *Paspalum jaguaense* León, *Paspalum kleinianum* J.Presl, *Paspalum littorale* R.Br., *Paspalum longiflorum* P. Beauv., *Paspalum reimarioides* Chapm., *Paspalum squamatum* Steud., *Paspalum tristachyum* J.Le Conte, *Paspalum vaginatum* var. *littorale* (R.Br.) Trin., *Paspalum vaginatum* f. *longipes* Lange, *Paspalum vaginatum* var. *longipes* Lange, *Paspalum vaginatum* subsp. *nanum* (Döll) Loxton, *Paspalum vaginatum* var. *nanum* Döll, *Paspalu vaginatum* var. *reimarioides* Chapm., *Sanguinaria vaginata* (Sw.) Bubani

69. *Pennisetum setosum* (Sw.) Rich.

Synonyms: *Cenchrus polystachios* (L.) Morrone, *Cenchrus ramosus* (Hochst.) Morrone, *Cenchrus retusus* Sw., *Cenchrus setosus* Sw., *Cenchrus subangustus* (Schumach.) Morrone, *Gymnotrix geniculata* Schult., *Gymnotrix polystachya* (L.) Sw. ex Trin., *Gymnotrix ramosa* Hochst., *Panicum alopecuros* Lam, *Panicum barbatum* Roxb, *Panicum cauda-ratti* Schumach., *Panicum cenchroides* Rich., *Panicum densispica* Poir., *Panicum dentispica* Kunth, *Panicum erubescens* Willd., *Panicum fuscescens* Willd. ex Nees, *Panicum holcoides* Roxb., *Panicum longisetum* Poir., *Panicum polystachion* L., *Panicum subangustum* Schumach., *Panicum triticoides* Poir., *Pennisetum alopecuroides* Ham., *Pennisetum arvense* Pilg., *Pennisetum barbatum* Schult., *Pennisetum borbonicum* Kunth, *Pennisetum breve* Nees, *Pennisetum cauda-ratti* (Schumach.) Franch., *Pennisetum ciliatum* Parl. ex Webb, *Pennisetum dasystachyum* Desv., *Pennisetum erubescens* (Willd.) Desv., *Pennisetum erubescens* (Willd.) Link, *Pennisetum flavescens* J.Presl, *Pennisetum gabonense* Franch., *Pennisetum gracile* Benth., *Pennisetum hamiltonii* Steud., *Pennisetum hirsutum* Nees, *Pennisetum holcoides* (Roxb.) Schult., *Pennisetum myurus* Parl., *Pennisetum nicaraguense* E.Fourn., *Pennisetum ovale* Rupr. ex Steud., *Pennisetum pallidum* Nees, *Pennisetum polystachion* subsp. *setosum* (Sw.) Brunken, *Pennisetum polystachion* f. *viviparum* Fosberg & Sachet, *Pennisetum polystachyum* (L.) Schult., *Pennisetum ramosum* (Hochst.) Schweinf., *Pennisetum reversum* Hack.,

Pennisetum richardii Kunth, *Pennisetum setosum* var. *breve* (Nees) Döll, *Pennisetum sieberi* Kunth, *Pennisetum stenostachyum* Peter, *Pennisetum subangustum* (Schumach.) Stapf & C.E.Hubb., *Pennisetum tenuispiculatum* Steud., *Pennisetum triticoides* (Poir.) Roem. & Schult., *Pennisetum uniflorum* Kunth, *Setaria cenchroides* (Rich.) Roem. & Schult., *Setaria erubescens* (Willd.) P.Beauv.

70. *Setaria glauca* (L.) P.Beauv.

Synonyms: *Alopecurus typhoides* Burm.f., *Cenchrus americanus* (L.) Morrone, *Cenchrus spicatus* (L.) Cav., *Cenchrus spicatus* (L.) Kuntze, *Chaetochloa glauca* (L.) Scribn., *Chaetochloa lutescens* (Weigel) Stuntz, *Chamaeraphis glauca* (L.) Kuntze, *Holcus racemosus* Forssk., *Holcus spicatus* L., *Ixophorus glaucus* (L.) Nash, *Panicum americanum* L., *Panicum coeruleum* Mill., *Panicum glaucum* L., *Panicum glaucum* Nees, *Panicum indicum* Mill., *Panicum involucratum* Roxb., *Panicum lutescens* Weigel, *Panicum sericeum* Aiton, *Panicum spicatum* (L.) Roxb., *Penicillaria arabica* A.Braun, *Penicillaria ciliata* Willd., *Penicillaria cylindrica* Roem. & H. Schult., *Penicillaria deflexa* Andersson ex A.Braun, *Penicillaria deflexa* Andersson, *Penicillaria indica* A. Braun, *Penicillaria involucrata* (Roxb.) Schult., *Penicillaria macrostachya* Klotzsch, *Penicillaria mossambicensis* Müll.Berol, *Penicillaria mossambicensis* Klotzsch ex A. Braun, *Penicillaria nigritarum* Schltdl., *Penicillaria plukenetii* Link, *Penicillaria roxburghii* Müll.Berol, *Penicillaria roxburghii* A. Braun, *Penicillaria solitaria* Stokes, *Penicillaria spicata* (L.) Willd., *Penicillaria spicata* P. Beauv., *Penicillaria typhoidea* (Burm.) Schltdl., *Penicillaria typhoides* Schltdl., *Penicillaria vulpina* Müll. Berol., *Penicillaria willdenowii* Klotzsch ex. A.Braun & C.D.Bouché, *Pennisetum albicauda* Stapf & C.E.Hubb., *Pennisetum americanum* (L.) Leeke, *Pennisetum americanum* (L.) K. Schum., *Pennisetum americanum* subsp. *americanum*, *Pennisetum americanum* f. *echinurus* (K.Schum.) Leeke, *Pennisetum americanum* subsp. *spicatum* (L.) Maire & Weiller, *Pennisetum americanum* subsp. *typhoideum* (Rich.) Maire & Weiller, *Pennisetum ancylochaete* Stapf & C.E.Hubb., *Pennisetum aureum* Link, *Pennisetum cereale* Trin., *Pennisetum cinereum* Stapf & C.E.Hubb., *Pennisetum echinurus* (K.Schum.) Stapf & C.E.Hubb., *Pennisetum gambiense* Stapf & C.E.Hubb., *Pennisetum gibbosum* Stapf & C.E.Hubb., *Pennisetum glaucum* (L.) R.Br., *Pennisetum leonis* Stapf & C.E.Hubb.,

Pennisetum linnaei Kunth, *Pennisetum maiwa* Stapf & C.E.Hubb., *Pennisetum malacochaete* Stapf & C.E.Hubb., *Pennisetum megastachyum* Steud., *Pennisetum nigritarum* (Schltdl.) T.Durand & Schinz, *Pennisetum nigritarum* var. *deflexum* (A.Braun) T.Durand & Schinz, *Pennisetum nigritarum* var. *macrostachyum* (A.Braun) T.Durand & Schinz, *Pennisetum plukenetii* (Link) T.Durand & Schinz, *Pennisetum pycnostachyum* Stapf & C.E.Hubb., *Pennisetum solitarium* Stokes, *Pennisetum spicatum* (L.) Körn., *Pennisetum spicatum* (L.) Roem. & Schult., *Pennisetum spicatum* var. *echinurus* K.Schum., *Pennisetum spicatum* var. *longipedunculata* K.Schum., *Pennisetum spicatum* var. *macrostachyum* (A.Braun) K.Schum., *Pennisetum spicatum* var. *typhoideum* (Rich.) T.Durand & Schinz, *Pennisetum spicatum* var. *typhoideum* Chiov., *Pennisetum spicatum* subsp. *willdenowii* K.Schum., *Pennisetum typhoides* (Burm.f.) Stapf & C.E.Hubb., *Pennisetum typhoideum* Rich., *Pennisetum typhoideum* Delile, *Pennisetum typhoideum* var. *echinurus* (K.Schum.) Rendle, *Pennisetum typhoideum* var. *plukenetii* (Link) Rendle, *Phleum africanum* Lour., *Setaria glauca* Hack., *Setaria humifusa* Dumort., *Setaria lutescens* (Weigel) F.T.Hubb., *Setaria rufa* Chevall., *Setaria sericea* (Sol.) P.Beauv., *Setariopsis glauca* (L.) Samp.

71. *Setaria tomentosa* (Roxb.) Kunth

Synonyms: *Chaetochloa intermedia* (Roem. & Schult.) Stuntz, *Oplismenus tomentosus* (Roxb.) Schult., *Panicum hookerianum* Balansa, *Panicum intermedium* (Roem. & Schult.) Roth, *Panicum tomentosum* Roxb., *Setaria intermedia* Roem. & Schult.

72. *Setaria verticillata* (L.) P.Beauv.

Synonyms: *Chaetochloa brevispica* Scribn. & Merr., *Chaetochloa brevispica* Scribn., *Chaetochloa verticillata* (L.) Scribn., *Chaetochloa verticillata* var. *breviseta* (Mutel) Farw., *Chamaeraphis italica* var. *aparine* (Steud.) Kuntze, *Chamaeraphis italica* var. *densa* Kuntze, *Chamaeraphis italica* var. *rottleri* (Spreng.) Kuntze, *Chamaeraphis italica* var. *verticillata* (L.) Kuntze, *Chamaeraphis verticillata* (L.) Porter, *Cynosurus paniceus* L., *Ixophorus verticillatus* (L.) Nash, *Panicum adhaerens* Forssk., *Panicum albospiculatum* Swallen, *Panicum aparine* Steud., *Panicum apricum* Swallen, *Panicum asperum* Lam., *Panicum* *bambusifolium* Desv., *Panicum italicum* Ucria, *Panicum kleinii* Swallen, *Panicum pompale* Swallen, *Panicum respiciens* (A.Rich.) Hochst. ex Steud., *Panicum respiciens* Hochst. ex A.Rich., *Panicum rottleri* (Spreng.) Nees, *Panicum rude* Nees, *Panicum secundum* Trin., *Panicum secundum* var. *inaequiglume* Döll, *Panicum secundum* var. *subaequiglume* Döll, *Panicum semitectum* Swallen, *Panicum vagum* Scop., *Panicum verticillatum* L., *Panicum verticillatum* Rottler ex Spreng., *Panicum verticillatum* subsp. *aparine* (Steud.) T.Durand & Schinz, *Panicum verticillatum* var. *aparine* (Steud.) Asch. & Schweinf., *Panicum verticillatum* var. *arenosum* (Schur) Asch. & Graebn., *Panicum verticillatum* var. *brevisetum* Mutel, *Panicum verticillatum* var. *parviflorum* Döll, *Panicum verticillatum* var. *retrorsum* Asch. & Schweinf., *Panicum viride* Desf., *Pennisetum respiciens* A.Rich., *Pennisetum verticillatum* (L.) R.Br., *Pennisetum verticillatum* R.Br. ex Sweet, *Setaria adhaerens* (Forssk.) Chiov., *Setaria adhaerens* var. *font-queri* Calduch, *Setaria adhaerens* subsp. *verticillata* (L.) Belo-Corr., *Setaria adhaerens* var. *verticillata* (L.) Belo-Corr., *Setaria adhaerens* subsp. *verticillata* Belo-Correira, *Setaria adhaerens* var. *vertillata* (L.) Belo-Corr., *Setaria ambigua* f. *major* Bujor., *Setaria ambigua* var. *major* Bujor., *Setaria ambigua* f. *ramiflora* Bujor., *Setaria aparine* (Steud.) Chiov., *Setaria brevispica* (Scribn. & Merr.) K.Schum., *Setaria carnei* Hitchc., *Setaria decipiens* f. *major* (Bujor.) Soó, *Setaria depauperata* Phil., *Setaria floribunda* Spreng., *Setaria italica* var. *aparine* (Steud.) Kuntze, *Setaria leiantha* f. *subhirsuta* Hack., *Setaria nubica* Link, *Setaria panicea* (L.) Schinz & Thell., *Setaria pratensis* Phil., *Setaria respiciens* (A.Rich.) Walp., *Setaria rottleri* Spreng., *Setaria teysmannii* Miq., *Setaria verticillata* f. *ambigua* (Guss.) T. Koyama, *Setaria verticillata* subsp. *aparine* (Steud.) T.Durand & Schinz, *Setaria verticillata* var. *aparine* (Steud.) Asch. & Graebn., *Setaria verticillata* var. *aparine* (Steud.) Asch. & Schweinf., *Setaria verticillata* f. *arenosa* (Schur) Morariu, *Setaria verticillata* var. *arenosa* Schur, *Setaria verticillata* var. *font-queri* (Calduch) O.Bolòs & Vigo, *Setaria verticillata* var. *pilifera* B.de Lesd., *Setaria verticillata* var. *respiciens* (A.Rich.) A.Braun, *Setaria verticillata* var. *respiciens* (Walp.) K. Schum., *Setaria verticillata* var. *verticillata*, *Setaria verticillata* f. *verticillata*, *Setaria verticilliformis* Dumort., *Setaria viridis* var. *insularis* N.Terracc., *Setariopsis verticillata* (L.) Samp.

GROUP: POOIDEAE

TRIBE: ALEUROPODEAE

73. *Aeluropus lagopoides* (L.) Thwaites

Synonyms: *Aeluropus bombycinus* Fig. & De Not., *Aeluropus brevifolius* (Willd.) Nees ex Steud., *Aeluropus brevifolius* var. *longifolius* Chiov., *Aeluropus brevifolius* var. *pygmaeus* Chiov., *Aeluropus concinnus* Fig. & De Not., *Aeluropus erythraeus* Mattei, *Aeluropus erythraeus* var. *scandens* Mattei, *Aeluropus laevis* Trin., *Aeluropus lagopodioides* Trin. ex Thwaites, *Aeluropus lagopoides* (L.) Chiov., *Aeluropus lagopoides* var. *glabrifolius* Czopanov, *Aeluropus lagopoides* var. *hispidulus* Halácsy, *Aeluropus lagopoides* subsp. *repens* (Desf.) Tzvelev, *Aeluropus littoralis* var. *mesopotamicus* (Nábelek) Bor, *Aeluropus littoralis* subsp. *repens* (Desf.) Trab., *Aeluropus littoralis* var. *repens* (Desf.) Coss. & Durieu, *Aeluropus longispicatus* Parsa, *Aeluropus massauensis* (Fresen.) Mattei, *Aeluropus mucronatus* var. *erythraeus* N.Terracc., *Aeluropus niliacus* (Spreng.) Steud., *Aeluropus pubescens* Trin. ex Steud., *Aeluropus repens* (Desf.) Parl., *Aeluropus sinaicus* Fig. & De Not., *Aeluropus villosus* Hook.f., *Aeluropus villosus* Trin., *Aira lagopoides* (L.) Scop., *Calotheca massauensis* (Fresen.) Hochst. & Steud., *Calotheca niliaca* Spreng., *Calotheca repens* (Desf.) Spreng., *Coelachyrum indicum* Hack., *Dactylis bombycina* (Fig. & De Not.) Steud., *Dactylis cynosuroides* Roth, *Dactylis lagopodioides* Dalzell & A.Gibson, *Dactylis lagopoides* L., *Dactylis massauensis* (Fresen.) Steud., *Dactylis repens* Desf., *Distichlis sudanensis* Beetle, *Eleusine brevifolia* (Willd.) R.Br. ex Hook.f., *Eragrostis brevifolia* (Willd.) Benth, *Koeleria brevifolia* (Willd.) Spreng., *Koeleria lagopoides* (L.) Panz. ex Spreng., *Melica reptans* Raspail, *Poa brevifolia* (Willd.) Kunth, *Poa lagopodoides* (L.) Kunth, *Poa lagopoides* (L.) Kunth, *Poa maritima* Cav., *Poa massauensis* Fresen., *Poa pungens* Georgi, *Poa ramosa* Savi, *Poa repens* (Desf.) M.Bieb., *Poa tunetana* Spreng., *Sesleria brevifolia* (Willd.) Panz., *Sesleria lagopoides* (L.) Spreng.

TRIBE: ISACHNEAE

74. *Isachne globosa* (Thunb.) Kuntze

Synonyms: *Agrostis globosa* (Thunb.) Poir., *Aira violacea* Willd. ex Steud., *Eriochloa globosa* (Thunb.) Kunth, *Helopus globosus* (Thunb.)

Steud., *Isachne adstans* (Steud.) Miq., *Isachne atrovirens* (Trin.) Trin., *Isachne australis* R.Br., *Isachne australis* var. *effusa* Trin. ex Hook.f., *Isachne dispar* Trin., *Isachne dispar* var. *villosa* C.E.C.Fisch., *Isachne geniculata* Griff., *Isachne globosa* var. *brevispicula* Ohwi, *Isachne globosa* var. *ciliaris* Ohwi, *Isachne globosa* var. *compacta* W.Z.Fang ex S.L.Chen, *Isachne globosa* var. *daviumbuensis* Jansen, *Isachne globosa* var. *effusa* (Trin. ex Hook.f.) Senaratna, *Isachne globosa* var. *obscura* (Buse) Henrard, *Isachne heterantha* Hayata, *Isachne javana* Nees ex Miq., *Isachne lepidota* (Steud.) Walp., *Isachne lutaria* Santos, *Isachne miliacea* var. *javanica* (Buse) Henrard, *Isachne miliacea* var. *madurensis* Jansen, *Isachne miliacea* var. *minutula* (Gaudich.) Fosberg & Sachet, *Isachne miliacea* var. *obscura* Buse, *Isachne miliacea* var. *ovalifolia* Jansen, *Isachne miliacea* var. *stricta* Ridl., *Isachne minutula* (Gaudich.) Kunth, *Isachne minutula* var. *javanica* Buse, *Isachne nodibarbata* (Hochst. ex Steud.) Henrard, *Isachne pangerangensis* var. *rhabdina* (Steud.) Jansen, *Isachne pangerangensis* var. *rhabdina* (Steud.) Henrard, *Isachne ponapensis* Hosok., *Isachne rhabdina* (Steud.) Henrard, *Isachne rhabdina* (Steud.) Ohwi, *Isachne stigmatosa* Griff., *Isachne subglobosa* Hatus. & T.Koyama, *Milium globosum* Thunb., *Panicum adstans* Steud., *Panicum antipodum* Spreng., *Panicum antipodum* var. *atrovirens* (Trin.) Trin., *Panicum atrovirens* Trin., *Panicum australe* (R.Br.) Raspail, *Panicum batavicum* Steud., *Panicum dispar* (Trin.) Steud., *Panicum lepidotum* Steud., *Panicum minutulum* Gaudich., *Panicum nodibarbatum* Hochst. ex Steud., *Panicum rhabdinum* Steud., *Panicum stigmatosum* Drury, *Panicum violaceum* Klein ex Thiele

TRIBE: PETORIDEAE

75. *Perotis indica* (L.) Kuntze

Synonyms: *Perotis indica* f. *glabra* Chiov., *Perotis indica* f. *hirtigluma* Chiov., *Perotis indica* var. *keelakaraiensis* P.Umam. & P.Daniel

TRIBE: ARISTIDEAE

76. *Aristida adscensionis* L.

Synonyms: *Aristida abyssinica* Trin. & Rupr., *Aristida abyssinica* (Trin & Rupr.) Henrard, *Aristida adscensionis* var. *abortiva* A.Beetle, *Aristida*

adscensionis var. *abyssinica* (Trin. & Rupr.) Engl., *Aristida adscensionis* var. *adscensionis*, *Aristida adscensionis* var. *aethiopica* (Trin. & Rupr.) T.Durand & Schinz, *Aristida adscensionis* var. *breviseta* Hack., *Aristida adscensionis* var. *bromoides* (Kunth) Henrard, *Aristida adscensionis* subsp. *caerulescens* (Bourreil & Trouin ex P. Auquier & J. Duvigneaud) Bourreil & Trouin, *Aristida adscensionis* var. *canariensis* (Willd.) T.Durand & Schinz, *Aristida adscensionis* var. *coarctata* (Kunth) Kuntze, *Aristida adscensionis* subsp. *coerulescens* (Desf.) Bourreil & Trouin ex Auquier & J.Duvign., *Aristida adscensionis* var. *coerulescens* (Desf.) Hack., *Aristida adscensionis* subsp. *coerulescens* (Desf.) R. Malagarr., *Aristida adscensionis* var. *condensata* (Hack.) Henrard, *Aristida adscensionis* var. *decolorata* (E.Fourn.) Beetle, *Aristida adscensionis* var. *ehrenbergii* (Trin. & Rupr.) Henrard, *Aristida adscensionis* var. *festucoides* (Poir.) Henrard, *Aristida adscensionis* var. *gigantea* (L.f.) Kuntze, *Aristida adscensionis* var. *glabricallis* Maire & Weiller, *Aristida adscensionis* subsp. *guineensis* (Trin. & Rupr.) Henrard, *Aristida adscensionis* var. *guineensis* (Trin. & Rupr.) Henrard, *Aristida adscensionis* subsp. *heymannii* (Regel) Tzvelev, *Aristida adscensionis* var. *humilis* (Kunth) Kuntze, *Aristida adscensionis* var. *interrupta* (Cav.) Beetle, *Aristida adscensionis* var. *modesta* Hack., *Aristida adscensionis* f. *modestina* Hack., *Aristida adscensionis* var. *nigrescens* (J.Presl) Beetle, *Aristida adscensionis* var. *normalis* Kuntze, *Aristida adscensionis* var. *pumila* (Decne.) Coss. & Durieu, *Aristida adscensionis* var. *pygmaea* (Trin. & Rupr.) T.Durand & Schinz, *Aristida adscensionis* var. *scabriflora* Hack., *Aristida adscensionis* var. *senegalensis* (Trin.) T.Durand & Schinz, *Aristida adscensionis* var. *spigera* (Trin. & Rupr.) Henrard, *Aristida adscensionis* var. *strictiflora* (Trin. & Rupr.) T.Durand & Schinz, *Aristida adscensionis* f. *viridis* Kuntze, *Aristida adscensionis* var. *vulpioides* (Hance) Hack. ex Henrard, *Aristida aethiopica* Trin. & Rupr., *Aristida aethiopica* (Trin. & Rupr.) Trin. & Rupr. ex Henrard, *Aristida americana* var. *bromoides* (Kunth) Scribn. & Merr., *Aristida arabica* Trin. & Rupr., *Aristida bromoides* Kunth, *Aristida caerulescens* var. *pumila* (Decne.) Post, *Aristida canariensis* Willd., *Aristida cardosoi* Cout., *Aristida coarctata* Kunth, *Aristida coerulescens* var. *arabica* Henrard, *Aristida coerulescens* var. *breviaristata* Schweinf., *Aristida coerulescens* var. *laevilemma* Maire, *Aristida coerulescens* var. *pumila* (Decne.) Döll, *Aristida confusa* Trin. & Rupr., *Aristida confusa* (Trin. & Rupr.) Trin. & Rupr. ex Henrard, *Aristida curvata* (Nees) Nees ex A.Rich., *Aristida curvata* var. *abyssinica* A.Rich., *Aristida curvata* var. *nana* (Nees) Henrard, *Aristida debilis* Mez, *Aristida depressa* Retz., *Aristida depressa* var. *bromoides* (Kunth) Trin. & Rupr., *Aristida depressa* var. *coarctata* (Kunth) Trin. & Rupr., *Aristida dispersa* Trin. & Rupr., *Aristida dispersa* var. *bromoides* (Kunth) Trin. & Rupr., *Aristida dispersa* var. *coarctata* (Kunth) Trin. & Rupr., *Aristida dispersa* var. *humilis* (Kunth) Trin. & Rupr., *Aristida dispersa* var. *nana* Trin. & Rupr., *Aristida dispersa* var. *nigrescens* (J.Presl) Trin. & Rupr., *Aristida ehrenbergii* Trin. & Rupr., *Aristida ehrenbergii* (Trin. & Rupr.) Trin. & Rupr. ex Henrard, *Aristida elatior* Cav., *Aristida fasciculata* Torr., *Aristida festucoides* Poir., *Aristida festucoides* Steud. & Hochst., *Aristida gigantea* L.f., *Aristida grisebachiana* E.Fourn., *Aristida grisebachiana* var. *decolorata* E.Fourn., *Aristida grisebachiana* var. *grisebachiana*, *Aristida guineensis* Trin. & Rupr., *Aristida heymannii* Regel, *Aristida humilis* Kunth, *Aristida interrupta* Cav., *Aristida luzoniensis* Cav., *Aristida macrochloa* Hochst., *Aristida maritima* Steud., *Aristida mauritiana* Kunth, *Aristida mauritiana* Hochst. ex A.Rich., *Aristida modatica* Steud., *Aristida mongholica* Trin., *Aristida mongholica* (Trin. & Rupr.) Henrard, *Aristida nana* Steud., *Aristida nigrescens* J.Presl, *Aristida paniculata* Forssk., *Aristida peruviana* Beetle, *Aristida pumila* Decne., *Aristida pusilla* Trin. & Rupr., *Aristida pygmaea* Trin. & Rupr., *Aristida racemosa* Spreng. *Aristida schaffneri* E.Fourn., *Aristida schaffneri* E. Fourn. ex Hemsl., *Aristida senegalensis* (Trin. & Rupr.) Henrard, *Aristida simplicissima* Steud., *Aristida spicigera* Trin. & Rupr., *Aristida stricta* var. *decolorata* E.Fourn. ex Dávila & Sánchez-Ken, *Aristida strictiflora* Trin. & Rupr., *Aristida submucronata* Schumach. & Thonn., *Aristida submucronata* var. *scabra* Henrard, *Aristida swartziana* Steud., *Aristida teneriffae* Steud., *Aristida thonningii* Trin. & Rupr., *Aristida viciosorum* Pau, *Aristida vulgaris* Trin. & Rupr, *Aristida vulgaris* var. *abyssinica* Trin. & Rupr., *Aristida vulgaris* var. *aethiopica* Trin. & Rupr., *Aristida vulgaris* var. *canariensis* (Willd.) Trin. & Rupr., *Aristida vulgaris* var. *coerulescens* (Desf.) Trin. & Rupr., *Aristida vulgaris* var. *confusa* Trin. & Rupr., *Aristida vulgaris* var. *depressa* (Retz.) Trin. & Rupr., *Aristida vulgaris* var. *ehrenbergii* Trin. & Rupr., *Aristida vulgaris* var. *mongholica* Trin. & Rupr., *Aristida vulgaris* var. *pumila* (Decne.) Trin. & Rupr., *Aristida vulgaris* var. *senegalensis* Trin. & Rupr.,

Aristida vulgaris var. *strictiflora* Trin. & Rupr., *Aristida vulpioides* Hance, *Arthratherum adscensionis* subsp. *heymannii* (Regel) Tzvelev, *Chaetaria adscensionis* (L.) P.Beauv., *Chaetaria bromoides* (Kunth) Roem. & Schult., *Chaetaria canariensis* (Willd.) P.Beauv., *Chaetaria canariensis* (Willd.) Nees, *Chaetaria coarctata* (Kunth) Roem. & Schult., *Chaetaria coerulescens* (Desf.) P.Beauv., *Chaetaria curvata* Nees, *Chaetaria depressa* (Retz.) P.Beauv., *Chaetaria elatior* (Cav.) P.Beauv., *Chaetaria fasciculata* (Torr.) Schult. & Schult.f., *Chaetaria festucoides* (Poir.) P.Beauv., *Chaetaria forskolii* Nees, *Chaetaria gigantea* (L.f.) P.Beauv., *Chaetaria humilis* (Humb. & Bonpl.) Roem. & Schult., *Chaetaria interrupta* (Cav.) P.Beauv., *Chaetaria luzoniensis* (Cav.) P.Beauv., *Chaetaria mauritiana* (Kunth) Nees, *Chaetaria mauritiana* var. *nana* Nees

77. *Aristida funiculata* Trin. & Rupr.

Synonyms: *Aristida foenicularis* Edgew., *Aristida funiculata* var. *brevis* Maire, *Aristida funiculata* var. *mallica* (Edgew.) Henrard, *Aristida funiculata* var. *paradoxa* (J.A.Schmidt) Henrard, *Aristida funiculata* var. *royleana* (Trin. & Rupr.) Hook.f., *Aristida macrathera* A.Rich., *Aristida mallica* Edgew., *Aristida paradoxa* J.A.Schmidt, *Aristida royleana* Trin. & Rupr., *Aristida stipacea* Ehrenb. & Hemprich ex Boiss., *Arthratherum royleanum* (Trin. & Rupr.) Edgew. ex Aitch.

TRIBE: CHLORIDEAE

78. *Chloris barbata* Sw.

Synonyms: *Andropogon barbatus* L., *Chloris barbata* var. *divaricata* Kuntze, *Chloris inflata* Link, *Chloris longifolia* Steud., *Chloris paraguaiensis* Steud., *Chloris rufescens* Steud.

79. *Chloris montana* Roxb.

Synonyms: *Chloris montana* var. *glauca* Hook.f.

80. *Chloris virgata* Sw.

Synonyms: *Agrostomia barbata* Cerv., *Agrostomia barbata* Cerv., *Chloris alba* var. *aristulata* Torr., *Chloris albertii* Regel, *Chloris barbata* var. *decora* (Steud.) Benth., *Chloris barbata* var. *meccana* (Hochst. ex Steud.) Asch. & Schweinf., *Chloris brachystachys* Andersson, *Chloris* *caudata* Trin., *Chloris compressa* DC., *Chloris decora* Nees ex Steud., *Chloris decora* Thwaites, *Chloris elegans* Kunth, *Chloris gabrielae* Domin, *Chloris madagascariensis* Steud., *Chloris meccana* Hochst. ex Steud., *Chloris meccana* Hochst. & Steud. ex Schltdl., *Chloris multiradiata* Hochst., *Chloris notocoma* Hochst., *Chloris penicillata* Jan ex Trin., *Chloris polydactyla* P.Durand, *Chloris polydactyla* subsp. *multiradiata* (Hochst.) Chiov., *Chloris pubescens* Lag., *Chloris rogeonii* A.Chev., *Chloris tibestica* Quézel, *Chlori virgata* var. *elegans* (Kunth) Stapf

81. *Cynadon dactylon* (L.) Pers.

Synonyms: *Agrostis linearis* Retz., *Agrostis stellata* Willd., *Capriola dactylon* (L.) Kuntze, *Capriola dactylon* (L.) Hitchc., *Capriola dactylon* var. *maritima* (Kunth) Hitchc., *Chloris cynodon* Trin., *Chloris maritima* Trin., *Chloris paytensis* Steud., *Cynodon affinis* Caro & E.A.Sánchez, *Cynodon aristiglumis* Caro & E.A.Sánchez, *Cynodon aristulatus* Caro & E.A.Sánchez, *Cynodon barberi* f. *longifolia* Join, *Cynodon dactylon* var. *affinis* (Caro & E.A.Sánchez) Romero Zarco, *Cynodon dactylon* var. *aridus* J.R.Harlan & de Wet, *Cynodon dactylon* var. *biflorus* Merino, *Cynodon dactylon* var. *densus* Hurcombe, *Cynodon dactylon* var. *elegans* Rendle, *Cynodon dactylon* subsp. *glabratus* (Steud.) A.Chev., *Cynodon dactylon* var. *glabratus* (Steud.) Chiov., *Cynodon dactylon* var. *hirsutissimus* (Litard. & Maire) Maire, *Cynodo dactylon* var. *longiglumis* Caro & E.A.Sánchez, *Cynodon dactylon* f. *major* (Beck) Soó, *Cynodon dactylon* var. *maritimus* (Kunth) Hack., *Cynodon dactylon* subsp. *nipponicus* (Ohwi) T.Koyama, *Cynodon dactylon* var. *nipponicus* Ohwi, *Cynodon dactylon* var. *pilosus* Caro & E.A.Sánchez, *Cynodon dactylon* var. *polevansii* (Stent) J.R. Harlan & de Wet, *Cynodon dactylon* var. *pulchellus* Benth., *Cynodon dactylon* var. *sarmentosus* Parodi, *Cynodon dactylon* var. *septentrionalis* (Asch. & Graebn.) Ravarut, *Cynodon dactylon* var. *stellatus* (Willd.) T. Durand & Schinz, *Cynodon dactylon* f. *villosus* (Grossh.) Regel ex Roshev., *Cynodon dactylon* var. *villosus* Regel, *Cynodon dactylon* var. *villosus* Grossh., *Cynodon dactylon* f. *viviparus* Beetle, *Cynodon decipiens* Caro & E.A.Sánchez, *Cynodon distichloides* Caro & E.A.Sánchez, *Cynodon erectus* J.Presl, *Cynodon glabratus* Steud., *Cynodon hirsutissimus* (Litard. & Maire) Caro & E.A.Sánchez, *Cynodon iraquensis* Caro, *Cynodon laeviglumis* Caro & E.A.Sánchez, *Cynodon* *linearis* Willd., *Cynodon maritimus* Kunth, *Cynodon maritimus* var. *breviglumis* Caro & E.A.Sánchez, *Cynodon maritimus* var. *grandispiculus* Caro & E.A.Sánchez, *Cynodon maritimus* var. *vaginiflorus* Caro, *Cynodon mucronatus* Caro & E.A.Sánchez, *Cynodon nitidus* Caro & E.A.Sánchez, *Cynodon pascuus* Nees, *Cynodon pedicellatus* Caro, *Cynodon polevansii* Stent, *Cynodon repens* Dulac, *Cynodon sarmentosus* Gray, *Cynodon scabrifolius* Caro, *Cynodon stellatus* Willd., *Cynodon tenuis* Trin., *Cynodon umbellatus* (Lam.) Caro, *Cynosurus dactylon* (L.) Pers., *Cynosurus uniflorus* Walter, *Dactilon officinale* Vill., *Digitaria ambigua* (Lapeyr. ex DC.) Mérat, *Digitaria dactylon* (L.) Scop., *Digitaria glumaepatula* (Steud.) Miq., *Digitaria glumipatula* (Steud.) Miq., *Digitaria linearis* (L.) Pers., *Digitaria linearis* (Retz.) Spreng., *Digitaria littoralis* Salisb., *Digitaria maritima* (Kunth) Spreng., *Digitaria stolonifera* Schrad., *Fibichia dactylon* (L.) Beck, *Fibichia umbellata* Koeler, *Fibichia umbellata* var. *biflora* (Merr.) Beck, *Fibichia umbellata* f. *glabrescens* Beck, *Fibichia umbellata* f. *major* Beck, *Milium dactylon* (L.) Moench, *Panicum ambiguum* (DC.) Le Turq., *Panicum dactylon* L., *Panicum glumipatulum* Steud., *Panicum lineare* L., *Paspalum ambiguum* DC., *Paspalum dactylon* (L.) Lam., *Paspalum umbellatum* Lam., *Phleum dactylon* (L.) Georgi, *Syntherisma linearis* (L.) Nash, *Vilfa linearis* (Retz.) P.Beauv., *Vilfa stellata* (Willd.) P.Beauv.

82. *Melanocenchris jaequemontii* Jaub. & Spach

Synonyms: *Amphipogon humilis* Buch.-Ham. ex Wall., *Gracilea royleana* Hook.f.

83. *Oropetium villosulum* Stapf ex Bor

No synonyms recorded

84. *Schoenefeldia gracilis* Kunth

Synonyms: *Chloris myosuroides* Hook.f., *Chloris pallida* (Edgew.) Hook.f., *Schoenefeldia nutans* Steud., *Schoenefeldia pallida* Edgew., *Schoenefeldia ramosa* Trin., *Schoenefeldia stricta* Steud.

85. *Tetrapogon tenellus* (Roxb.) Chiov.

Synonyms: *Chloris macrantha* Desv., *Chloris macrantha* Jaub. & Spach, *Chloris tenella* J.Koenig ex Roxb., *Chloris triangulata* Hochst. ex A.Rich., *Ctenium indicum* Spreng., *Lepidopironia triangulata* (Hochst. ex A.Rich.)

Hochst. ex Schumach., *Tetrapogon macranthus* (Desv.) Benth., *Tetrapogon macranthus* (Desv.) Chiov., *Tetrapogon triangularis* (Hochst. ex A.Rich.) Hochst., *Tetrapogon triangulatus* (A.Rich.) Schweinf., *Tetrapogon triangulatus* var. *sericatus* Chiov.

86. *Tetrapogon villosus* Desf.

Synonyms: *Chloris tetrapogon* P.Beauv., *Chloris villosa* (Desf.) Pers., *Chloris villosa* var. *sinaica* Decne., *Tetrapogon villosus* var. *monostachyus* Batt. & Trab.

TRIBE: ERAGROSTEAE

87. *Dactyloctenium aegyptium* (L.) Willd.

Synonyms: *Aegilops saccharina* Walter, *Chloris guineensis* Schumach. & Thonn., *Chloris mucronata* Michx., *Chloris prostrata* (Willd.) Poir., *Cynosurus aegyptius* L., *Dactyloctenium aegyptiacum* Willd., *Dactyloctenium aegyptiacus* Willd., *Dactyloctenium aegyptium* (L.) Richt., *Dactyloctenium aegyptium* (L.) P. Beauv., *Dactyloctenium aegyptium* var. *mucronatum* (Michx.) Schweinf., *Dactyloctenium aegyptium* f. *viviparum* Beetle, *Dactyloctenium aegyptius* var. *mucronatum* (Michx.) Lanza & Mattei, *Dactyloctenium figarii* De Not., *Dactyloctenium meridionale* Ham., *Dactyloctenium mpuetense* De Wild., *Dactyloctenium mucronatum* (Michx.) Willd., *Dactyloctenium mucronatum* var. *erectum* E.Fourn., *Dactyloctenium prostratum* Willd., *Eleusine aegyptia* (L.) Roxb., *Eleusine aegyptia* (L.) Pers., *Eleusine aegyptia* Raf., *Eleusine aegyptia* (L.) Roberty, *Eleusine aegyptia* (L.) Desf., *Eleusine aegyptiaca* (L.) Desf., *Eleusine cruciata* Lam., *Eleusine egyptia* Raf., *Eleusine mucronata* Stokes, *Eleusine mucronata* (Michx.) Hornem., *Eleusine pectinata* Moench, *Eleusine prostrata* Spreng., *Rabdochloa mucronata* (Michx.) P.Beauv.

88. *Dactyloctenium sindicum* Boiss.

Synonyms: *Dactyloctenium glaucophyllum* Courbai, *Dactyloctenium glaucophyllum* var. *elongatior* Courbai, *Dactyloctenium glaucophyllum* var. *robustior* Courbai, *Eleusine glaucophylla* (Courbon) Munro ex Benth., *Eleusine scindica* (Boiss.) Duthie

89. *Dactyloctenium giganteum* Fischer and Schweick.

Synonyms: No synonyms

90. *Desmostachya bipinnata* (L.) Stapf

Synonyms: *Briza bipinnata* L., *Cynosurus durus* Forssk., *Desmostachya cynosuroides* (Retz.) Stapf ex Massey, *Desmostachya cynosuroides* (Retz.) Haines, *Desmostachys bipinnata* (L.) Stapf, *Dinebra dura* (L.) Lag., *Eragrostis bipinnata* (L.) K.Schum., *Eragrostis bipinnata* (L.) Muschl., *Eragrostis cynosuroides* (Retz.) P.Beauv., *Eragrostis thunbergii* Baill., *Leptochloa bipinnata* (L.) Hochst., *Megastachya bipinnata* (L.) P.Beauv., *Poa cynosuroides* Retz., *Pogonarthria bipinnata* (L.) Chiov., *Rabdochloa bipinnata* (L.) Kuntze, *Stapfiola bipinnata* (L.) Kuntze, *Uniola bipinnata* (L.) L.

91. *Dinebra retroflexa* (Vahl) Panz.

Synonyms: *Cynosurus retroflexus* Vahl, *Dactylis paspaloides* Willd., *Dinebra aegyptiaca* Delile, *Dinebra aegyptiaca* Kunth, *Dinebra arabica* Jacq., *Dinebra brevifolia* Steud., *Dinebra paspaloides* P.Beauv., *Dinebra retroflexa* var. *brevifolia* (Steud.) T.Durand & Schinz, *Eleusine calycina* Roxb., *Leptochloa arabica* (Jacq.) Steud., *Leptochloa arabica* (Jacq.) Kunth, *Leptochloa calycina* (Roxb.) Kunth, *Leptochloa coromandelina* Steud.

92. *Eleusine indica* (L.) Gaertn.

Synonyms: *Agropyron geminatum* Schult. & Schult.f., *Cynodon indicus* (L.) Raspail, *Cynosurus indicus* L., *Cynosurus pectinatus* Lam., *Eleusine distachya* Nees, *Eleusine distans* Moench, *Eleusine distans* Link, *Eleusine glabra* Schumach., *Eleusine gonantha* Schrank, *Eleusine gouinii* E.Fourn., *Eleusine gracilis* Salisb., *Eleusine inaequalis* E.Fourn., *Eleusine inaequalis* E.Fourn. ex Hemsl., *Eleusine indica* var. *major* E.Fourn., *Eleusine indica* var. *monostachya* F.M.Bailey, *Eleusine indica* var. *oligostachya* Honda, *Eleusine japonica* Steud., *Eleusine macrosperma* Stokes, *Eleusine marginata* Lindl., *Eleusine polydactyla* Steud., *Eleusine rigidifolia* E.Fourn., *Eleusine rigidifolia* E. Fourn. ex Hemsl., *Eleusine scabra* E.Fourn., *Eleusine scabra* E.Fourn. ex Hemsl., *Juncus loureiroana* Schult. & Schult.f., *Leptochloa pectinata* (Lam.) Kunth, *Paspalum dissectum* Kniph., *Triticum geminatum* Spreng.

93. *Eleusine verticillata* Roxb.

Synonyms: *Acrachne racemosa* (Heyne ex Roth) Ohwi, *Acrachne verticillata* (Roxb.) Wight & Arn. ex Lindl., *Acrachne verticillata* (Roxb.)

Wight & Arn. ex Chiov., *Acrachne verticillata* (Roxb.) B.D. Jacks., *Dactyloctenium verticillatum* (Roxb.) Munro, *Eleusine racemosa* B.Heyne ex Roth, *Leptochloa racemosa* (Roem. & Schult.) Kunth, *Leptochloa schimperiana* Hochst., *Leptochloa verticillata* (Roxb.) Kunth, *Sclerodactylon micrandrum* Keng f. & L.Liou

94.*Eragrostiella bifaria* (Vahl) Bor

Synonyms: *Catapodium bifarium* (Vahl) Link, *Catapodium coromandelianum* (J.Koenig ex Rottler) Link, *Eragrostiella bifaria* var. *australiana* F.M.Bailey, *Eragrostiella coromandeliana* (K.D.Koenig ex Rottler) Keng f. & L.Liou, *Eragrostis bifaria* (Vahl) Wight, *Eragrostis bifaria* (Vahl) Wight ex Steud., *Eragrostis coromandeliana* (J.Koenig ex Rottler) Trin., *Eragrostis cretacea* Nees, *Poa bifaria* Vahl, *Poa coromandeliana* J.Koenig ex Rottler, *Triticum bifarium* (Vahl) Kuntze

95.*Eragrostiella bachyphylla* (Stapf) Bor

Synonyms: *Eragrostis brachyphylla* Stapf

96.*Eragrostis cilianensis* (All.) Janch.

Synonyms: *Briza eragrostis* L., *Briza major* L. ex Kunth, *Briza megastachya* (Koeler) hort. ex Roem. & Schult., *Briza oblonga* Moench, *Briza purpurascens* Muhl., *Calotheca purpurascens* (Muhl.) Spreng., *Eragrostis argentina* Jedwabn., *Eragrostis articulata* De Wild., *Eragrostis cilianensis* (Bellardi) Link ex Vignolo, *Eragrostis cilianensis* (Bellardi) Mosher, *Eragrostis cilianensis* (Bellardi) F.T. Hubb., *Eragrostis cilianensis* subsp. *major* (Host) Maire & Weiller, *Eragrostis cilianensis* f. *megastachya* (Koeler) Maire & Weiller, *Eragrostis cilianensis* subsp. *megastachya* (Koeler) Maire & Weiller, *Eragrostis cilianensis* subsp. *starosselskyi* (Grossh.) Tzvelev, *Eragrostis cilianensis* var. *starosselskyi* (Grossh.) Dobignard & Portal, *Eragrostis cilianensis* var. *subbiloba* (Chiov.) Dobignard & Portal, *Eragrostis cilianensis* var. *thyrsiflora* (Willk. & Lange) Dobignard & Portal, *Eragrostis cilianensis* subsp. *thyrsiflora* (Willk.) H.Scholz & Valdés, *Eragrostis costata* B.L.Turner, *Eragrostis eragrostis* (L.) Blatt. & McCann, *Eragrostis flexuosa* Steud., *Eragrostis leersioides* (J.Presl) Steud., *Eragrostis leersioides* (C. Presl) Guss., *Eragrostis leersioides* var. *leersioides* (C. Presl) Richter, *Eragrostis major* Host, *Eragrostis major* var. *subbiloba* (Chiov.)

Chiov., *Eragrostis megalostachya* (Koeler) St.-Lag., *Eragrostis megastachya* (Koeler) Link, *Eragrostis megastachya* var. *cilianensis* (All.) Asch. & Graebn., *Eragrostis megastachya* var. *compacta* (Regel) Krylov, *Eragrostis megastachya* f. *nana* Lorentz & Niederl., *Eragrostis megastachya* var. *thyrsiflora* Willk. & Lange, *Eragrostis minor* var. *megastachya* (Koeler) Burtt Davy, *Eragrostis monodii* A.Camus, *Eragrostis multiflora* var. *cilianensis* (All.) Maire, *Eragrostis multiflora* var. *glandulifera* Chiov., *Eragrostis multiflora* var. *insularis* Chiov., *Eragrostis multiflora* var. *leersioides* (Guss.) Richt., *Eragrostis multiflora* var. *subbiloba* Chiov., *Eragrostis multiflora* var. *triticea* (C. Presl) Richt., *Eragrostis multiflora* f. *violacea* N.Terrac. ex Chiov., *Eragrostis oblonga* (Moench) Baumg., *Eragrostis pappiana* var. *insularis* (Chiov.) Mattei, *Eragrostis pappiana* var. *insularis* Terracciano ex Chiov., *Eragrostis pappii* Gand., *Eragrostis poaeoides* var. *megastachya* P. Beauv. ex Vasey, *Eragrostis polyadenia* Mattei, *Eragrostis polymorpha* Roem. & Schult., *Eragrostis polysperma* Peter, *Eragrostis pooides* var. *compacta* Regel, *Eragrostis pooides* var. *megastachya* (Koeler) A.Gray, *Eragrostis schweinfurthiana* Jedwabn., *Eragrostis starosselskyi* Grossh., *Eragrostis triticea* (C.Presl) Steud., *Eragrostis virletii* E.Fourn., *Eragrostis vulgaris* C.Presl ex Steud., *Eragrostis vulgaris* subsp. *megastachya* (Koeler) Douin, *Eragrostis vulgaris* var. *megastachya* (Koeler) Coss. & Germ., *Erosion cilianense* (Bellardi) Lunell, *Erosion cilianensis* (All.) Lunell, *Megastachya eragrostis* (L.) Roem. & Schult., *Megastachya leersioides* C.Presl, *Megastachya oblonga* (Moench) P.Beauv., *Megastachya obtusa* Schult., *Megastachya purpurascens* (Muhl.) Schult., *Megastachya triticea* C.Presl, *Poa cachectica* Schumach., *Poa cilianensis* All., *Poa eragrostis* (L.) Brot., *Poa flexuosa* Roxb., *Poa leersioides* (J.Presl) Guss., *Poa megastachya* Koeler, *Poa nuttallii* Spreng., *Poa oblonga* (Moench) Baumg., *Poa obtusa* Nutt., *Poa pennsylvanica* Nutt., *Poa philadelphica* W.P.C.Barton, *Poa polymorpha* J.Koenig ex Rottb., *Poa roxburghiana* Schult., *Poa tortuosa* Spreng., *Poa triticea* (J.Presl) Kunth, *Poa triticea* (C. Presl) Guss.

97. *Eragrostis ciliaris* (L.) R.Br.

Synonyms: *Cynodon ciliaris* (L.) Raspail, *Eragrostis arabica* Jaub. & Spach, *Eragrostis boryana* (Willd.) Steud., *Eragrostis ciliaris* (L.) Link,

Eragrostis ciliaris (L.) Nees, *Eragrostis ciliaris* subsp. *brachystachya* (Boiss.) H.Scholz, *Eragrostis ciliaris* var. *brachystachya* Boiss., *Eragrostis ciliaris* var. *clarkei* Stapf ex Hook.f., *Eragrostis ciliaris* var. *compta* (Link) Schrad., *Eragrostis ciliaris* var. *latifolia* Hack., *Eragrostis ciliaris* var. *laxa* Kunth, *Eragrostis compta* Link, *Eragrostis lapida* Hochst., *Eragrostis lobata* Trin., *Eragrostis pulchella* Parl., *Eragrostis villosa* Trin., *Erosion ciliare* (L.) Lunell, *Macroblepharus contractus* Phil., *Megastachya boryana* (Willd.) Roem. & Schult., *Megastachya ciliaris* (L.) P.Beauv., *Poa boryana* Willd., *Poa ciliaris* L., *Poa compta* (Link) Kunth, *Poa elegans* Poir., *Poa lobata* (Trin.) Kunth

98. *Eragrostis japonica* (Thunb.) Trin.

Synonyms: *Eragrostis japonica* var. *interrupta* (Lam.) Henrard

99. *Eragrostis nutans* (Retz.) Nees ex Steud.

Synonyms: *Eragrostis nutans* (Retz.) Nees

100.*Eragrostis pilosa* (L.) P.Beauv.

Synonyms: *Catabrosa verticillata* (Cav.) P.Beauv., *Eragrostis afghanica* Gand., *Eragrostis amurensis* Prob., *Eragrostis bagdadensis* Boiss., *Eragrostis baguirmensis* Chev., *Eragrostis baguirmiensis* A.Chev., *Eragrostis collocarpa* K.Schum., *Eragrostis damiensiana* var. *laxior* Thell., *Eragrostis filiformis* Link, *Eragrostis gracilis* Schrad., *Eragrostis gracilis* Velen., *Eragrostis gracillima* Hack., *Eragrostis imberbis* (Franch.) Prob., *Eragrostis inberbis* (Franch.) Prob., *Eragrostis indica* (J.Koenig ex Rottler) Willd. ex Steud., *Eragrostis jeholensis* Honda, *Eragrostis linkii* (Kunth) Steud., *Eragrostis multispicula* Kitag., *Eragrostis petersii* Trin., *Eragrostis pilosa* var. *amurensis* (Prob.) Vorosch., *Eragrostis pilosa* f. *imberbis* Franch., *Eragrostis pilosa* subsp. *imberbis* (Franch.) Tzvelev, *Eragrostis pilosa* var. *imberbis* Franch., *Eragrostis pilosa* var. *jeholensis* (Honda) Ohwi, *Eragrostis pilosa* var. *major* Litv., *Eragrostis pilosa* subsp. *neglecta* H.Scholz, *Eragrostis pilosa* var. *verticillata* (Cav.) Rchb., *Eragrostis punctata* (L.) Link ex Steud., *Eragrostis senegalensis* (Desv.) A.Chev., *Eragrostis tenuiflora* Rupr. ex Steud., *Eragrostis verticillata* (Cav.) P. Beauv., *Eragrostis verticillata* (Cav.) Roem. & Schult., *Eragrostis verticillata* Coss. & Lange, *Eragrostis verticillata* var. *indica* (Rottler)

Wight & Arn. ex Nees, *Poa bohemica* J.C.Mayer ex Mert. & W.D.J.Koch, *Poa delicatior* Steud., *Poa eragrostis* Walter, *Poa indica* J.Koenig ex Rottler, *Poa linkii* Kunth, *Poa mexicana* Link, *Poa pilosa* L., *Poa pilosa* Larrañaga, *Poa pilosa* var. *tenuis* Regel, *Poa plumosa* Schrad., *Poa poiretii* Roem. & Schult., *Poa punctata* L.f., *Poa senegalensis* Desv., *Poa tenella* Pall., *Poa tenella* Willd. (L. misappl.) Pursh, *Poa tenuiflora* Steud., *Poa verticillata* Cav.

101. *Eragrostis tenella* (Linn.) P.Beauv. ex Roem.

Synonyms: *Poa tenella* Linn., *Poa amabilis* Linn., *Poa plumose* Retz., *Eragrostis plumose* (Retz.) Link., *Megastachya tenella* (Linn.) Bojer, *Eragrostis amabilis* (Linn.) Wight et Arn. ex Hook. et Arn., *Eragrostis tenella* Var. *plumose* (Retz.) Stapf, *Eragrostis tenella* var. *tenella* (Linn.) Hook. f., *Poa amabilis* Linn.

102. *Eragrostis tremula* Hochst. ex Steud.

Synonyms: *Eragrostis lamarckii* Steud., *Eragrostis multiflora* Trin., *Eragrostis rhachitricha* Hochst. ex Miq., *Eragrostis serpula* Chiov., *Poa multiflora* Roxb., *Poa tremula* Lam.

103. *Eragrostis unioloides* (Retz.) Nees ex Steud.

Synonyms: *Eragrostis amabilis* var. *contracta* Buse, *Eragrostis amabilis* var. *effusa* Buse, *Eragrostis amabilis* var. *prostrata* Buse, *Eragrostis amabilis* var. *scabriuscula* Buse, *Eragrostis euchroa* Steud., *Eragrostis formosana* Hayata, *Eragrostis rubens* (Lam.) Hochst. ex Miq., *Eragrostis rubens* (Lam.) Steud., *Eragrostis rubens* var. *contracta* (Buse) Miq., *Eragrostis rubens* var. *effusa* (Buse) Miq., *Eragrostis rubens* f. *linearis* Boerl., *Eragrostis rubens* var. *scabriuscula* (Buse) Miq., *Eragrostis tenella* Nees, *Eragrostis unioloides* var. *ongiemensis* A.Camus, *Poa rubens* Lam., *Poa unioloides* Retz., *Uniola indica* Spreng.

104. *Eragrostis viscosa* (Retz.) Trin.

Synonyms: *Eragrostis hirsutissima* Peter, *Eragrostis mangalorica* Hochst. ex Miq., *Eragrostis retinorrhoea* Steud., *Eragrostis strigosa* Andersson, *Eragrostis tenella* var. *viscosa* (Retz.) Stapf, *Eragrostis transvaalensis* Gand., *Eragrostis viscosa* var. *pilosissima* (Hochst. ex A.Rich.) Hochst., *Poa viscosa* Retz., *Poa viscosa* var. *pilosissima* Hochst. ex A.Rich.

TRIBE: SPOROBOLEAE

105. *Sporobolus coromardelianus* (Retz.) Kunth

Synonyms: *Agrostis coromandeliana* Retz., *Agrostis rura* Buch.-Ham. ex Wall., *Sporobolus commutatus* (Trin.) Kunth, *Sporobolus coromandelianus* (Retz.) R. Br., *Sporobolus javensis* Ohwi, *Sporobolus parvulus* Stent, *Sporobolus violascens* Mez, *Vilfa commutata* Trin., *Vilfa coromandeliana* (Retz.) P.Beauv., *Vilfa roxburghiana* Nees ex Wight

106. *Sporobolus diander* (Retz.) P. Beauv.

Synonyms: *Sporobolus diandrus* (Retz.) P.Beauv., *Agrostis diandra* Retz., *Agrostis elongata* var. *flaccida* Roem. & Schult., *Spermachiton involutum* Llanos, *Sporobolus diandrus* var. *diandrus*, *Sporobolus diandrus* var. *nanus* Hook.f., *Sporobolus indicus* var. *diandrus* (Retz.) Jovet & Guédès, *Sporobolus indicus* var. *flaccidus* (Roem. & Schult.) Veldkamp, *Sporobolus trimenii* Senaratna, *Vilfa diandra* (Retz.) Trin., *Vilfa erosa* Trin., *Vilfa retzii* Steud.

107. *Sporobolus indicus* (L.) R.Br.

Synonyms: *Agrostis elongata* Lam., *Agrostis indica* L., *Agrostis orientalis* Nees, *Agrostis tenacissima* L.f., *Agrostis tenacissima* Jacq., *Agrostis tenuissima* Spreng., *Agrostis tenuissima* Larrañaga, *Andropogon intortum* Crantz, *Paspalum lanceaefolium* Desv., *Sporobolus angustus* Buckley, *Sporobolus berteroanus* (Trin.) Hitchc. & Chase, *Sporobolus exilis* (Trin.) Balansa, *Sporobolus indicus* f. *africanoides* Jovet & Guédès, *Sporobolus indicus* var. *andinus* Renvoize, *Sporobolus indicus* var. *exilis* (Trin.) T.Koyama, *Sporobolus indicus* var. *indicus*, *Sporobolus indicus* f. *indicus*, *Sporobolus indicus* subsp. *indicus*, *Sporobolus indicus* f. *microspiculus* Jovet & Guédès, *Sporobolus indicus* var. *tenacissimus* (L.f.) Peter, *Sporobolus jacquemontii* Borhidi, *Sporobolus lamarckii* Desv., *Sporobolus orientalis* (Nees) Kunth, *Sporobolus tenacissimus* (L.f.) P.Beauv., *Vilfa angusta* Buckley, *Vilfa berteroana* Trin., *Vilfa elongata* (Lam.) P.Beauv., *Vilfa elongata* (R. Br.) Trin., *Vilfa exilis* Trin., *Vilfa indica* (L.) Trin. ex Steud., *Vilfa indica* Trin., *Vilfa orientalis* Nees ex Trin., *Vilfa tenacissima* (L.f.) Kunth, *Vilfa tenacissima* var. *exilis* (Trin.) E.Fourn., *Vilfa tenacissima* var. *intermedia* E. Fourn., *Vilfa tenacissima* var. *robusta* E. Fourn.

108. *Urochondra setulosa* (Trin.) C.E.Hubb.

Synonyms: *Crypsis dura* Boiss., *Crypsis setulosa* (Trin.) Mez, *Heleochloa dura* (Boiss.) Boiss., *Heleochloa dura* subsp. *kuriensis* Vierh., *Heleochloa setulosa* (Trin.) Blatt. & McCann, *Heterochloa dura* (Boiss.) Balf.f., *Sporobolus setulosus* (Trin.) N.Terracc., *Vilfa setulosa* Trin.

TRIBE: ZOYSIEAE

109. *Tragus biflorus* (Roxb.) Schult. (illegimate name)

Synonyms: *Tragus racemosus* (L.) All., *Aira malatrina* Buch.-Ham. ex Wall., *Cenchrus linearis* Lam., *Cenchrus racemosus* L., *Lappago biflora* Roxb., *Lappago decipiens* Fig. & De Not., *Lappago racemosa* (L.) Honck., *Nazia racemosa* (L.) Kuntze, *Phalaris muricata* Forssk., *Tragus adriaticus* Gand., *Tragus arenarius* Bremek. & Oberm., *Tragus brevicaulis* Boiss., *Tragus decipiens* (Fig. & De Not.) Boiss., *Tragus echinatus* Cav., *Tragus muricatus* (Forssk.) Moench, *Tragus pallens* Gand., *Tragus paucispinus* Hack., *Tragus racemosus* var. *decipiens* (Fig. & De Not.) T.Durand & Schinz, *Tragus racemosus* f. *divaricatus* Döll, *Tragus racemosus* f. *erectus* Döll, *Tragus racemosus* var. *erectus* Mutel, *Tragus racemosus* var. *paucispinus* (Hack.) Maire, *Tragus racemosus* var. *remotus* Suess.

110. *Zoysia matrella* (L.) Merr.

Synonyms: *Agrostis matrella* L., *Matrella juncea* Pers., *Osterdamia matrella* (L.) Kuntze, *Osterdamia tenuifolia* (Trin.) Kuntze, *Osterdamia zoysia* Honda, *Osterdamia zoysia* var. *tenuifolia* (Willd. ex Trin.) Honda, *Zoysia aristata* C.Muell., *Zoysia griffithiana* C.Muell., *Zoysia malaccensis* Gand., *Zoysia matrella* subsp. *tenuifolia* (Willd. ex Trin.) T.Koyama, *Zoysia matrella* var. *tenuifolia* (Willd. ex Trin.) T.Koyama, *Zoysia matrella* var. *tenuifolia* (Thiele) Sasaki, *Zoysia pungens* Willd., *Zoysia pungens* var. *tenuifolia* (Willd. ex Trin.) T.Durand & Schinz, *Zoysia serrulata* Mez, *Zoysia tenuifolia* Willd. ex Thiele

INDEX

A

Acrachne racemosa, 33, 37, 38, 47, 76, 130, 145, 156, 243, 252, 296
Acuminate termination point, 252–257
Acute termination point, 253, 254
Aeluropus lagopoides (L.) Thwaites, 14, 34, 46, 72, 123, 146, 164, 234, 256, 288
Aleuropoideae, 34
Alloteropis, 217, 255, 258
Alloteropsis cimicina (L.) Stapf, 13, 43, 62, 104, 149, 163, 217, 255, 270
Alopecurus typhoides Burm.f, 285
Anastrophus paspalodes (Michx.) Nash, 283
Andropogon pachyarthrus Hack, 261
Andropogon pumilus Roxb, 12, 40, 51, 85, 149, 162, 201, 255, 261
Andropogon serrulatus, 262
Andropogoneae, 3, 30–36, 156, 166
angiosperms, 12, 160
  fruit, 17
angled broad undulation, 110, 123, 124, 131
Angled nodal region, 253, 254
animal husbandry, 2
anticlinal bands, 113, 124
anticlinal wall, 89–119, 122–125, 128, 139, 142, 143
Aplaaauda mutica L, 12, 36, 40, 51, 85, 147, 174, 201, 254, 261, 262
Apluda, 12, 33, 36, 40, 51, 85, 147, 174, 201, 254, 261, 262, 266
Arctic regions, 1
Aristida adscendionsis, 256
Aristida adscensionis *L*, 14, 33–35, 37, 46, 72, 124, 144, 164, 235, 289, 290
Aristida funiculata Trin. & Rupr, 14, 33–37, 73, 124, 144, 164, 236, 257, 258, 292
Arthraxon lanceolatus (Roxb.) Hochst, 12, 40, 52, 86, 147, 156, 162, 202, 251, 262
Arthrolophis pumilus, 261
Asteraceae, 4

B

Bamboos, 2
barley, 2
biennial plants, 3
blisters, 130, 131, 134, 143
Bothriochloa pertusa (L.) A. Camus, 12, 36, 40, 52, 86, 150, 162, 202, 253, 262
Bracharia cruciformis, 43, 63, 148, 255
Brachiaria distachya (L.) Stapf, 13, 43, 63, 105, 146, 163, 218, 256, 271
Brachiaria eruciformis (Sm.) Griseb, 13, 105, 163, 217, 270
Brachiaria ramosa (L.) Stapf, 13, 43, 63, 148, 163, 255, 271
Brachiaria reptans (L.), 13, 107, 163, 219, 271
breeding programs, 2
broad depressed channels, 142

C

callus, 15
  shape, 15
Capillipedium hugelii (Hack.) A. Camus, 12, 40, 52, 87, 150, 163, 262
Caryopses divided, 18
  embryo, 18
  endosperm, 18
caryopsis, 3, 10, 11, 17, 32, 85, 141

Cenchrus, 13, 15, 44, 149, 150, 163, 164, 174, 253, 256, 264, 272, 273, 285, 302
Cenchrus biflorus Roxb, 13, 64, 107, 163, 174, 219, 272
Cenchrus ciliaris *L*, 13, 44, 64, 108, 164, 220, 258, 272
Cenchrus polystachios (L.) Morrone, 284
Cenchrus preurii (Kunth) Maire, 13, 44, 65, 109, 273
Cenchrus setigerus *Vahl*, 13, 44, 109, 164, 220, 273
central axis, 4, 15
Chaetochloa brevispica Scribn. & Merr, 286
characteristic features of caryopses, 17
Chionachne koenigii (Spr.) Thw, 12, 40, 51, 84, 149, 261
Chlorideae, 10, 33–36, 39, 156
Chloris barbata *Sw*, 14, 39, 46, 73, 126, 149, 164, 236, 252, 292
Chloris gayana, 161
Chloris montana Roxb, 14, 126, 149, 292
Chloris virgata *Sw*, 14, 47, 74, 127, 145, 164, 237, 252, 292
Chryosopogon, 30
Chrysopogon fulvus (Spreng.) Chiov, 12, 40, 53, 88, 147, 163, 174, 203, 253, 263
Coix, 12, 33–37, 40, 50, 84, 146, 162, 200, 254, 261
Coix lachryma-jobi *L.*, 12, 34, 36, 84, 162, 200, 261
compact panicles, 4
compound reticulate, 38
concave interspace, 85–92, 97, 99–109, 112–123, 127, 129, 130, 133–141, 144
Cymbopogon martini (Roxb.) W. Watson, 12, 33, 40, 53, 88, 147, 163, 174, 203, 253, 263
Cynadon, 14, 47, 74, 147, 164, 237, 256, 258, 293
Cynadon dactylon (L.) Pers, 14, 47, 74, 127, 147, 164, 237, 293

D

Dactyloctenium, 14, 37, 38, 47, 48, 77, 131, 132, 145, 165, 240, 241, 253, 255, 295, 297
Dactyloctenium aegyptium (L.) Willd, 14, 47, 77, 131, 145, 165, 240, 253, 295
Dactyloctenium and Eleusine, 37
Dactyloctenium giganteum Fischer and Schweick, 14, 48, 77, 132, 145, 295
Dactyloctenium sindicum *Boiss*, 14, 48, 77, 132, 145, 165, 241, 255, 295
dendrogram, 150, 151, 154, 156, 166, 174, 199, 257, 258
desert, 2
Desmostachya bipinnata (L.) Stapf, 14, 48, 78, 133, 146, 165, 241, 252, 296
Dicanthium annulatum, 40, 53, 150, 163, 252
Dicanthium caricosum, 41, 54, 150, 163, 253
Dichanthium annulatum (Forssk.) Stapf, 12, 89, 204, 263
Dichanthium caricosum, 89, 204, 263
Dichanthium caricosum (L.) A.Camus, 263
Digitaria adpressa Bojer, 65, 273, 281
Digitaria ciliaris (Retz.) Koeler, 13, 44, 66, 110, 150, 164, 221, 254, 273
Digitaria eriantha, 161
Digitaria foliosa Lag, 283
Digitaria granularis (Trin.) Henrard, 13, 30, 34, 44, 66, 111, 147, 155, 164, 222, 254, 273
Digitaria longiflora (Retz.) Pers, 164, 222, 254, 274
Digitaria microbachne (Persl) Hern, 13, 111, 150, 164, 221, 254, 273
Digitaria stircta *Roth*, 164, 223, 254, 274
digitate raceme, 4
digitate spikes, 4
Dihcanthium caricosum (L.) A.Camus, 12
Dimeria, 12, 30, 35, 41, 145, 263

Dimeria orinthopoda *Trin*, 12, 36, 54, 90, 263
Dinebra retroflexa (Vahl) Panz, 14, 33, 48, 78, 133, 145, 165, 242, 254, 258, 296
Dorsal surface of lamina, 199
Dorsal/lateral striations, 146–149

E

E. cilianensis, 18, 38
E. japonica, 38
E. pilosa, 38
E. tremula, 18, 38
E. unioloides, 38
Echinochloa, 13, 33, 34, 38, 39, 44, 45, 66, 67, 112, 113, 148, 149, 164, 256, 257, 270, 271, 274, 276–279, 281
Echinochloa colona (L.) Link, 13, 112, 164, 274
Echinochloa colonum, 38, 39, 44, 66, 149, 257, 274
Echinochloa crus-galli, 259, 274–277
Echinochloa crusgalli (L.) P. Beauv, 13, 38, 39, 44, 67, 113, 148, 164, 224, 275, 276
Echinochloa stagnina (Retz.) P. Beauv, 13, 34, 39, 45, 67, 113, 148, 164, 256, 277
ecosystem, 2, 160
Electron Microscopy, 84–144
Eleusine indica (L.) Gaertn, 14, 33, 34, 37, 38, 48, 78, 134, 145, 156, 165, 242, 252, 296
Eleusine verticillata *Roxb*, 14, 165, 296
embryo axis, 30, 91, 136
embryo proportion, 11
embryo percentage caryopses, 36
  embryo 45% or under, 36
    short embryo class
  embryo 46% or over, 36
    large embryo class
embryo types, 30
  L type: 'blob-like' embryo without clear embryo axis, 30
  N type: clearly defined embryo axis, 30
endosperm, 18
endosperm boundary, 18
epidermal features, 5
Eragrosteae, 31–34, 36–38, 156
Eragrosteae members, 31, 37
Eragrosti pilosa, 245
Eragrostideae, 34
Eragrostiella bachyphylla (Stapf) Bor, 165, 244, 253, 297
Eragrostiella bifaria (Vahl) Bor, 14, 48, 79, 135, 146, 165, 243, 253, 258, 297
Eragrostis, 14, 15, 18, 34–38, 49, 145, 147, 165, 244, 257, 258, 288, 296–300
Eragrostis cilianensis (All.) Janch, 14, 48, 79, 135, 146, 165, 297
Eragrostis ciliaris (L.) R.Br, 14, 48, 79, 136, 145, 165, 245, 298, 299
Eragrostis japonica (Thunb.) Trin, 14, 48, 80, 137, 299
Eragrostis nutans (Retz.), 14, 48, 80, 137, 299
Eragrostis pilosa (L.) P. Beauv, 14, 38, 80, 138, 299
Eragrostis tenella (Linn.) P.Beauv. ex Roem, 49, 81, 138, 165, 246, 257, 300
Eragrostis tremula Hochst. ex Steud, 15, 49, 81, 139, 165, 246, 257, 300
Eragrostis unioloides (Retz.) Nees ex Steud, 15, 49, 81, 140, 165, 247, 256, 300
Eragrostis viscosa (Retz.) Trin., 15, 49, 82, 140, 165, 247, 300
Eragrostreae, 30
Eremopogon, 164, 225, 258, 278
Eremopogon foveolatus (Delile) Stapf, 164, 225, 255, 278
Eriochloa procera (Retz.) C. E. Hubb, 13, 33, 45, 67, 114, 150, 164, 226, 256, 258, 278
evolutionary tendency, 11

F

family, 1, 4–6, 17

Fan shaped hilum, 145, 148–150
fertile offspring, 6
Flat termination point, 252–256
Flattened leaf sheath type, 251–256
Folded vernation, 251, 256
foveate interspace, 104, 123, 131, 132, 136
fruit shape, 10
funicle, 32

G

genus, 6, 11, 12, 16, 18, 34, 38
germination, 3, 30
Glabrous
    angled nodal region, 254
    internode, 254, 255
    leaf blade surface, 252–255
    nodal region, 251–255
    surface, 170, 200, 201, 203, 204, 209–212, 214–218, 221, 223–233, 235–237, 239, 241, 243, 245–248, 255
Globular slimy glands, 139
Gramineae, 1, 4, 5, 10, 161
grass fruit, 17
grass seedlings, 161, 162, 175–199

H

habit, 3
Hackelochloa, 163, 205, 251, 258, 264
Hackelochloa granularis (L.) Kuntze, 163, 205, 251, 258 264
Herbarium specimens, 166
Heteropogon contortus var. *contortus* sub var. *genuinus* Blatt. & McCann, 12, 35, 41, 54, 55, 91, 149, 163, 205, 206, 252, 264
Heteropogon contortus var. *contortus* sub var. *typicus* Blatt. & McCann, 12, 35, 41, 54, 90, 149, 163, 264
Heteropogon ritcheii (Hook.f.) Blatt. & McCann, 12, 35, 41, 55, 92, 149, 163, 207, 252, 264
Heteropogon triticeus (R.Br.) Stapf ex Craib, 12, 34, 41, 55, 92, 150, 163, 174, 207, 264
hexagonal reticulum, 98, 110, 134, 139
Hilum, 18, 32, 37, 40–144, 149, 150, 155
hilum. Chlorideae, 33
hilum position, 32
    basal, 32
    linear, 32
    sub basal, 32
hilum shape, 32
    circular, 32
    fan shaped, 32
    linear, 32
    oval, 32
    'V' shaped, 32
hilum type, 32
    basal–circular, 32
    basal–fan shaped, 32
    basal–linear, 32
    basal–oval type, 32
    basal–unclear, 32
    basal–'V' shaped, 32
    linear (long type), 32
    linear (short type), 32

I

identical species, 11
Illustrated terminologies, 159
    auricle, 174
    collar, 174
    growth habit, 170
        annual, 170
        perennial, 171
    leaf blade, 172
        amplexicaul, 172
        leaf lamina, 172
        leaf margin, 172
        midvein, 172
        scabrous, 172
        surface, 172
        tuberculated hairs, 172
    leaf sheath, 173

close, 173
open, 173
splitted, 173
ligule, 173
acuminate, 173
acute, 173
ciliate, 173
cirrhose, 174
truncate, 173
villous, 174
internode, 171
glabrous, 171
pubescent, 171
node, 171
bearded, 171
glabrous, 171
pubescent, 171
termination point, 172
acuminate, 172
acute, 172
vernation, 171
folded, 171
rolled, 171
Imperata cylindrica (L.) Raeusch, 208, 252, 264, 265
inflorescence, 3, 4, 15
inflorescences type, 2
interbreeding and producing, 6
irregular crimples, 142
Isachne globosa (Thunb.) Kuntze, 14, 33, 34, 46, 72, 123, 144, 164, 234, 254, 288, 289
Ischaemum indicum (Houtt.) Merr, 12, 41, 56, 93, 146, 163, 208, 254, 265
Ischaemum molle Hook.f, 12, 41, 56, 93, 146, 163, 209, 255, 258, 265
Ischaemum pilosum (Willd.) Wight, 12, 41, 56, 94, 150, 163, 209, 254, 265
Ischaemum rugosum Salisb, 12, 41, 57, 94, 150, 163, 210, 255, 266
Iseilema, 13, 42, 57, 95, 150, 163, 210, 258, 266
Iseilema laxum Hack, 13, 42, 57, 95, 150, 163, 210, 252, 266

K

Keeled termination point, 252–256

L

lachryma-jobi, 12, 36, 40, 50, 146, 162, 200, 254, 261
lachrymaljobi, 37
lanceolate, 3, 98, 147, 149, 202, 216, 218
Leaf
sheath margin, 170, 201, 202–257
close, 253, 255, 257
Leguminosae, 4
Lemma/palea, 2, 3, 17
length, 3
Leptoloma miliacea, 281
Light microscopic study, 40–49
lodicules, 3

M

macroand microstructure, 11
maize, 2
material/plumbing, 2
Megathyrsus maximus, 280
Melanocenchris, 14, 36, 47, 75, 128, 146, 164, 238, 254, 258, 294
Melanocenchris jaequemontii Jaub. & Spach, 14, 36, 47, 75, 128, 146, 164, 238, 254, 294
Melastomataceae, 12
Membranous, 170, 201–216, 220–223, 228, 230–240, 244, 251
Mentzelia, Seavey, 12
micromorphological characters, 11
microscopic data, 17
microstructural features, 11
microstructural pattern analysis, 11
millets, 2
monocotyledons, 5
monogeneric tribe, 10
Mordechai Kislev's computerized identification, 11
more practical value, 160
morphological and molecular data, 10

morphological and taxonomic features., 166
Morphometric features, 15
callus, 15
characteristic features of caryopses, 15
caryopses length, 34
caryopsis breadth, 34
compression and thickness: breadth ratio, 35
dorsal/lateral striations, 33
embryo size (%), 36
embryo size, 36
embryo type, 30
embryo, 18
hilum visibility, 33
hilum, 32
measurements, 17
narrowness and length: breadth ratio, 35
presentation of Data, 17
scutellum shape, 30
shape, 18
structure, 17
ventral furrow, 33
spikelet and floret, 15
morphological
features, 5, 10, 11, 17, 37, 166
structure, 10
morphometric features, 18
multidirectional
crimples, 141, 142
elevations, 125, 128, 133, 134
multiple
racemes, 4
spikes, 4

N

Natural grasslands, 2
nomenclatural hierarchy, 6
northeast tropical, 197, 200, 201, 204–206, 208, 209, 212–214, 216–220, 222, 225–227, 229–233, 235–238, 248, 250
numerical/scanning electron microscopic, 156

O

oats, 2
Oklahoma grassland, 161
Ophiuros exaltatus (L.) Kuntze, 13, 42, 57, 96, 146, 163, 211, 253, 266
Oplismenous, 258
Oplismenus burmanii (Retz.) P. Beauv, 13, 30, 33, 38, 45, 68, 115, 146, 278, 279
Oplismenus composites (L.) P. Beauv, 13, 33, 45, 150, 164, 226, 255, 278
Oplismenus compositus, 68, 115, 279
Orchidaceae and Rubiaceae, 4
Oropetium villosulum Stapf ex Bor, 14, 47, 75, 128, 145, 164, 238, 254, 294
Orthopogon junghuhnii, 279
Oryza sativa, 10
Oval shaped hilum, 145–150
overlapping margins, 173, 204, 206, 207, 209, 211–221, 223–226, 228–235, 238–244, 246- 249, 251–253, 255–257

P

Panicum antidotalePanicum antidotale Retz, 13, 45, 68, 116, 148, 164, 227, 280
Panicum coloratum, 161
Panicum curvatum var. tenellum, 281
Panicum maximum Jacq, 161, 164, 227, 252, 280
Panicum miliaceum, 13, 45, 69, 118, 164, 228, 255, 281
Panicum miliare, 33, 148, 280
Panicum trypheron Schult, 14, 45, 69, 117, 149, 164, 228, 255, 258, 281
parallel veins, 201, 202, 204–228, 230–239, 241, 245, 246, 250
Paspalidium flavidum (Retz.) A. Camus, 14, 45, 69, 118, 148, 164, 229, 256, 258, 281

Paspalidium geminatum (Forssk.) Stapf, 14, 45, 70, 119, 148, 164, 229, 256, 258, 281, 282
Paspalum adelogaeum Steud, 282
Paspalum distichum *L*, 164, 231, 253, 283, 284
Paspalum scrobiculatum *L*, 14, 35, 45, 70, 119, 148, 164, 230, 254, 282, 283
Paspalum vaginatum *Sw*, 164, 231, 253, 283, 284
pasture species, 2
pasture-improvement program, 160
Pennisetum dandestinum, 161
Pennisetum setosum (Sw.) Rich, 14, 46, 70, 120, 164, 232, 256, 284, 285
pericarp, 17
periclinal wall, 89–97, 99–110, 113–119, 122–125, 128, 135, 139, 143
Perotis indica (L.) Kuntze, 14, 46, 73, 125, 145, 289
phenetic relationships, 11
Photographs, 17, 162, 174
photosynthesis, 3, 172
phylogenetic relationship, 10
physiological groups, 3
plant
  kingdoms, 2
  taxonomists, 10
Poaceae, 3–5, 261
Pointed termination point, 253–257
Pooideae, 5, 30, 34–37, 166
Pubescent
  internode, 254, 255
  leaf blade surface, 252–255
  nodal region, 251–255
  surface, 170, 182, 202, 205–207, 211–214, 217–223, 225, 226, 230, 232–236, 238–242, 245, 248–251, 255

Q

Qualitative features, 175–199

R

rachis of spikelets, 4
reproductive
  parts, 2
  unit, 15
reticulate pattern, 37–39, 145, 149, 150
Reticulate-scalariform rugose, 97
rice, 2, 10
Rolled vernation, 251, 256
Rottbelia, 15, 174
Rottboellia exaltata Linn. f, 13, 42, 58, 96, 146, 163, 211, 253, 266
Round leaf sheath type, 251–256
Rugose pattern, 104
Ruminate rugose, 84–88, 90, 93, 95, 96, 114, 127, 128, 133, 134

S

S. splendid, 161
Saccharum spontanum *L*, 13, 33, 42, 58, 97, 147, 266
Sachoenefeldia gracilis, 36, 47, 75, 147, 239, 254, 258
Scanning electron microscope (SEM), 11
  studies, 37, 50–83
Schoenefeldia gracilis *Kunth*, 14, 129, 164, 294
sculpturing pattern, 11
scutellum, 18, 30, 31, 37, 38, 91, 136, 138, 144–148
scutellum shape, 30
  sickle shaped, 30
  'V' shaped, 30
Sehima ischaemoides Forssk, 13, 33, 42, 58, 97, 149, 163, 212, 255, 267
Sehima nervosum (Rottler) Stapf, 13, 33, 42, 59, 98, 146, 163, 212, 253, 267
Sehima sulcatum (Hack.) A.Camus, 13, 33, 42, 59, 98, 149, 163, 213, 253, 258, 267
Setaria, 14, 33, 38, 46, 71, 121, 122, 148, 164, 232, 233, 254, 271, 272, 274, 275, 278, 281, 282, 285–287
Setaria glauca, 14, 46, 71, 121, 148, 164, 232, 285, 286
Setaria glauca (L.) P.Beauv, 14, 164, 285

Setaria sphacelata, 161
Setaria tomentosa (Roxb.) Kunth, 14, 46, 71, 121, 148, 164, 233, 254, 286
Setaria verticillata (L.) P. Beauv, 14, 46, 71, 122, 148, 164, 233, 254, 286, 287
short panicle braches, 4
Sickle shaped scutellum, 30, 31, 148
Simple nodal region, 254–257
single
  midvein, 200–202, 256
  prominent midvein, 202–256
smaller inner palea, 15
solitary spikes, 4
Sorghum group, 15
Sorghum halepense (L.) Pers, 13, 42, 59, 99, 147, 163, 213, 252, 267, 268
Sorghum purpureo-sericeum (A.Rich.) Schweinf. and Asch, 13, 42, 60, 99, 147, 163, 252, 268
species, 3–6, 10–12, 17, 18, 30–39, 156, 161, 162, 166, 174, 257–259
  identification, 3
Spikelet, 3–5, 15
  arrangement, 4
  floret, 15
Sporoboleae, 30–36, 156
Sporobolus coromandelianus, 35, 141, 301
Sporobolus coromandliens, 256
Sporobolus coromardelianus (Retz.) Kunth, 15, 36, 49, 82, 147, 165, 248, 301
Sporobolus diander (Retz.) P. Beauv, 15, 49, 82, 141, 147, 165, 248, 256, 301
Sporobolus indicus (L.) R.Br, 15, 49, 83, 142, 146, 165, 249, 256, 301
spreading panicles, 4
stamens and pistil, 3
starchy endosperm, 17, 18
stereoscopic, 11
stolons/rhizomes, 2
subbasal hila, 18
subfamily, 5
subtle bran and hilum characters, 11
sugarcane, 2

T

taxonomic
  problems, 11
  value, 10, 11
Termination point, 170, 175–199, 201–252
Tetrapogon tenellus (Roxb.) Chiov, 14, 36, 47, 76, 129, 146, 165, 239, 253, 294
Tetrapogon villosus *Desf*, 14, 47, 76, 130, 146, 165, 240, 253, 295
Thelepogn elegans *Roth*, 13, 42, 60, 100, 146, 163, 214, 254, 268
Themeda, 11, 13, 33, 43, 60, 61, 101–103, 146, 149, 163, 174, 215, 216, 252–258, 268–270
Themeda cymbaria Hack, 13, 43, 60, 101, 146, 163, 215, 252, 258, 268
Themeda laxa A. Camus, 13, 43, 61, 101, 149, 163, 215, 255, 268
Themeda quadrivalvis (L.) Ketz, 13, 43, 61, 103, 149, 270
Themeda triandra Forssk, 13, 43, 61, 102, 149, 163, 174, 216, 253, 268, 269
topography, 10
Tragus, 15, 165, 250, 255, 258, 302
Tragus biflorus (Roxb.) Schult., 15, 49, 83, 143, 149, 165, 250, 255, 302
tribe, 5, 10, 34–38, 156, 166
Triplopogon ramosissimus (Hack.) Bor, 13, 43, 62, 103, 147, 163, 216, 252, 270

U

Undulating
  margin, 140
  periclinal wall, 95
Urochloa brachiaria, 161
Urochondra setulosa (Trin.) C. E. Hubb, 15, 49, 83, 143, 146, 165, 174, 250, 255, 302

V

vegetative
  reproduction, 160

stage, 160
ventral
groove, 37–39, 84–146, 149, 155
surface of lamina, 199
surface, 40–145, 150, 172, 175–198
Vetivaria zizanoides (L.) Nash, 13, 43, 62, 104, 147, 270
Vetiveria, 30, 270

W

west-central tropical, 200, 205–209, 212–214, 216–222, 225–233, 235–247, 250
wheat, 2

Z

Zoysia matrella (L.) Merr, 165, 251, 253, 302